STOCHASTIC DYNAMICS

STOCHASTIC DYNAMICS

Søren R. K. Nielsen
Zili Zhang

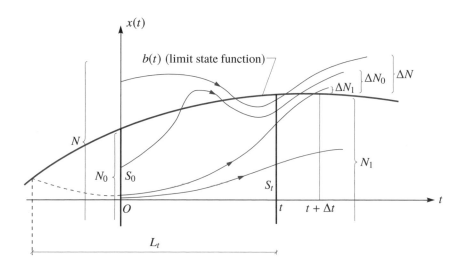

Aarhus University Press

Stochastic Dynamics

© Søren R.K. Nielsen, Zili Zhang and Aarhus University Press 2017
Cover: Nethe Ellinge Nielsen, Trefold
Publishing editor: Simon Olling Rebsdorf

Printed on Amber Graphic by Narayana Press

Printed in Denmark 2017
ISBN 978 87 7184 232 6

AARHUS UNIVERSITY PRESS

Finlandsgade 29
DK-8200 Aarhus N
Denmark
www.unipress.dk

Weblinks were active when the book was printed. They may no longer be active.

INTERNATIONAL DISTRIBUTION

UK & Eire
Gazelle Book Services Ltd.
White Cross Mills, Hightown
Lancaster LA1 4XS, England
www.gazellebooks.com

North America
ISD
70 Enterprise Drive
Bristol, CT 06010
USA
www.isddistribution.com

Contents

PREFACE

The present textbook has been written based on previous lecture notes for a course on stochastic vibration theory given in the autumn semester at Aarhus University for M.Sc. students in structural engineering.

In chapter 1, the basic assumptions of the random vibration theory are emphasized. In chapters 2 and 3, pertinent results of stochastic variables and stochastic processes have been indicated. Chapter 4 deals with the stochastic response analysis of single-degree-of-freedom, multi-degree-of-freedom and continuous linear structural systems. In principle, an introductory course on linear structural dynamics is presupposed. However, in order to make this textbook self-contained, short reviews of the most important results of linear deterministic vibration theory have been included in the start of the relevant sub-sections. Chapter 5 outlines the reliability theory for dynamically excited civil engineering structures, i.e., reliability theory for narrow-banded response processes. Finally, Chapter 6 gives an introduction to Monte Carlo simulation techniques, which become increasingly important and useful as the computers become more and more powerful.

Aarhus University, December 2016

Søren R. K. Nielsen
Zili Zhang

CHAPTER 1
INTRODUCTION

A structural system may be considered as an ensemble of mass particles, each of which is exposed to external loading, and to internal forces from neighbouring mass particles. If the initial conditions, and the external and internal forces are perfectly described for all mass particles, the motion of the system can in principle be determined from *Newton's 2nd law of motion*, leaving no room for any indeterminism. This is the principle of *causality* or *determinism*, postulated in classical physics.

In practice, neither the initial conditions, nor the external and internal forces can be perfectly determined for the mass particles. Usually, the design loads of the structure are some future extreme loadings, which cannot be specified in space and time in the sense presumed above. Even if this was the case, the internal forces cannot be observed or specified. Instead, these internal forces have to be theoretically determined by a mathematical model. In the continuum mechanics approach, the internal forces are expressed in terms of stresses. Significant modelling errors may stem from the constitutive equation which relates the stress components to the motions of the mass particles, and also from certain kinematical approximations rendering analytical solutions possible (e.g., classical beam-, plate-, and shell theories). These errors or approximations introduce uncertainty into the predicted motions of the mass particles. The accuracy of the prediction depends on the level of uncertainty with which the *initial values*, and the external and internal forces can be specified.

Generally, uncertainties are classified as either aleatoric or epistemic.

Aleatoric uncertainties are uncertainties that cannot be removed or reduced by any means. Typical examples are external dynamic natural loads from winds, waves and earthquakes, but also some man made loads such as traffic loads.

Epistemic uncertainties can be removed or reduced to a certain extent. An example is the measurement uncertainties in active feedback vibration control problems, which can be reduced by better measurement equipments or filtering procedures. Another example is the structural modelling uncertainty, which can be reduced by better numerical or analytical models. Epistemic uncertainties can only be reduced to a certain extent, after which they should be considered as aleatoric.

Both aleatoric and epistemic uncertainties are quantified by stochastic models.

The subject of *stochastic mechanics* is the quantification of uncertainty of the structural response based on the quantified uncertainty of the initial values and the external and internal forces. The word "stochastic" (from greek *stochas'mos* = presumerable), with the usual meaning that something is happening by chance, is in a way misleading, because nothing in nature is considered to happen by chance. Stochastic mechanics should merely be interpreted as a tool for quantification of uncertainty of some quantities in the mechanical problem. If these quantities are observed or controlled, the uncertainty is removed or reduced. This would not be possible if an inherent indeterminism is present in nature.

As stated, a mathematical model must be adapted for the determination of internal forces. The quantification of the errors associated with this choice is one of the unsolved problems in stochastic mechanics. The frequently mentioned suggestion to evaluate the performance of the model against a more elaborated model provides insight into the accuracy of the selected model, but does not carry the uncertainty of the chosen model into an operable scheme, where the modelling uncertainty can be weighted along with the uncertainties associated with the initial values and the external loadings.

Even for a well-defined mathematical model for the internal forces, additional uncertainty can be introduced if the parameters of the model cannot be properly calibrated. As an example, the beam in Fig. 1-1 is assumed to be modelled using *Bernoulli-Euler beam theory*. However, the bending stiffness $EI(x)$ in this model may still be uncertain. This is the case for reinforced concrete members under dynamic loading, where cracked and uncracked sections may alternate in an uncontrolled way along the beam. In contrary to the more fundamental modelling problem for the internal forces, parameter uncertainty can easily be treated within the realm of stochastic mechanics.

In stochastic mechanics, the uncertainties of the external loadings and possible parameter fields of the mathematical model for the internal forces are usually modelled as *stochastic variables* or *stochastic processes*. The stochastic structure of these stochastic variables and stochastic processes is estimated from available measurements, statistical inference and engineering

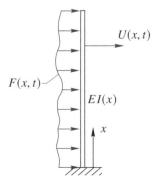

Fig. 1–1 Bernoulli-Euler beam with uncertain bending stiffness.

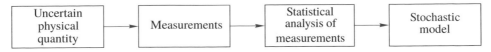

Fig. 1–2 Stochastic modelling of uncertain physical quantity.

judgement. However, this statistical calibration process is not considered a subject of the stochastic response analysis, which assumes that such a calibration has already been performed. The various steps in a stochastic modelling problem have been illustrated in Fig. 1-2. Then, the stochastic analysis merely concentrates on the determination of the stochastic structure of the response processes in terms of the corresponding stochastic structure of the external loading processes and of possible parameter fields associated with the adapted mathematical model.

Dynamic loadings on civil engineering structures are all characterized as highly uncertain. For most structures, the uncertainty of these loadings is much larger than the uncertainty associated with conventional mathematical models for structural analysis, even though this is not always true. For example, the modelling of the dissipation mechanism of mechanical energy into heat may be related with large uncertainty. Nevertheless, in a large number of cases, the modelling uncertainty can be disregarded on condition that a conventional and well-tested mathematical model for structural analysis is adopted. This is the basic assumption in classical stochastic dynamic theory, which is also the basis of the present book. The adopted structural model may be linear or non-linear. However, in the following, only linear structural models are considered for which the *superposition theorem* is applicable. Then, the structural response is determined as a linear functional of the external loading in terms of a memory integral known as *Duhamel's integral*. Since the structure is considered deterministic, no uncertainty is related to the kernel of the memory integral (the so-called *impulse response function*).

The ultimate aim of any engineering analysis is to make decisions on the ability of a certain proposed structural design to fulfil its purpose. Within the context of stochastic mechanics, the decision parameter of interest is given by the *probability of failure*. If it is too high, the draft proposal must be changed. In the present book, various approximate methods have been indicated for the determination of this quantity for dynamically excited structures. In any case, the reliability analysis assumes a preceding stochastic dynamic analysis which provides the necessary parameters needed.

CHAPTER 2
STOCHASTIC VARIABLES

2.1 Introduction to stochastic variables

A *stochastic variable* X is a real function defined on a *sample space* Ω, with the property that any subset of Ω of the following type

$$A_x = \{\omega \in \Omega \mid X(\omega) \le x\} \tag{2-1}$$

is an *event*. This means that a probability $P(A_x)$ can be assigned to the event A_x. $P(\cdot)$ is the *probability function* mapping all events $A \subseteq \Omega$ into the interval $[0, 1]$.

Next, the *probability distribution function* $F_X(x)$ of the stochastic variable X is defined as:

$$F_X(x) = P(A_x) \tag{2-2}$$

The following abbreviated notation will be used in the text: $F_X(x) = P(X \le x)$.

It is noticed that $A_{-\infty} = \emptyset$ and $A_{+\infty} = \Omega$, and $P(\emptyset) = 0$, $P(\Omega) = 1$, where $\emptyset$ signifies an *impossible event*. Hence the following results hold:

$$F_X(-\infty) = 0, \quad F_X(\infty) = 1 \tag{2-3}$$

Since $x_1 \le x_2 \Rightarrow A_{x_1} \subseteq A_{x_2}$, one has $x_1 \le x_2 \Rightarrow F_X(x_1) \le F_X(x_2)$. This establishes the probability distribution function as a non-decreasing function of x.

In what follows, it is assumed that $F_X(x)$ is continuous and piecewise differentiable. The derivative of the probability distribution function, called the *probability density function*, is defined as:

$$f_X(x) = \frac{d}{dx} F_X(x) \tag{2-4}$$

Then, the probability of the event $\{x_1 < X(\omega) \le x_2\}$ becomes:

$$P(x_1 < X \le x_2) = F_X(x_2) - F_X(x_1) = \int_{x_1}^{x_2} f_X(x)\, dx \tag{2-5}$$

Table 2-1 Probability density functions for some continuous stochastic variables.

Distribution	Probability density function	Short notation
Uniform	$f_X(x) = \begin{cases} 0, & x < a \\ \frac{1}{b-a}, & a \leq x \leq b \\ 0, & x > b \end{cases}$	$X \sim U(a, b)$
Normal	$f_X(x) = \frac{1}{\sqrt{2\pi}\sigma} \exp\left(-\frac{1}{2}\frac{(x-\mu)^2}{\sigma^2}\right)$	$X \sim N(\mu, \sigma^2)$
Gamma	$f_X(x) = \begin{cases} 0, & x < 0 \\ \frac{\beta^{\alpha+1}}{\Gamma(\alpha+1)} x^\alpha e^{-\beta x}, & x \geq 0 \end{cases}$	$X \sim G(\alpha, \beta)$
Exponential	$f_X(x) = \begin{cases} 0, & x < 0 \\ \beta e^{-\beta x}, & x \geq 0 \end{cases}$	$X \sim E(\beta)$
Rayleigh	$f_X(x) = \begin{cases} 0, & x < 0 \\ \frac{x}{\sigma^2} \exp\left(-\frac{x^2}{2\sigma^2}\right), & x \geq 0 \end{cases}$	$X \sim R(\sigma^2)$
Weibull	$f_X(x) = \begin{cases} 0, & x < x_0 \\ \frac{\alpha}{x_1}(\frac{x-x_0}{x_1})^{\alpha-1} \exp\left(-(\frac{x-x_0}{x_1})^\alpha\right), & x \geq x_0 \end{cases}$	$X \sim W(x_0, x_1, \alpha)$

The probability density function is nonnegative, and due to Eq. (2-3) the following normalization condition prevails:

$$\int_{-\infty}^{\infty} f_X(x)\, dx = 1 \tag{2–6}$$

A stochastic variable is completely specified by its probability distribution function or its probability density function. In Table 2-1, some well-known probability density functions are listed.

Let X and Y be stochastic variables defined on the same sample space Ω. The column vector $[X, Y]^T$ is termed a *two-dimensional stochastic vector*. Considering the events $A_x = \{X \leq x\}$ and $B_y = \{Y \leq y\}$, the intersection $A_x \cap B_y$ is also an event, and consequently the probability $P(A_x \cap B_y)$ can be defined as well. Obviously, this is a function of the pair (x, y), i.e., $F_{XY}(x, y)$, which is termed the *joint probability distribution function* of the two-dimensional stochastic vector $[X, Y]^T$. The following short notation is applied:

$$F_{XY}(x, y) = P(X \leq x \wedge Y \leq y) \tag{2–7}$$

Since $A_\infty = B_\infty = \Omega$ and $A_{-\infty} = B_{-\infty} = \emptyset$, it follows that:

$$
\left.
\begin{aligned}
F_{XY}(x, \infty) &= P(A_x \cap \Omega) = P(A_x) = F_X(x) \\
F_{XY}(\infty, y) &= P(\Omega \cap B_y) = P(B_y) = F_Y(y) \\
F_{XY}(\infty, \infty) &= P(\Omega \cap \Omega) = P(\Omega) = 1 \\
F_{XY}(-\infty, y) &= P(\emptyset \cap B_y) = P(\emptyset) = 0 \\
F_{XY}(x, -\infty) &= P(A_x \cap \emptyset) = P(\emptyset) = 0
\end{aligned}
\right\}
\tag{2-8}
$$

where $F_X(x) = F_{XY}(x, \infty)$ and $F_Y(y) = F_{XY}(\infty, y)$ are termed the marginal probability distribution functions of the stochastic vector $[X, Y]^T$.

$F_{XY}(x, y)$ is assumed to be continuous with piecewise continuous mixed derivatives of 2nd order. Then, the *joint probability density function* of $[X, Y]^T$ is defined as:

$$
f_{XY}(x, y) = \frac{\partial^2}{\partial x \partial y} F_{XY}(x, y)
\tag{2-9}
$$

From Eqs. (2-8) and (2-9), it follows that:

$$
F_{XY}(x, y) = \int_{-\infty}^{x} \int_{-\infty}^{y} f_{XY}(x, y) \, dx \, dy
\tag{2-10}
$$

Since $f_{XY}(x, y) \, dx \, dy$ represents the probability of $[X, Y] \in [x, x + dx] \times [y, y + dy]$, $f_{XY}(x, y)$ is a nonnegative function of x and y.

X and Y are said to be *independent stochastic variables*, if:

$$
f_{XY}(x, y) = f_X(x) f_Y(y)
\tag{2-11}
$$

Let $X_1, \ldots, X_n$ be a set of stochastic variables all defined on the same sample space Ω. Then, the column vector $\mathbf{X} = [X_1, \ldots, X_n]^T$ is termed an *n-dimensional stochastic vector*.

As a straightforward generalization of the two-dimensional case, the joint probability distribution function of $\mathbf{X}$ is defined as:

$$
\begin{aligned}
F_{\mathbf{X}}(\mathbf{x}) &= F_{X_1 \cdots X_n}(x_1, \ldots, x_n) = P(\{X_1 \le x_1\} \cap \cdots \cap \{X_n \le x_n\}) \\
&= P(X_1 \le x_1 \wedge \cdots \wedge X_n \le x_n) = P(\mathbf{X} \le \mathbf{x})
\end{aligned}
\tag{2-12}
$$

where $\mathbf{x} = [x_1, \ldots, x_n]^T$ is a vector of sample values of the stochastic vector $\mathbf{X}$. The joint probability density function of $\mathbf{X}$ is defined as:

$$
f_{\mathbf{X}}(\mathbf{x}) = f_{X_1 \cdots X_n}(x_1, \ldots, x_n) = \frac{\partial^n}{\partial x_1 \ldots \partial x_n} F_{X_1 \cdots X_n}(x_1, \ldots, x_n)
\tag{2-13}
$$

The probability that $\mathbf{X}$ attains values in an arbitrary domain $D \subseteq R^n$ can be written as:

$$
P(\mathbf{X} \in D) = \int_D f_{\mathbf{X}}(\mathbf{x}) \, d\mathbf{x}
\tag{2-14}
$$

$\mathbf{X}^T = [X_1, \ldots, X_n]$ is said to be a *normally distributed* stochastic vector, in short written as $\mathbf{X} \sim N(\boldsymbol{\mu}, \mathbf{C})$, if:

$$f_{\mathbf{X}}(\mathbf{x}) = \frac{1}{(2\pi)^{\frac{n}{2}} (\det(\mathbf{C}))^{\frac{1}{2}}} \exp\left(-\tfrac{1}{2} (\mathbf{x} - \boldsymbol{\mu})^T \mathbf{C}^{-1} (\mathbf{x} - \boldsymbol{\mu})\right) \tag{2–15}$$

where $\boldsymbol{\mu}$ is an n-dimensional vector, termed the *mean value vector*. $\mathbf{C}$ is a symmetric *positive definite matrix*, termed the *covariance matrix*. $\det(\mathbf{C})$ is the determinant of $\mathbf{C}$, which is always positive. Definitions of $\boldsymbol{\mu}$ and $\mathbf{C}$ will be given in Section 2.3.

Example 2-1: Two-dimensional normal stochastic vector

For a *two-dimensional normally distributed stochastic vector*, we may write:

$$\boldsymbol{\mu} = \begin{bmatrix} \mu_1 \\ \mu_2 \end{bmatrix}, \quad \mathbf{C} = \begin{bmatrix} \sigma_1^2 & \sigma_1 \sigma_2 \rho \\ \sigma_1 \sigma_2 \rho & \sigma_2^2 \end{bmatrix} \tag{2–16}$$

$\mathbf{C}$ will be positive definite if $\rho^2 < 1$. Then, Eq. (2-15) attains the form:

$$f_{\mathbf{X}}(\mathbf{x}) = f_{X_1 X_2}(x_1, x_2) = \frac{1}{2\pi \sigma_1 \sigma_2 \sqrt{1 - \rho^2}} \exp\left(-\frac{1}{2} \frac{\xi_1^2 - 2\rho \xi_1 \xi_2 + \xi_2^2}{1 - \rho^2}\right) \tag{2–17}$$

$$\xi_j = \frac{x_j - \mu_j}{\sigma_j}, \quad j = 1, 2 \tag{2–18}$$

2.2   Combined stochastic variables

Let us consider a continuous function $g(x)$. If x is replaced by the random variable $X : \Omega \to R$, a new stochastic variable $Y : \Omega \to R$ is defined by the combined mapping:

$$\forall \omega \in \Omega : Y(\omega) = g(X(\omega)) \tag{2–19}$$

Further, $g(x)$ is assumed to be a differentiable monotonously increasing or decreasing function of x, corresponding to a one-to-one (bijective) mapping. Then the probability of $Y \in [y, y + dy]$ is equal to the probability of $X \in [x, x + dx]$, where $y = g(x)$. The following expression is thus obtained:

$$f_Y(y) \, dy = f_X(x) \, dx \qquad \Rightarrow$$

$$f_Y(y) = f_X(x) \frac{1}{|g'(x)|} = f_X(g^{-1}(y)) \frac{1}{|g'(g^{-1}(y))|} \tag{2–20}$$

where $g'(x) = \frac{dy}{dx} = \frac{dg(x)}{dx}$, and $x = g^{-1}(y)$ signifies the inverse mapping.

Example 2-2: Lognormally distributed stochastic variable

Let $X \sim N(\mu, \sigma^2)$ and $y = g(x) = e^x$. Then, $g'(x) = e^x$, $x = g^{-1}(y) = \ln y$, $g'(g^{-1}(y)) = y$. Using Eq. (2-20), we have the following probability density function of Y:

$$f_Y(y) = \frac{1}{\sqrt{2\pi}\sigma} \exp\left(-\frac{(\ln y - \mu)^2}{2\sigma^2}\right) \frac{1}{y}, \quad y > 0 \tag{2–21}$$

A stochastic variable with the probability density function as given by Eq. (2-21) is termed a *lognormally distributed* stochastic variable, $Y \sim LN(\mu, \sigma^2)$.

Example 2-3: Square of a normally distributed stochastic variable

Let $X \sim N(\mu, \sigma^2)$ and $y = g(x) = x^2$. Then, $g'(x) = 2x$, $x = \pm\sqrt{y}$, $|\frac{dx}{dy}| = \frac{1}{2\sqrt{y}}$. This is an example of a mapping which is not one-to-one. The probability of $Y \in [y, y + dy]$ is equal to the probability of $X \in [-x - dx, -x] \cup [x, x + dx]$, i.e.:

$$f_Y(y)\, dy \;=\; \frac{1}{\sqrt{2\pi}\sigma} \exp\left(-\frac{(-x-\mu)^2}{2\sigma^2}\right) dx \;+\; \frac{1}{\sqrt{2\pi}\sigma} \exp\left(-\frac{(x-\mu)^2}{2\sigma^2}\right) dx \qquad \Rightarrow$$

$$f_Y(y) \;=\; \frac{1}{\sqrt{8\pi}\sigma} \exp\left(-\frac{(-\sqrt{y}-\mu)^2}{2\sigma^2}\right) \frac{1}{\sqrt{y}} \;+\; \frac{1}{\sqrt{8\pi}\sigma} \exp\left(-\frac{(\sqrt{y}-\mu)^2}{2\sigma^2}\right) \frac{1}{\sqrt{y}} \qquad (2\text{–}22)$$

Alternatively, the probability density function may be determined from the following basic definition of the probability distribution function:

$$F_Y(y) \;=\; P(Y \leq y) \;=\; P(-\sqrt{y} < X \leq \sqrt{y}) \;=\; \Phi\left(\frac{\sqrt{y}-\mu}{\sigma}\right) - \Phi\left(\frac{-\sqrt{y}-\mu}{\sigma}\right) \qquad (2\text{–}23)$$

Next, $f_Y(y)$ can be obtained by taking the derivative of $F_Y(y)$ with respect to y. Here and in the following, $\Phi(x)$ indicates the probability distribution function of a *standardized normal stochastic variable* with zero mean value and unit variance, $X \sim N(0, 1)$, defined as:

$$\Phi(x) = \int_{-\infty}^{x} \varphi(u)\, du, \quad \varphi(x) = \frac{1}{\sqrt{2\pi}} \exp\left(-\tfrac{1}{2} x^2\right) \qquad (2\text{–}24)$$

where $\varphi(x) = \frac{d}{dx}\Phi(x)$ indicates the corresponding probability density function.

Example 2-4: Inverse method for generating samples of a stochastic variable

Especially, we may choose the mapping function $g(x)$ as the probability distribution function $F_X(x)$ of a given stochastic variable X, so $Y = g(X) = F_X(X)$. Based on the relationship between the probability density function and the probability distribution function in Eq. (2-4), we have:

$$dy \;=\; \frac{d}{dx} F_X(x)\, dx \;=\; f_X(x)\, dx \qquad (2\text{–}25)$$

At the same time, we have the following equality from Eq. (2-20):

$$f_Y(y)\, dy \;=\; f_X(x)\, dx \qquad (2\text{–}26)$$

In combination of Eqs. (2-25) and (2-26) yields:

$$f_Y(y) \equiv 1 \qquad (2\text{–}27)$$

Hence $Y \sim U(0, 1)$. A sample y of the uniformly distributed random variable Y can be obtained by a standard pseudo random generator. The related sample x of X is then obtained as:

$$x \;=\; F_X^{-1}(y) \qquad (2\text{–}28)$$

Eq. (2-28) is referred to as the *inverse method* for generating samples of a stochastic variable with given probability distribution function in *Monte Carlo simulation*.

Next, let us consider the multi-dimensional continuous nonlinear bijective mapping of the n-dimensional stochastic vector $\mathbf{X}$ onto the n-dimensional stochastic vector $\mathbf{Y}$:

$$\mathbf{Y} = \mathbf{g}(\mathbf{X}) \tag{2-29}$$

A differential volume element $d\mathbf{x}$ placed at $\mathbf{x}$ is considered. $d\mathbf{x}$ is mapped onto $d\mathbf{y}$ at $\mathbf{y} = \mathbf{g}(\mathbf{x})$ by the mapping in Eq. (2-29). These two volume elements are related as $d\mathbf{y} = |\det(\mathbf{J})| \, d\mathbf{x}$, where $\mathbf{J} = \nabla \mathbf{g}(\mathbf{x})$ indicates the *gradient matrix* and $|\det(\mathbf{J})|$ is the *Jacobian* of the transformation.

The probability of $\mathbf{Y} \in d\mathbf{y}$ must be equal to the probability of $\mathbf{X} \in d\mathbf{x}$, leading to:

$$f_{\mathbf{Y}}(\mathbf{y}) \, d\mathbf{y} = f_{\mathbf{X}}(\mathbf{x}) \, d\mathbf{x} \qquad \Rightarrow$$

$$f_{\mathbf{Y}}(\mathbf{y}) = f_{\mathbf{X}}\big(\mathbf{g}^{-1}(\mathbf{y})\big) \frac{1}{\left| \det \big(\nabla \mathbf{g} \, (\mathbf{g}^{-1}(\mathbf{y})) \big) \right|} \tag{2-30}$$

Eq. (2-30), which is a multi-dimensional generalization of Eq. (2-20), is only applicable for one-to-one mappings of stochastic vectors of equal dimension n. Generalizations to other mappings and to cases where the dimensions of the stochastic vectors are of different dimensions require special attention along the line illustrated in Example 2-3.

Example 2-5: Linear mapping of a normally distributed stochastic vector

The n-dimensional stochastic vectors $\mathbf{X}$ and $\mathbf{Y}$ are assumed to be related by the linear transformation:

$$\mathbf{Y} = \mathbf{g}(\mathbf{X}) = \mathbf{a} + \mathbf{B}\mathbf{X} \tag{2-31}$$

$\mathbf{X}$ is normally distributed with the probability density function given by Eq. (2-15), i.e., $\mathbf{X} \sim N(\boldsymbol{\mu}, \mathbf{C})$. $\mathbf{a}$ is an n-dimensional deterministic vector, and $\mathbf{B}$ is an $n \times n$-dimensional deterministic, non-singular matrix, so $\det(\mathbf{B}) \neq 0$. $\nabla \mathbf{g}(\mathbf{x}) = \mathbf{B}$, and $\mathbf{g}^{-1}(\mathbf{y}) = \mathbf{B}^{-1}(\mathbf{y} - \mathbf{a})$. Then, Eq. (2-30) attains the form:

$$f_{\mathbf{Y}}(\mathbf{y}) = f_{\mathbf{X}}\big(\mathbf{B}^{-1}(\mathbf{y} - \mathbf{a})\big) \frac{1}{|\det(\mathbf{B})|}$$

$$= \frac{1}{(2\pi)^{\frac{n}{2}} \left(\det(\mathbf{C}) \right)^{\frac{1}{2}} |\det(\mathbf{B})|} \exp\left(-\tfrac{1}{2} \big(\mathbf{B}^{-1}(\mathbf{y} - \mathbf{a}) - \boldsymbol{\mu} \big)^T \mathbf{C}^{-1} \big(\mathbf{B}^{-1}(\mathbf{y} - \mathbf{a}) - \boldsymbol{\mu} \big) \right)$$

$$= \frac{1}{(2\pi)^{\frac{n}{2}} \left(\det(\mathbf{B}\mathbf{C}\mathbf{B}^T) \right)^{\frac{1}{2}}} \exp\left(-\tfrac{1}{2} \big(\mathbf{y} - \mathbf{a} - \mathbf{B}\boldsymbol{\mu} \big)^T \big(\mathbf{B}\mathbf{C}\mathbf{B}^T \big)^{-1} \big(\mathbf{y} - \mathbf{a} - \mathbf{B}\boldsymbol{\mu} \big) \right) \tag{2-32}$$

where it has been used that $\det(\mathbf{B}\mathbf{C}\mathbf{B}^T) = \det(\mathbf{B})\det(\mathbf{C})\det(\mathbf{B}^T) = \det(\mathbf{C})\big(\det(\mathbf{B})\big)^2$ and $\big(\mathbf{B}\mathbf{C}\mathbf{B}^T\big)^{-1} = \big(\mathbf{B}^T\big)^{-1}\mathbf{C}^{-1}\mathbf{B}^{-1} = \big(\mathbf{B}^{-1}\big)^T\mathbf{C}^{-1}\mathbf{B}^{-1}$.

Eq. (2-32) shows that $\mathbf{Y} \sim N(\mathbf{a}+\mathbf{B}\boldsymbol{\mu}, \mathbf{B}\mathbf{C}\mathbf{B}^T)$. Hence, normality is preserved under linear transformations as in Eq. (2-31). This result can be shown to hold even when $\mathbf{B}$ is a non-quadratic matrix, i.e., when $\mathbf{X}$ and $\mathbf{Y}$ have different dimensions m and n, respectively.

Example 2-6: Transformation of a normally distributed stochastic vector to a vector of mutually independent standardized normal stochastic variables

Let $\mathbf{X} \sim N(\boldsymbol{\mu}, \mathbf{C})$ be a normally distributed stochastic vector of dimension n. We shall indicate a linear transformation of the type in Eq. (2-31), where the components Y_j of $\mathbf{Y}$ become mutually independent standardized normal stochastic variables, i.e., $Y_j \sim N(0, 1)$.

The following *eigenvalue problem* for the covariance matrix $\mathbf{C}$ is considered:

$$\mathbf{C}\mathbf{v}_j = \lambda_j \mathbf{v}_j, \quad j = 1, \ldots, n \tag{2–33}$$

Since the covariance matrix is symmetric, i.e., $\mathbf{C} = \mathbf{C}^T$, the eigenvalues λ_j and eigenvectors $\mathbf{v}_j$ are real. Further, since $\mathbf{C}$ is positive definite, all eigenvalues are positive. By a suitable normalization, the eigenvalues fulfill the *orthonormality condition*:

$$\mathbf{v}_j^T \mathbf{v}_k = \begin{cases} 0, & j \neq k \\ 1, & j = k \end{cases} \tag{2–34}$$

The eigenvalue problem in Eq. (2-33) may be assembled into the following matrix formulation:

$$\mathbf{C}\mathbf{V} = \mathbf{V}\boldsymbol{\Lambda} \tag{2–35}$$

where:

$$\mathbf{V} = \begin{bmatrix} \mathbf{v}_1 \mathbf{v}_2 \cdots \mathbf{v}_n \end{bmatrix}, \quad \boldsymbol{\Lambda} = \begin{bmatrix} \lambda_1 & 0 & \cdots & 0 \\ 0 & \lambda_2 & \cdots & 0 \\ \vdots & \vdots & \ddots & \vdots \\ 0 & 0 & \cdots & \lambda_n \end{bmatrix} \tag{2–36}$$

Because of the orthogonality condition in Eq. (2-34), the *modal matrix* $\mathbf{V}$ is orthonormal, thus $\mathbf{V}^T = \mathbf{V}^{-1}$. Then, Eq. (2-35) may be written as:

$$\mathbf{V}^{-1}\mathbf{C}\mathbf{V} = \boldsymbol{\Lambda} \quad \Rightarrow \quad \mathbf{B}^T \mathbf{C}\mathbf{B} = \mathbf{I} \tag{2–37}$$

where $\mathbf{I}$ indicates the *identity matrix*, and:

$$\mathbf{B} = \mathbf{V}\boldsymbol{\Lambda}^{-\frac{1}{2}}, \quad \boldsymbol{\Lambda}^{-\frac{1}{2}} = \begin{bmatrix} \lambda_1^{-\frac{1}{2}} & 0 & \cdots & 0 \\ 0 & \lambda_2^{-\frac{1}{2}} & \cdots & 0 \\ \vdots & \vdots & \ddots & \vdots \\ 0 & 0 & \cdots & \lambda_n^{-\frac{1}{2}} \end{bmatrix} \tag{2–38}$$

Further, the parameter vector $\mathbf{a}$ in Eq. (2-31) may be chosen to fulfill the following equation:

$$\mathbf{a} + \mathbf{B}\boldsymbol{\mu} = \mathbf{0} \tag{2–39}$$

With the choices of $\mathbf{B}$ and $\mathbf{a}$ as given by Eqs. (2-38) and (2-39), Eq. (2-32) attains the form:

$$f_{\mathbf{Y}}(\mathbf{y}) = \prod_{i=1}^{n} \frac{1}{\sqrt{2\pi}} \exp\left(-\tfrac{1}{2} y_i^2\right) \tag{2-40}$$

Eq. (2-40) shows that all components in $\mathbf{Y}$ are mutually independent stochastic variables, i.e., $Y_j \sim N(0, 1)$.

Assuming that $\mathbf{X} \sim N(\boldsymbol{\mu}, \mathbf{C})$, and $\mathbf{a}$ and $\mathbf{B}$ are chosen as indicated by Eqs. (2-38) and (2-39), we have:

$$\int_{R^n} f_{\mathbf{X}}(\mathbf{x})\, d\mathbf{x} = \int_{R^n} f_{\mathbf{Y}}(\mathbf{y})\, d\mathbf{y} = \prod_{j=1}^{n} \int_{-\infty}^{\infty} \frac{1}{\sqrt{2\pi}} \exp\left(-\tfrac{1}{2} y_j^2\right) dy_j = 1 \tag{2-41}$$

Eq. (2-41) shows that Eq. (2-15) is an acceptable probability density function in the sense that its integration over the sample domain amounts to 1.

It should be noticed that infinite many other transformation matrices $\mathbf{B}$ exist than the one indicated in Eq. (2-38), which fulfill the property in Eq. (2-37).

Example 2-7: Transformation of a pair of stochastic variables to a pair of independent normally distributed stochastic variables

Let $\mathbf{X} = [R, \Phi]^T$, where $R \sim R(\sigma^2)$ and $\Phi \sim U(0, 2\pi)$ are assumed to be stochastically independent. The following non-linear one-to-one transformation is considered:

$$\begin{bmatrix} Y_1 \\ Y_2 \end{bmatrix} = \mathbf{g}(\mathbf{X}) = \begin{bmatrix} R\cos(\alpha + \Phi) \\ R\sin(\alpha + \Phi) \end{bmatrix} \tag{2-42}$$

where α is an arbitrary deterministic constant.

The Jacobian of the transformation in Eq. (2-42) becomes:

$$\left|\det\left(\nabla\mathbf{g}(\mathbf{x})\right)\right| = \left|\det\left(\begin{bmatrix} \cos(\alpha + \varphi) & -r\sin(\alpha + \varphi) \\ \sin(\alpha + \varphi) & r\cos(\alpha + \varphi) \end{bmatrix}\right)\right| = r \tag{2-43}$$

From Eq. (2-42), we have:

$$r = \sqrt{y_1^2 + y_2^2} \tag{2-44}$$

The joint probability density function of $[Y_1, Y_2]^T$ follows from Eq. (2-30):

$$f_{Y_1 Y_2}(y_1, y_2) = \frac{1}{r} f_{R\Phi}(r, \varphi) = \frac{1}{r} \frac{r}{\sigma^2} \exp\left(-\frac{r^2}{2\sigma^2}\right) \cdot \frac{1}{2\pi} = \frac{1}{2\pi\sigma^2} \exp\left(-\frac{y_1^2 + y_2^2}{2\sigma^2}\right)$$

$$= \frac{1}{\sqrt{2\pi}\sigma} \exp\left(-\frac{y_1^2}{2\sigma^2}\right) \cdot \frac{1}{\sqrt{2\pi}\sigma} \exp\left(-\frac{y_2^2}{2\sigma^2}\right) = f_{Y_1}(y_1) \cdot f_{Y_2}(y_2) \tag{2-45}$$

where it has been used that R and Φ are stochastically independent variables with probability density functions indicated in Table 2-1.

Eq. (2-45) shows that $Y_1 \sim N(0, \sigma^2)$ and $Y_2 \sim N(0, \sigma^2)$, and that Y_1 and Y_2 are stochastically independent variables.

Next, let $\mathbf{X} = [U_1, U_2]^T$, where $U_j \sim U(0, 1)$, $j = 1, 2$, and U_1 and U_2 are stochastically independent variables. The non-linear one-to-one transformation is considered:

$$\begin{bmatrix} Y_1 \\ Y_2 \end{bmatrix} = \begin{bmatrix} \sqrt{-2 \ln U_2} \, \cos(2\pi U_1) \\ \sqrt{-2 \ln U_2} \, \sin(2\pi U_1) \end{bmatrix} \tag{2-46}$$

The Jacobian of the transformation in Eq. (2-46) becomes:

$$\left| \det\left(\nabla \mathbf{g}(\mathbf{x}) \right) \right| = \left| \det\left(\begin{bmatrix} -2\pi\sqrt{-2 \ln u_2} \, \sin(2\pi u_1) & -\frac{1}{u_2\sqrt{-2\ln u_2}} \cos(2\pi u_1) \\ 2\pi\sqrt{-2 \ln u_2} \, \cos(2\pi u_1) & -\frac{1}{u_2\sqrt{-2\ln u_2}} \sin(2\pi u_1) \end{bmatrix} \right) \right| = \frac{2\pi}{u_2} \tag{2-47}$$

From Eq. (2-46), we have:

$$-2 \ln u_2 = y_1^2 + y_2^2 \quad \Rightarrow \quad u_2 = \exp\left(-\frac{y_1^2 + y_2^2}{2} \right) \tag{2-48}$$

The joint probability density function of $[Y_1, Y_2]$ follows from Eq. (2-30):

$$f_{Y_1 Y_2}(y_1, y_2) = \frac{1}{\frac{2\pi}{u_2}} \cdot f_{U_1}(u_1) \cdot f_{U_2}(u_2) = \frac{u_2}{2\pi} \cdot 1 \cdot 1 = \frac{1}{2\pi} \exp\left(-\frac{y_1^2 + y_2^2}{2} \right)$$

$$= \frac{1}{\sqrt{2\pi}} \exp\left(-\frac{y_1^2}{2} \right) \cdot \frac{1}{\sqrt{2\pi}} \exp\left(-\frac{y_2^2}{2} \right) = f_{Y_1}(y_1) \cdot f_{Y_2}(y_2) \tag{2-49}$$

Eq. (2-49) shows that $Y_1 \sim N(0, 1)$ and $Y_2 \sim N(0, 1)$, and that Y_1 and Y_2 are stochastically independent variables.

The transformation in Eq. (2-46) is known as the *Box-Müller transformation*. This transformation is often used for generating samples of normally distributed random variables in Monte Carlo simulation procedures.

2.3 Expectations

The *expected value (mean value)* $E[X]$ of the stochastic variable X is defined as:

$$E[X] = \int_{-\infty}^{\infty} x \, f_X(x) \, dx \tag{2-50}$$

In the following, the symbol μ_X is reserved for $E[X]$.

The *variance* σ_X^2 of X is defined as the expected value of the combined stochastic variable $Y = (X - \mu_X)^2$, i.e.:

$$\sigma_X^2 = E\left[(X - \mu_X)^2 \right] = \int_{-\infty}^{\infty} (x - \mu_X)^2 \, f_X(x) \, dx = E[X^2] - \mu_X^2 \tag{2-51}$$

The square root of the variance, σ_X, is termed the *standard deviation*. In the case of $\mu_X \neq 0$, the *variational coefficient* V_X is defined as:

$$V_X = \frac{\sigma_X}{\mu_X} \tag{2–52}$$

Example 2-8: Statistical moments of a normal stochastic variable

Let $X \sim N(\mu, \sigma^2)$. According to Table 2-1, we have:

$$\mu_X = \int_{-\infty}^{\infty} x \frac{1}{\sqrt{2\pi}\sigma} \exp\left(-\frac{1}{2}\left(\frac{x-\mu}{\sigma}\right)^2\right) dx = \mu \tag{2–53}$$

$$\sigma_X^2 = \int_{-\infty}^{\infty} (x-\mu)^2 \frac{1}{\sqrt{2\pi}\sigma} \exp\left(-\frac{1}{2}\left(\frac{x-\mu}{\sigma}\right)^2\right) dx = \sigma^2 \tag{2–54}$$

Hence, the mean value and the variance are given by the parameters μ and σ entering the distribution. Higher order moments of a normally distributed stochastic variable have been indicated in Appendix C.

The results in Eqs. (2-53) and (2-54) can be generalized to an n-dimensional normally distributed stochastic vector $\mathbf{X}$ with the probability density function given by Eq. (2-15). The mean value vector $\boldsymbol{\mu_X}$ and the covariance matrix $\mathbf{C_{XX}}$ of $\mathbf{X}$ become:

$$\boldsymbol{\mu_X} = E[\mathbf{X}] = \int_{R^n} \mathbf{x} \frac{1}{(2\pi)^{\frac{n}{2}}(\det(\mathbf{C}))^{\frac{1}{2}}} \exp\left(-\frac{1}{2}(\mathbf{x}-\boldsymbol{\mu})^T \mathbf{C}^{-1}(\mathbf{x}-\boldsymbol{\mu})\right) d\mathbf{x} = \boldsymbol{\mu} \tag{2–55}$$

$$\mathbf{C_{XX}} = E\left[(\mathbf{X}-\boldsymbol{\mu_X})(\mathbf{X}-\boldsymbol{\mu_X})^T\right]$$

$$= \int_{R^n} (\mathbf{x}-\boldsymbol{\mu_X})(\mathbf{x}-\boldsymbol{\mu_X})^T \frac{1}{(2\pi)^{\frac{n}{2}}(\det(\mathbf{C}))^{\frac{1}{2}}} \exp\left(-\frac{1}{2}(\mathbf{x}-\boldsymbol{\mu})^T \mathbf{C}^{-1}(\mathbf{x}-\boldsymbol{\mu})\right) d\mathbf{x} = \mathbf{C} \tag{2–56}$$

More generally, let us consider a combined stochastic variable $Y = g(\mathbf{X}) = g(X_1, \ldots, X_n)$. The mean value of Y is computed as:

$$E[Y] = E\left[g(X_1, \ldots, X_n)\right] = \int_{R^n} g(x_1, \ldots, x_n) f_{X_1 \cdots X_n}(x_1, \ldots, x_n) \, dx_1 \cdots dx_n \tag{2–57}$$

$E[Y]$ exists if and only if $g f_{X_1 \cdots X_n}$ is integrable over R^n. For the two special cases of $g(\mathbf{X}) = X_j$ and $g(\mathbf{X}) = (X_j - \mu_{X_j})(X_k - \mu_{X_k})$, the mean value μ_{X_j} of the component X_j, and the *covariance* $\kappa_{X_j X_k}$ of the components X_j and X_k can be respectively obtained as:

$$\mu_{X_j} = \int_{R^n} x_j \, f_{X_1 \cdots X_j \cdots X_n}(x_1, \ldots, x_j, \ldots, x_n) \, dx_1 \cdots dx_j \cdots dx_n = \int_{-\infty}^{\infty} x_j \, f_{X_j}(x_j) \, dx_j \tag{2–58}$$

$$\kappa_{X_j X_k} = \int_{R^n} (x_j - \mu_{X_j})(x_k - \mu_{X_k}) f_{X_1 \cdots X_j \cdots X_k \cdots X_n}(x_1, \ldots, x_j, \ldots, x_k, \ldots, x_n) dx_1 \cdots dx_j \cdots dx_k \cdots dx_n$$

$$= \int_{-\infty}^{\infty} \int_{-\infty}^{\infty} (x_j - \mu_{X_j})(x_k - \mu_{X_k}) \, f_{X_j X_k}(x_j, x_k) \, dx_j dx_k \tag{2–59}$$

where $f_{X_j}(x_j)$ indicates the probability density function of X_j, and $f_{X_j X_k}(x_j, x_k)$ indicates the joint probability density function of $[X_j, X_k]^T$.

The variance $\sigma^2_{X_j}$ of the component X_j is obtained from Eq. (2-59) for the case of $k = j$.

The *correlation coefficient* of the components X_j and X_k is defined as:

$$\rho_{X_j X_k} = \frac{\kappa_{X_j X_k}}{\sigma_{X_j} \sigma_{X_k}} \tag{2–60}$$

It can be shown that $|\rho_{X_j X_k}| \leq 1$. X_j and X_k are said to be *uncorrelated* if $\rho_{X_j X_k} = 0$.

If X_j and X_k are independent stochastic variables, the covariance $\kappa_{X_j X_k}$ becomes:

$$
\begin{aligned}
\kappa_{X_j X_k} &= \int_{-\infty}^{\infty} \int_{-\infty}^{\infty} (x_j - \mu_{X_j})(x_k - \mu_{X_k}) f_{X_j}(x_j) f_{X_k}(x_k) \, dx_j dx_k \\
&= \int_{-\infty}^{\infty} (x_j - \mu_{X_j}) f_{X_j}(x_j) \, dx_j \int_{-\infty}^{\infty} (x_k - \mu_{X_k}) f_{X_k}(x_j) \, dx_k = 0
\end{aligned}
\tag{2–61}
$$

Hence, independent stochastic variables are always uncorrelated. On the other hand, if all component stochastic variables in $\mathbf{X}$ are mutually uncorrelated, $\mathbf{C_{XX}}$ becomes a diagonal matrix. If $\mathbf{X}$ is further assumed to be a normal vector, it follows from Eq. (2-15) that $f_{\mathbf{X}}(\mathbf{x}) = \prod_{j=1}^{n} f_{X_j}(x_j)$. Therefore, uncorrelated normal stochastic variables will also be mutually independent. This result is unique for normally distributed stochastic variables.

2.4   Conditional distributions

Let X and Y be arbitrary stochastic variables. The *conditional probability density function* of X on $Y = y$ is defined as:

$$f_{X|Y}(x \mid y) = \frac{f_{XY}(x, y)}{f_Y(y)} \tag{2–62}$$

Obviously, the conditional probability density function can only be defined for samples of Y, where the marginal probability density function $f_Y(y) > 0$.

$f_{X|Y}(x \mid y)$ has the properties of a usual probability density function. It is nonnegative, and the following result applies:

$$\int_{-\infty}^{\infty} f_{X|Y}(x \mid y) \, dx = \frac{1}{f_Y(y)} \int_{-\infty}^{\infty} f_{XY}(x, y) \, dx = \frac{f_Y(y)}{f_Y(y)} = 1 \tag{2–63}$$

If X and Y are independent stochastic variables, $f_{X|Y}(x \mid y)$ is equal to the marginal probability density function $f_X(x)$, as follows from:

$$f_{X|Y}(x \mid y) = \frac{f_{XY}(x, y)}{f_Y(y)} = \frac{f_X(x) f_Y(y)}{f_Y(y)} = f_X(x) \tag{2–64}$$

The expected value of the combined stochastic variable $g(X)$ on condition of $Y = y$ is defined as:

$$E[g(X)|y] = \int_{-\infty}^{\infty} g(x)f_{X|Y}(x|y)\,dx \tag{2-65}$$

If $g(X) = X$, Eq. (2-65) determines the *conditional mean value* $\mu_{X|Y} = E[X|Y = y]$, and if $g(X) = (X - \mu_{X|Y})^2$, the *conditional variance* $\sigma_{X|Y}^2 = E[(X - \mu_{X|Y})^2|Y = y]$ is obtained.

$E[g(X)|y]$ is a function of y. Hence, if y is replaced with the stochastic variable Y, a combined stochastic variable is obtained. The expected value of $E[g(X)|Y]$ becomes:

$$E\Big[E[g(X)|Y]\Big] = \int_{-\infty}^{\infty} E[g(X)|y]\,f_Y(y)\,dy = \int_{-\infty}^{\infty} \left(\int_{-\infty}^{\infty} g(x)\,f_{X|Y}(x|y)\,dx \right) f_Y(y)\,dy$$

$$= \int_{-\infty}^{\infty} \int_{-\infty}^{\infty} g(x)\,f_{XY}(x,y)\,dx\,dy = \int_{-\infty}^{\infty} g(x)\,f_X(x)\,dx = E[g(X)] \tag{2-66}$$

Next, let $\mathbf{X}$ and $\mathbf{Y}$ be stochastic vectors of dimensions m and n, respectively. The conditional probability density function of $\mathbf{X}$ on $\mathbf{Y} = \mathbf{y}$ is defined as:

$$f_{\mathbf{X}|\mathbf{Y}}(\mathbf{x}|\mathbf{y}) = \frac{f_{\mathbf{XY}}(\mathbf{x},\mathbf{y})}{f_{\mathbf{Y}}(\mathbf{y})} \tag{2-67}$$

where $f_{\mathbf{XY}}(\mathbf{x},\mathbf{y})$ is the joint probability density function of $\mathbf{X}$ and $\mathbf{Y}$, and $f_{\mathbf{Y}}(\mathbf{y})$ is the marginal joint probability density function of $\mathbf{Y}$.

Let us consider the combined stochastic vector $\mathbf{g}(\mathbf{X})$, which can be either a linear or non-linear function of $\mathbf{X}$. Generalizing the results from Eqs. (2-65) and (2-66), we have:

$$E[\mathbf{g}(\mathbf{X})|\mathbf{y}] = \int_{R^m} \mathbf{g}(\mathbf{x})f_{\mathbf{X}|\mathbf{Y}}(\mathbf{x}|\mathbf{y})\,d\mathbf{x} \tag{2-68}$$

$$E\Big[E[\mathbf{g}(\mathbf{X})|\mathbf{Y}]\Big] = E[\mathbf{g}(\mathbf{X})] \tag{2-69}$$

The probability density function of a normally distributed stochastic vector $\mathbf{Z} \sim N(\boldsymbol{\mu}_{\mathbf{Z}}, \mathbf{C}_{\mathbf{ZZ}})$ of dimension $m + n$, is given as, cf. Eq. (2-15):

$$f_{\mathbf{Z}}(\mathbf{z}) = \frac{1}{(2\pi)^{\frac{m+n}{2}} \left(\det(\mathbf{C}_{\mathbf{ZZ}}) \right)^{\frac{1}{2}}} \exp\left(-\tfrac{1}{2}(\mathbf{z} - \boldsymbol{\mu}_{\mathbf{Z}})^T \mathbf{C}_{\mathbf{ZZ}}^{-1}(\mathbf{z} - \boldsymbol{\mu}_{\mathbf{Z}}) \right) \tag{2-70}$$

where $\mathbf{Z}$ is partitioned into stochastic subvectors $\mathbf{X}$ and $\mathbf{Y}$ of dimension m and n:

$$\mathbf{Z} = \begin{bmatrix} \mathbf{X} \\ \mathbf{Y} \end{bmatrix} \tag{2-71}$$

Correspondingly, the mean value vector $\mu_{\mathbf{Z}}$ and the covariance matrix $\mathbf{C_{ZZ}}$ are partitioned as:

$$\mu_{\mathbf{Z}} = \begin{bmatrix} \mu_{\mathbf{X}} \\ \mu_{\mathbf{Y}} \end{bmatrix}, \quad \mathbf{C_{ZZ}} = \begin{bmatrix} \mathbf{C_{XX}} & \mathbf{C_{XY}} \\ \mathbf{C_{XY}^T} & \mathbf{C_{YY}} \end{bmatrix} \tag{2–72}$$

where $\mu_{\mathbf{X}}$, $\mu_{\mathbf{Y}}$ and $\mathbf{C_{XX}}$, $\mathbf{C_{YY}}$ signify the mean value vectors and the covariance matrices of the stochastic subvectors $\mathbf{X}$ and $\mathbf{Y}$, and $\mathbf{C_{XY}} = E\left[(\mathbf{X} - \mu_{\mathbf{X}})(\mathbf{Y} - \mu_{\mathbf{Y}})^T \right]$ is the *cross-covariance matrix*.

The following identities can be proved:[1]

$$\det\left(\mathbf{C_{ZZ}}\right) = \det\left(\mathbf{C_{XX|Y}}\right) \det\left(\mathbf{C_{YY}}\right) \tag{2–73}$$

$$(\mathbf{z} - \mu_{\mathbf{Z}})^T \mathbf{C_{ZZ}^{-1}} (\mathbf{z} - \mu_{\mathbf{Z}}) = (\mathbf{x} - \mu_{\mathbf{X|Y}})^T \mathbf{C_{XX|Y}^{-1}} (\mathbf{x} - \mu_{\mathbf{X|Y}}) + (\mathbf{y} - \mu_{\mathbf{Y}})^T \mathbf{C_{YY}^{-1}} (\mathbf{y} - \mu_{\mathbf{Y}}) \tag{2–74}$$

where:

$$\mu_{\mathbf{X|Y}} = \mu_{\mathbf{X}} + \mathbf{C_{XY}} \mathbf{C_{YY}^{-1}} (\mathbf{y} - \mu_{\mathbf{Y}}) \tag{2–75}$$

$$\mathbf{C_{XX|Y}} = \mathbf{C_{XX}} - \mathbf{C_{XY}} \mathbf{C_{YY}^{-1}} \mathbf{C_{XY}^T} \tag{2–76}$$

Applying Eqs. (2-70), (2-73) and (2-74), the following equations for the conditional probability density function of $\mathbf{X}$ on $\mathbf{Y} = \mathbf{y}$ can be obtained:

$$f_{\mathbf{Z}}(\mathbf{z}) = f_{\mathbf{X|Y}}(\mathbf{x} \mid \mathbf{y}) f_{\mathbf{Y}}(\mathbf{y}) \tag{2–77}$$

$$f_{\mathbf{X|Y}}(\mathbf{x} \mid \mathbf{y}) = \frac{1}{(2\pi)^{\frac{m}{2}} \left(\det(\mathbf{C_{XX|Y}}) \right)^{\frac{1}{2}}} \exp\left(-\tfrac{1}{2}(\mathbf{x} - \mu_{\mathbf{X|Y}})^T \mathbf{C_{XX|Y}^{-1}} (\mathbf{x} - \mu_{\mathbf{X|Y}}) \right) \tag{2–78}$$

Eqs. (2-77) and (2-78) imply that the conditional distribution of $\mathbf{X}$ on $\mathbf{Y} = \mathbf{y}$ is normal with the conditional mean value vector $\mu_{\mathbf{X|Y}}$ given by Eq. (2-75), and the conditional covariance matrix $\mathbf{C_{XX|Y}}$ given by Eq. (2-76). Further, the marginal distribution of the stochastic subvector $\mathbf{Y}$ is normal as well. Hence, any conditional or marginal distribution of a normal stochastic vector is normal. It should also be noticed that the conditional covariance matrix $\mathbf{C_{XX|Y}}$ is independent of $\mathbf{y}$.

Example 2-9: Conditional probability density function of a 2-dimensional stochastic vector

Let $\mathbf{Z} = [X, Y]^T \sim N(\mu_{\mathbf{Z}}, \mathbf{C_{ZZ}})$, with its mean value vector and covariance matrix given by:

$$\mu_{\mathbf{Z}} = \begin{bmatrix} \mu_X \\ \mu_Y \end{bmatrix} \tag{2–79}$$

---

[1]A.H. Jazwinski. *Stochastic Processes and Filtering Theory*. Dover Publications, Inc., New York, 2007.

$$\mathbf{C_{ZZ}} = \begin{bmatrix} \sigma_X^2 & \sigma_X \sigma_Y \rho \\ \sigma_X \sigma_Y \rho & \sigma_Y^2 \end{bmatrix} \tag{2–80}$$

where ρ indicates the correlation coefficient between the stochastic variables X and Y.

The conditional mean value $\mu_{X|Y}$ and the conditional variance $\sigma_{X|Y}^2$ follow from Eqs. (2-75) and (2-76):

$$\mu_{X|Y} = \mu_X + \sigma_X \sigma_Y \rho \frac{1}{\sigma_Y^2} (y - \mu_Y) = \mu_X + \rho \frac{\sigma_X}{\sigma_Y} (y - \mu_Y) \tag{2–81}$$

$$\sigma_{X|Y}^2 = \sigma_X^2 - \rho \sigma_X \sigma_Y \frac{1}{\sigma_Y^2} \rho \sigma_X \sigma_Y = \sigma_X^2 (1 - \rho^2) \tag{2–82}$$

Finally, the conditional probability density function $f_{X|Y}(x \mid y)$ becomes, cf. Eq.(2-78):

$$f_{X|Y}(x \mid y) = \frac{1}{\sqrt{2\pi} \sigma_X \sqrt{1 - \rho^2}} \exp\left(-\frac{1}{2} \left(\frac{x - \mu_X - \rho \frac{\sigma_X}{\sigma_Y} (y - \mu_Y)}{\sigma_X \sqrt{1 - \rho^2}} \right)^2 \right) \tag{2–83}$$

CHAPTER 3
STOCHASTIC PROCESSES

3.1 Introduction to stochastic processes

A *stochastic process* is defined as an indexed set $\{X(t), t \in T\}$ of stochastic variables $X(t)$ defined on the same sample space Ω. t is designated the *index parameter* and T is the *index set*. The index parameter identifies a specific stochastic variable in the process. The stochastic process is characterized based on the structure of the index set:

$$T = \{1, 2, \ldots, n\} \qquad : \{X(t),\, t \in T\} \text{ is an } n\text{-dimensional stochastic vector.}$$
$$T = \{1, 2, \ldots, n, \ldots\} : \{X(t),\, t \in T\} \text{ is a } discrete\ parameter\ stochastic\ process.$$
$$T \subseteq R^n \qquad\qquad : \{X(t),\, t \in T\} \text{ is a } continuous\ parameter\ stochastic\ process.$$

In what follows, only continuous parameter stochastic processes will be considered.

As in the case of n-dimensional stochastic vectors, the probability structure of a stochastic process is specified by the joint probability distribution functions of the stochastic variables in the process. The decisive difference is that the number of stochastic variables may be infinite. With ever increasing accuracy, the probability structure of the process can be described by a sequence of *joint probability distribution functions*:

$$\left. \begin{aligned} & F_{X(t_1)}(x_1) \\ & F_{X(t_1)X(t_2)}(x_1, x_2) \\ & \quad \vdots \\ & F_{X(t_1)\cdots X(t_n)}(x_1, \ldots, x_n) \\ & \quad \vdots \end{aligned} \right\} \tag{3-1}$$

for any index parameters $t_1, \ldots, t_n, \ldots \in T$. These functions are named the joint probability distribution functions of the 1st order, the 2nd order, etc. Of course, the joint probability distribution function $F_{X(t_1)\cdots X(t_n)}(x_1, \ldots, x_n)$ of the nth order depends on which stochastic variables are considered, and consequently may be considered a function of both the index parameters $t_1, \ldots, t_n$ and the *state variables* $x_1, \ldots, x_n$. For ease, the following notation will be introduced:

$$F_X(x_1, t_1; \ldots ; x_n, t_n) = F_{X(t_1)\cdots X(t_n)}(x_1, \ldots, x_n) \tag{3-2}$$

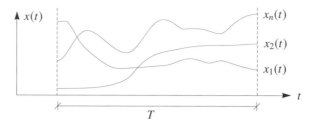

Fig. 3–1 Realizations of a stochastic process.

The nth order *joint probability density function* of the process is defined as, cf. Eq. (2-13):

$$f_X(x_1, t_1 ; \ldots ; x_n, t_n) = \frac{\partial^n}{\partial x_1 \cdots \partial x_n} F_X(x_1, t_1 ; \ldots ; x_n, t_n) \tag{3–3}$$

When a fixed $\omega_j \in \Omega$ is considered, the mapping points $x_j(t) = X(t, \omega_j)$ form a real function defined on T, which is termed a *realization*. When the index parameter represents time, we shall synonymously use the designation *time-series*. If all $\omega_j \in \Omega$ are considered, a sequence of realizations $x_1(t), x_2(t), \ldots, x_n(t), \ldots$ can be obtained as demonstrated in Fig. 3-1, from which the probability structure equally well may be obtained by statistical analysis.

Alternatively, the probability structure of the process may be described by a hierarchy of joint statistical moments of increasing order. Only the moments of the lowest order will be considered in the following.

The *mean value function* $\mu_X(t)$ specifies the expected value of the stochastic variable $X(t)$:

$$\mu_X(t) = E[X(t)] \tag{3–4}$$

In general, the expected value depends on the considered random variable in the process as indicated by the index parameter t, so the expected value becomes a function of t as well.

The *auto-covariance function* $\kappa_{XX}(t_1, t_2)$ specifies the covariance between the stochastic variables $X(t_1)$ and $X(t_2)$:

$$\kappa_{XX}(t_1, t_2) = E\left[\left(X(t_1) - \mu_X(t_1)\right)\left(X(t_2) - \mu_X(t_2)\right)\right]$$

$$= E[X(t_1)X(t_2)] - \mu_X(t_1)\mu_X(t_2) \tag{3–5}$$

The auto-covariance function fulfills the symmetry property:

$$\kappa_{XX}(t_1, t_2) = \kappa_{XX}(t_2, t_1) \tag{3–6}$$

Eq. (3-6) merely gives the trivial statement that the covariance of $X(t_1)$ and $X(t_2)$ is equal to the covariance of $X(t_2)$ and $X(t_1)$.

The *variance function* $\sigma_X^2(t)$ is obtained from Eq. (3-5) for $t_1 = t_2 = t$:

$$\sigma_X^2(t) = E\left[(X(t) - \mu_X(t))^2\right] = E\left[X^2(t)\right] - \mu_X^2(t) \tag{3-7}$$

The *auto-correlation coefficient function* $\rho_{XX}(t_1, t_2)$ specifies the correlation coefficient of $X(t_1)$ and $X(t_2)$, cf. Eq. (2-60):

$$\rho_{XX}(t_1, t_2) = \frac{\kappa_{XX}(t_1, t_2)}{\sigma_X(t_1)\,\sigma_X(t_2)} \tag{3-8}$$

Example 3-1: Rectangular beam with stochastic height and width

Fig. 3-2 shows a collection of beams all of the length L with rectangular cross-sections, where the height and width are related with uncertainties. At a distance x from one of the beam ends, the height and the width are modelled as stochastic variables $H(x)$ and $W(x)$, respectively. The set of stochastic variables from all beam sections form the stochastic processes $\{H(x),\ x \in [0, L]\}$ and $\{W(x),\ x \in [0, L]\}$. The sample space Ω is formed from all produced rectangular beams of the length L.

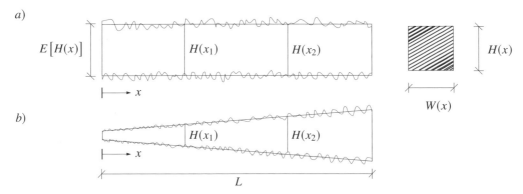

Fig. 3–2 Rectangular beam with stochastic height and width. a) Realization of the homogeneous height process. b) Realization of the non-homogeneous height process.

$\{X(t),\ t \in T\}$ is called a *Gaussian process*, if the n-dimensional stochastic vector $\mathbf{X} = [X(t_1), \ldots, X(t_n)]^T$ is normally distributed for any order n and arbitrary index parameters $t_1, \ldots, t_n \in T$. The mean value vector $\boldsymbol{\mu}_\mathbf{X}$ and the covariance matrix $\mathbf{C_{XX}}$ of $\mathbf{X}$ can be specified from the mean value function $\mu_X(t)$ and the auto-covariance function $\kappa_{XX}(t_1, t_2)$ of the process. Hence, a Gaussian process is completely described by its mean value function and auto-covariance function.

Example 3-2: Harmonic process

Let the Rayleigh-distributed stochastic variable $R \sim R(\sigma^2)$ and the *uniformly distributed* stochastic variable $\Phi \sim U(0, 2\pi)$ be stochastically independent. A *harmonic process* $\{X(t),\ t \in R\}$ is defined in the following form:

$$X(t) = R\cos(\omega_0 t + \Phi) \tag{3-9}$$

where ω_0 is a deterministic constant. Obviously, the randomness of the process is generated from merely two stochastic variables R and Φ. It follows from Example 2-7 that $X(t)$ is normally distributed for all index parameter values, i.e., $X(t) \sim N(0, \sigma_X^2)$. The mean value function and the auto-covariance function of this process become:

$$\mu_X(t) = E[X(t)] = 0 \tag{3-10}$$

$$
\begin{aligned}
\kappa_{XX}(t_1, t_2) &= E\left[R^2 \cos(\omega_0 t_1 + \Phi)\cos(\omega_0 t_2 + \Phi)\right] \\
&= E\left[R^2\right] E\left[\tfrac{1}{2}\cos\left(\omega_0(t_2 - t_1)\right) + \tfrac{1}{2}\cos\left(\omega_0(t_1 + t_2) + 2\Phi\right)\right] = \sigma^2 \cos\left(\omega_0(t_2 - t_1)\right)
\end{aligned}
\tag{3-11}
$$

where it has been used that:

$$E\left[\cos(\alpha + 2\Phi)\right] = \int_0^{2\pi} \cos(\alpha + 2x)\frac{1}{2\pi}\,dx = 0 \tag{3-12}$$

Setting $t_1 = t_2$ in Eq. (3-11) provides $\sigma_X^2 = \sigma^2$.

Even though $X(t)$ is normally distributed for all t, $\{X(t), t \in R\}$ is not a Gaussian process. Actually, none of the higher order joint probability distribution functions are normal.

A *stochastic vector process* is a set of N-dimensional stochastic vectors $\{\mathbf{X}(t), t \in T\}$, $\mathbf{X}(t) = [X_1(t), \ldots, X_N(t)]^T$, where all the component stochastic variables are defined on the same sample space Ω. A vector process can be considered a double indexed set of stochastic variables, where the elements of T as well as the component indices form the index parameters. In the same manner as for scalar processes, the stochastic structure of a vector process is specified by a sequence of joint probability distribution functions:

$$\left.\begin{aligned}
&F_{X_{i_1}(t_1)}(x_1) \\
&F_{X_{i_1}(t_1)X_{i_2}(t_2)}(x_1, x_2) \\
&\quad\vdots \\
&F_{X_{i_1}(t_1)\cdots X_{i_n}(t_n)}(x_1, \ldots, x_n) \\
&\quad\vdots
\end{aligned}\right\} \tag{3-13}$$

The joint probability distribution functions of the nth order depend on the considered stochastic variables $X_{i_1}(t_1), \ldots, X_{i_n}(t_n)$. Additional to the state variables $x_1, \ldots, x_n$, these functions depend on the index parameters $t_1, \ldots, t_n$ and on the coordinate indices $i_1, \ldots, i_n$. The following notation will be applied:

$$F_{\mathbf{X}}(x_1, t_1, i_1 ; \ldots ; x_n, t_n, i_n) = F_{X_{i_1}(t_1)\cdots X_{i_n}(t_n)}(x_1, \ldots, x_n) \tag{3-14}$$

Example 3-3: Depth and width of a rectangular beam organized as a stochastic vector process

The stochastic height $H(x)$ and width $W(x)$ at the position x are organized into a 2-dimensional stochastic vector $\mathbf{X}(x) = [H(x), W(x)]^T$. Hence, $\{\mathbf{X}(x), x \in [0, L]\}$ becomes a 2-dimensional vector process defined on the sample space Ω that is formed by all rectangular beams of the length L.

For vector processes, the prefix "cross" is usually applied to characterize the statistical moments when stochastic variables from different coordinate processes are considered. The prefix "auto" is reserved to cases where statistical moments for the same coordinate process $\{X_j(t), \in T\}$ are considered.

The *cross-covariance function* $\kappa_{X_{i_1} X_{i_2}}(t_1, t_2)$ specifies the covariance between the two stochastic variables $X_{i_1}(t_1)$ and $X_{i_2}(t_2)$, i.e.:

$$\kappa_{X_{i_1} X_{i_2}}(t_1, t_2) = E\left[\left(X_{i_1}(t_1) - \mu_{X_{i_1}}(t_1)\right)\left(X_{i_2}(t_2) - \mu_{X_{i_2}}(t_2)\right)\right]$$

$$= E\left[X_{i_1}(t_1) X_{i_2}(t_2)\right] - \mu_{X_{i_1}}(t_1)\mu_{X_{i_2}}(t_2) \tag{3-15}$$

Similar to Eq. (3-6), the following symmetry property applies:

$$\kappa_{X_{i_1} X_{i_2}}(t_1, t_2) = \kappa_{X_{i_2} X_{i_1}}(t_2, t_1) \tag{3-16}$$

3.2   Homogeneous processes

A stochastic process $\{X(t), t \in T\}$ is said to be *homogeneous* in the strict sense (or *strictly homogeneous*) if the set of finite dimensional joint probability distribution functions of the process is invariant under a linear translation of the index parameters, $t \rightarrow t + a$ for every $a \in T$. This means:

$$\left.\begin{array}{rl} F_X(x_1, t_1) & = F_X(x_1, t_1 + a) \\ F_X(x_1, t_1 ; x_2, t_2) & = F_X(x_1, t_1 + a; x_2, t_2 + a) \\ \vdots & \\ F_X(x_1, t_1 ; \ldots ; x_n, t_n) & = F_X(x_1, t_1 + a; \ldots ; x_n, t_n + a) \\ \vdots & \end{array}\right\} \tag{3-17}$$

If Eq. (3-17) only holds for $n = 1$ and $n = 2$, the process is termed homogeneous in the weak sense (or *weakly homogeneous*).

Especially, if $a = -t$ and $\{X(t), t \in R\}$ is weakly homogeneous, it is seen from Eq. (3-17) that joint probability distribution functions of the 1st order become independent of t, i.e., $F_X(x, t) = F_X(x)$. For $a = -t_1$ the joint probability distribution functions of the 2nd order will only depend on the index parameters t_1 and t_2 through the difference $t_2 - t_1$, i.e., $F_X(x_1, t_1 ; x_2, t_2) = F_X(x_1, x_2 ; t_2 - t_1)$. For a strictly homogeneous process, the higher order joint probability distribution functions depend on $t_1, \ldots, t_n$ through the differences $t_2 - t_1, t_3 - t_1, \ldots, t_n - t_1$, i.e., $F_X(x_1, t_1 ; x_2, t_2 ; \ldots ; x_n, t_n) = F_X(x_1, x_2, \ldots, x_n ; t_2 - t_1, \ldots, t_n - t_1)$.

Since $F_X(x)$ is independent of t, for a weakly homogeneous process, the mean value function and the variance function become independent of t as well, i.e., $\mu_X(t) = \mu_X$ and $\sigma_X^2(t) = \sigma_X^2$. As

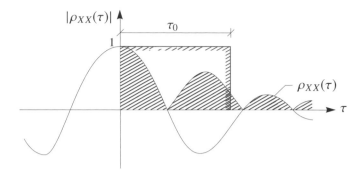

Fig. 3–3 Definition of the correlation length of a weakly homogeneous stochastic process.

$F_X(x_1, t_1 ; x_2, t_2)$ is merely a function of $(t_2 - t_1)$, it follows from Eq. (3-6) that the auto-covariance function of a weakly homogeneous process is an even function of the index difference:

$$\kappa_{XX}(\tau) = \kappa_{XX}(-\tau), \quad \tau = t_2 - t_1 \tag{3–18}$$

If the index parameter t denotes time, the terms strictly and weakly *stationary processes* are used synonymously for strictly and weakly homogeneous processes.

The realization of the beam height process shown in Fig. 3-2a may be considered a part of a weakly homogeneous process. The mean value function $\mu_X = E[H(x)]$ and the standard deviation σ_X measuring the magnitude of the irregularities seem to be constant along the beam. Further, the intervals between zero up-crossings of the mean value level are not changing systematically. In contrast, the height of the beam in Fig. 3-2b should be modelled as an inhomogeneous process. μ_X, σ_X and the intervals between zero up-crossings of μ_X all seem to increase with x.

For weakly homogeneous processes, the auto-correlation coefficient function becomes, cf. Eq. (3-8):

$$\rho_{XX}(\tau) = \frac{\kappa_{XX}(\tau)}{\sigma_X^2} \tag{3–19}$$

Based on the auto-correlation coefficient function, the *correlation length* is defined as:

$$\tau_0 = \int_0^\infty |\rho_{XX}(\tau)| \, d\tau \tag{3–20}$$

Eq. (3-20) has been illustrated in Fig. 3-3. τ_0 indicates the separation in the index parameters for which the stochastic variables become effectively uncorrelated.

Eq. (3-17) states that the joint probability distribution functions of the n-dimensional stochastic vectors $[X(t_1), \ldots, X(t_n)]^T$ and $[X(t_1 + a), \ldots, X(t_n + a)]^T$ are alike for any order n, any index values $t_1, \ldots, t_n \in T$ and any $a \in T$. Eq. (3-17) only makes sense if T is un-bounded, i.e., $T = R^n$. Actually, if T is bounded and t is placed on the boundary, then $t + a \notin T$, in which case Eq. (3-17) becomes meaningless. Since all natural processes are limited in time and space, the concept of homogeneity (stationarity) should be considered as an approximate mathematical modelling assumption. The concept may be applied to dynamic wave, wind and traffic load processes where the change of the probabilistic structure of the loads take place with a timescale much larger than the fundamental eigenperiod of the structure. This is not the case for earthquakes, which consequently must be modelled as a non-stationary load process.

The previous definitions can immediately be extended to N-dimensional stochastic vector processes $\{\mathbf{X}(t), t \in T\}$, $\mathbf{X}(t) = [X_1(t), \ldots, X_N(t)]^T$. The vector process is said to be strictly homogeneous, if for any order n, any combination of index parameter $(t_1, i_1), \ldots, (t_n, i_n) \in T \times \{1, \ldots, N\}$ and any $a \in T$:

$$
\left.
\begin{aligned}
F_{\mathbf{X}}(x_1, t_1, i_1) &= F_{\mathbf{X}}(x_1, t_1 + a, i_1) \\
F_{\mathbf{X}}(x_1, t_1, i_1 ; x_2, t_2, i_2) &= F_{\mathbf{X}}(x_1, t_1 + a, i_1 ; x_2, t_2 + a, i_2) \\
&\vdots \\
F_{\mathbf{X}}(x_1, t_1, i_1 ; \ldots ; x_n, t_n, i_n) &= F_{\mathbf{X}}(x_1, t_1 + a, i_1 ; \ldots ; x_n, t_n + a, i_n) \\
&\vdots
\end{aligned}
\right\}
\tag{3–21}
$$

Vector processes are called weakly homogeneous, if Eq. (3-21) only holds for $n = 1$ and $n = 2$. If $\{\mathbf{X}(t), t \in T\}$ is homogeneous, it implies that all coordinate processes are homogeneous. The opposite statement is not necessarily true.

For a *weakly homogeneous vector process*, the mean value functions become constant as a function of t, i.e., $\mu_{X_i}(t) = \mu_{X_i}$. The cross-covariance functions become a function of the index difference $t_2 - t_1$, i.e., $\kappa_{X_{i_1} X_{i_2}}(t_1, t_2) = \kappa_{X_{i_1} X_{i_2}}(t_2 - t_1)$. Then, the symmetry property in Eq. (3-16) attains the form:

$$
\kappa_{X_{i_1} X_{i_2}}(\tau) = \kappa_{X_{i_2} X_{i_1}}(-\tau), \quad \tau = t_2 - t_1
\tag{3–22}
$$

Assuming $|\kappa_{X_{i_1} X_{i_2}}(\tau)|$ to be integrable over the interval $]-\infty, \infty[$, the following *Fourier transforms* may be defined:

$$
\left.
\begin{aligned}
\kappa_{X_{i_1} X_{i_2}}(\tau) &= \int_{-\infty}^{\infty} e^{i\omega\tau} S_{X_{i_1} X_{i_2}}(\omega)\, d\omega \\
S_{X_{i_1} X_{i_2}}(\omega) &= \frac{1}{2\pi} \int_{-\infty}^{\infty} e^{-i\omega\tau} \kappa_{X_{i_1} X_{i_2}}(\tau)\, d\tau
\end{aligned}
\right\}
\tag{3–23}
$$

Eq. (3-23), which establishes a relation between $\kappa_{X_{i_1} X_{i_2}}(\tau)$ and its Fourier transform $S_{X_{i_1} X_{i_2}}(\omega)$, represents the so-called *Wiener-Khintchine relation*. $S_{X_{i_1} X_{i_2}}(\omega)$ is designated the *cross-spectral density function* (*auto-spectral density function* when $i_1 = i_2$). $\kappa_{X_{i_1} X_{i_2}}(\tau)$ is always real, whereas $S_{X_{i_1} X_{i_2}}(\omega)$ is generally complex. We shall later give a physical interpretation of $S_{X_{i_1} X_{i_2}}(\omega)$. At present, it should merely be considered as the Fourier transform of $\kappa_{X_{i_1} X_{i_2}}(\tau)$, and hence an alternative way of representing the correlation structure of the stochastic process.

Substitution of $u = -\tau$ into Eq. (3-23) and use of Eq. (3-22) provides:

$$S_{X_{i_1} X_{i_2}}(\omega) = \frac{1}{2\pi} \int_{-\infty}^{\infty} e^{i\omega u} \, \kappa_{X_{i_1} X_{i_2}}(-u) \, du = \frac{1}{2\pi} \int_{-\infty}^{\infty} e^{i\omega u} \, \kappa_{X_{i_2} X_{i_1}}(u) \, du$$

$$= \left(\frac{1}{2\pi} \int_{-\infty}^{\infty} e^{-i\omega u} \, \kappa_{X_{i_2} X_{i_1}}(u) \, du \right)^* = S_{X_{i_2} X_{i_1}}^*(\omega) \tag{3–24}$$

where $S_{X_{i_2} X_{i_1}}^*(\omega)$ denotes the *complex conjugate* of $S_{X_{i_2} X_{i_1}}(\omega)$. Further, from Eq. (3-23) it follows that:

$$S_{X_{i_1} X_{i_2}}(\omega) = \frac{1}{2\pi} \left(\int_{-\infty}^{\infty} e^{i\omega\tau} \, \kappa_{X_{i_1} X_{i_2}}(\tau) \, d\tau \right)^*$$

$$= \frac{1}{2\pi} \left(\int_{-\infty}^{\infty} e^{-i(-\omega)\tau} \, \kappa_{X_{i_1} X_{i_2}}(\tau) \, d\tau \right)^* = S_{X_{i_1} X_{i_2}}^*(-\omega) \tag{3–25}$$

The corresponding results for the auto-spectral density function is obtained by setting $i_1 = i_2 = i$ in Eqs. (3-24) and (3-25), i.e.:

$$S_{X_i X_i}(\omega) = S_{X_i X_i}^*(\omega) \tag{3–26}$$

$$S_{X_i X_i}(\omega) = S_{X_i X_i}^*(-\omega) = S_{X_i X_i}(-\omega) \tag{3–27}$$

From Eqs. (3-26) and (3-27), it is seen that the auto-spectral density function is a real and symmetric function of ω.

Since the auto-covariance function $\kappa_{X_i X_i}(\tau)$ and the auto-spectral density function $S_{X_i X_i}(\omega)$ are symmetric functions, the Wiener-Khintchine relation for these quantities can be recast as the following cosine transforms:

$$\left. \begin{aligned} \kappa_{X_i X_i}(\tau) &= 2 \int_0^{\infty} \cos(\omega\tau) \, S_{X_i X_i}(\omega) \, d\omega \\ S_{X_i X_i}(\omega) &= \frac{1}{\pi} \int_0^{\infty} \cos(\omega\tau) \, \kappa_{X_i X_i}(\tau) \, d\tau \end{aligned} \right\} \tag{3–28}$$

The variance function $\sigma_{X_i}^2 = \kappa_{X_i X_i}(0)$ is obtained by setting $\tau = 0$ in Eq (3-23) or Eq (3-28):

$$\sigma_{X_i}^2 = \int_{-\infty}^{\infty} S_{X_i X_i}(\omega) \, d\omega = 2 \int_0^{\infty} S_{X_i X_i}(\omega) \, d\omega \tag{3–29}$$

Positive values of ω may be interpreted as an angular frequency if τ represents time, whereas ω is the *wave number* if τ represents a space variable. Although ω will be referred to as the "angular frequency" in the following, this distinction should be kept in mind.

Eq. (3-29) shows that the auto-spectral density function can be interpreted as the distribution of the variance $\sigma_{X_i}^2$ in the frequency space. Therefore, $2S_{X_i X_i}(\omega)d\omega$ represents the variance contribution from harmonic components in the interval $[\omega, \omega + d\omega[$. Since these variance

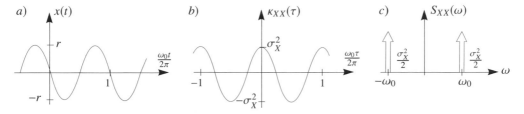

Fig. 3–4 Harmonic process. a) Realization. b) Auto-covariance function. c) Auto-spectral density function.

contributions need to be nonnegative, the auto-spectral density function must be nonnegative as well, i.e.:

$$S_{X_i X_i}(\omega) \geq 0 \tag{3–30}$$

Example 3-4: Auto-spectral density function of a harmonic process

The harmonic process considered in Example 3-2 can be shown to be homogeneous in the strict sense. The auto-spectral density function of the harmonic process follows from Eqs. (3-11) and (3-23):

$$S_{XX}(\omega) = \frac{1}{2\pi} \int_{-\infty}^{\infty} e^{-i\omega\tau} \, \sigma^2 \cos(\omega_0 \tau) \, d\tau = \frac{\sigma^2}{2\pi} \int_{-\infty}^{\infty} e^{-i\omega\tau} \tfrac{1}{2} \left(e^{i\omega_0\tau} + e^{-i\omega_0\tau} \right) d\tau$$

$$= \frac{\sigma^2}{4\pi} \int_{-\infty}^{\infty} \left(e^{-i(\omega-\omega_0)\tau} + e^{-i(\omega+\omega_0)\tau} \right) d\tau = \frac{\sigma^2}{2} \Big(\delta(\omega - \omega_0) + \delta(\omega + \omega_0) \Big) \tag{3–31}$$

where $\delta(\cdot)$ indicates the *Dirach delta function*. In the last statement of Eq. (3-31), the following formally mutual Fourier transforms of $f(\tau) = 1$ and $\delta(\omega)$ have been applied:

$$\left. \begin{aligned} 1 &= \int_{-\infty}^{\infty} \delta(\omega) \, e^{i\omega\tau} \, d\omega \\ \delta(\omega) &= \frac{1}{2\pi} \int_{-\infty}^{\infty} 1 \cdot e^{-i\omega\tau} \, d\tau \end{aligned} \right\} \tag{3–32}$$

A realization of the harmonic process with the amplitude $R = r$ is shown in Fig. 3-4, along with the auto-covariance function as given by Eq. (3-11), and the auto-spectral density function as given by Eq. (3-31). It should be noted that the correlation length of the harmonic process is $\tau_0 = \infty$, cf. Eq. (3-20).

Example 3-5: Representation of a zero-mean stationary Gaussian process with a continuous auto-spectral density function by a process made up of a sum of harmonic processes

A process $\{\tilde{X}(t), \, t \in R\}$ is defined as a sum of harmonic processes:

$$\tilde{X}(t) = \sum_{j=1}^{J} \tilde{X}_j(t) \tag{3–33}$$

where $\{\tilde{X}_j(t), \, t \in R\}$ are stochastically independent harmonic processes given by:

$$\tilde{X}_j(t) = R_j \cos(\omega_j t + \Phi_j), \quad j = 1, \ldots, J \tag{3–34}$$

R_j are Rayleigh-distributed random variables, $R_j \sim R(\sigma_j)$. Φ_j are uniformly distributed random variables, $\Phi_j \sim U(0, 2\pi)$. ω_j are deterministic constants. Further, R_j and Φ_j $(j = 1, \ldots, J)$ are stochastically independent.

It follows from Example 2-6 that $\tilde{X}_j(t) \sim N(0, \sigma_j^2)$. Hence, the mean value function of $\{\tilde{X}(t), t \in R\}$ becomes:

$$\mu_{\tilde{X}}(t) = \sum_{j=1}^{J} E\left[\tilde{X}_j(t)\right] = 0 \tag{3–35}$$

The auto-covariance function becomes, cf. Eq. (3-11):

$$
\begin{aligned}
\kappa_{\tilde{X}\tilde{X}}(t_1, t_2) &= \sum_{j=1}^{J} \sum_{k=1}^{J} E\left[\tilde{X}_j(t_1)\tilde{X}_k(t_2)\right] = \sum_{j=1}^{J} E\left[\tilde{X}_j(t_1)\tilde{X}_j(t_2)\right] \\
&= \sum_{j=1}^{J} \sigma_j^2 \cos\left(\omega_j(t_2 - t_1)\right) = \kappa_{\tilde{X}\tilde{X}}(\tau), \quad \tau = t_2 - t_1
\end{aligned}
\tag{3–36}
$$

where it has been used that $E\left[\tilde{X}_j(t_1)\tilde{X}_k(t_2)\right] = E\left[\tilde{X}_j(t_1)\right]E\left[\tilde{X}_k(t_2)\right] = 0$ for $j \neq k$.

The auto-spectral density function follows from Eq. (3-31):

$$S_{\tilde{X}\tilde{X}}(\omega) = \sum_{j=1}^{J} \frac{\sigma_j^2}{2} \left(\delta(\omega - \omega_j) + \delta(\omega + \omega_j)\right) \tag{3–37}$$

A sum of normal stochastic variables becomes normal itself, implying that $\tilde{X}(t) \sim N\left(0, \sum_{j=1}^{J} \sigma_j^2\right)$. However, higher order probability distribution functions of $\tilde{X}(t)$ are not normal, so $\{\tilde{X}(t), t \in R\}$ is not a Gaussian process. Nevertheless, based on the *central limit theorem* in probability theory and its multi-dimensional generalization, it can be argued that $\{\tilde{X}(t), t \in R\}$ converges to a zero-mean Gaussian (normal) process as $J \to \infty$ under rather mild conditions, merely specifying that the angular frequencies $\omega_1, \ldots \omega_J$ should be selected on the frequency axis, so the variances σ_j^2 become of comparable magnitudes.

Fig. 3-5 shows the procedure of representing a zero-mean stationary Gaussian process $\{X(t), t \in R\}$ with a continuous auto-spectral density function $S_{XX}(\omega)$, by an equivalent stochastic process $\{\tilde{X}(t), t \in R\}$ made up of a sum of harmonic processes. The continuous auto-spectral density function $S_{XX}(\omega)$ is illustrated in Fig. 3-5a, and the auto-spectral density function of the equivalent process as given by Eq. (3-37) is illustrated in Fig. 3-5b. The calibration of the harmonic component processes $\{\tilde{X}_j(t), t \in R\}$ is as follows. Let $\omega'_{j-1} = \frac{1}{2}(\omega_{j-1} + \omega_j)$ and $\omega'_j = \frac{1}{2}(\omega_j + \omega_{j+1})$ denote the angular frequencies at the mid-points of the frequency intervals adjacent to ω_j. Then, the variance σ_j^2 attributed to $\{\tilde{X}_j(t), t \in R\}$ is calculated as:

$$\frac{1}{2}\sigma_j^2 = \int_{\omega'_{j-1}}^{\omega'_j} S_{XX}(\omega)\, d\omega \quad \Rightarrow$$

$$\sigma_j^2 = 2 \int_{\omega'_{j-1}}^{\omega'_j} S_{XX}(\omega)\, d\omega, \quad j = 1, 2, \ldots, J \tag{3–38}$$

Eq. (3-38) indicates that the variance σ_j^2 of the harmonic process $\{\tilde{X}_j(t), t \in R\}$ is obtained by calculating the red hatched areas in Fig. 3-5a. With the calibrated variances σ_j^2, the equivalent process $\{\tilde{X}(t), t \in R\}$

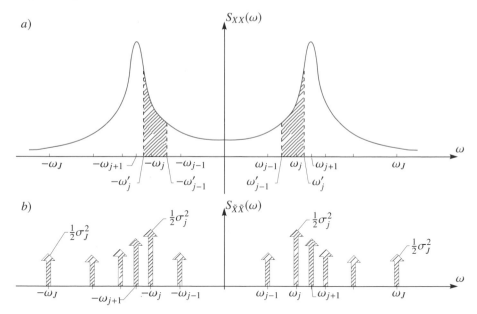

Fig. 3–5 a) Auto-spectral density function of a given stationary Gaussian process. b) Auto-spectral density function of the equivalent process made up of a sum of harmonic processes.

as in Eqs. (3-33) and (3-34) may be used to generate realizations of the underlying stochastic process $\{X(t), t \in R\}$.

A homogeneous (stationary) Gaussian process $\{F(t), t \in R\}$ with mean value function $\mu_F = 0$ is termed a *white noise*, if its auto-spectral density function is constant for all angular frequencies, i.e.:

$$S_{FF}(\omega) = S_0, \quad \omega \in R \tag{3–39}$$

Based on Eqs. (3-23) and (3-32), the auto-covariance function of a white noise process can be evaluated as:

$$\kappa_{FF}(\tau) = 2\pi S_0 \delta(\tau) \tag{3–40}$$

Especially, if $S_0 = \frac{1}{2\pi} \Leftrightarrow \kappa_{FF}(\tau) = \delta(\tau)$, the process is named a *unit white noise*. The auto-covariance function and the auto-spectral density function of a white noise process are shown in Fig. 3-6.

From Eq. (3-40), it follows that $F(t)$ and $F(t+\tau)$ for a white noise are stochastically independent no matter how small a value of $|\tau|$ is considered. Hence, the correlation length $\tau_0 = 0$, and the realizations are discontinuous at any index value t.

Insertion of Eq. (3-39) into Eq. (3-29) provides $\sigma_F^2 = \infty$. Assuming $a < b$, we have:

$$P(a < F(t) \le b) = \lim_{\sigma_F \to \infty} \left(\Phi\left(\frac{b}{\sigma_F}\right) - \Phi\left(\frac{a}{\sigma_F}\right) \right) = 0.5 - 0.5 = 0 \tag{3–41}$$

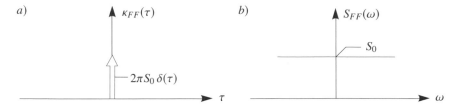

Fig. 3–6 White noise process. a) Auto-covariance function. b) Auto-spectral density function.

where $\Phi(\cdot)$ is the probability distribution function of a standardized normal stochastic variable, cf. Eq. (2-24).

In combination, Eqs. (3-40) and (3-41) imply that the realizations of a white noise process are discontinuous for all $t \in T$, and the sample values will be beyond of any finite interval with probability 1. Clearly, such a process is a mathematical abstraction. It is not even possible to sketch typical realizations of the process. The designation "white noise" originates from optics, where the auto-spectral density function of sunlight is almost constant over the visible frequency range of the electromagnetic spectrum.

The harmonic process dealt with in Example 3-5 and the white noise process can be considered as two limiting extreme stochastic processes, where the variance is concentrated at a single angular frequency ω_0 and is uniformly distributed over all frequencies, respectively. Both cases are idealizations, which are only approximately met in practice.

A *narrow-banded stochastic process* has a continuous distribution of the variance over a range of frequencies, but there will be a dominating contribution from a narrow interval close to $\pm\, \omega_0$ as shown in Fig. 3-7c. In this respect, a harmonic process may be considered a limiting case of a narrow-banded stochastic process.

The realization $x(t)$ of a narrow-banded process, as shown in Fig. 3-7a, is approximately harmonic with the angular frequency ω_0, although the amplitude is slowly varying in a random manner. Further, it is observed that each out-crossing of the t-axis by $x(t)$ is succeeded by exactly one local maximum or local minimum. The axis has been normalized with respect to $T_0 = \frac{2\pi}{\omega_0}$, which indicates the interval between succeeding up- or down-crossings of the t- axis.

The auto-covariance function shown in Fig. 3-7b appears like a cosine function with angular frequency ω_0 and with monotonously decreasing amplitudes. As seen, this decrease takes place over many periods T_0. Consequently, the correlation length τ_0, indicating the value of τ after which the local maxima of $\kappa_{XX}(\tau)$ have diminished to insignificant magnitudes, fulfills $\tau_0 \gg T_0$.

The dynamic response of a civil engineering structure dominated by a single vibration mode may typically be modelled as a narrow-banded process, due to the low structural damping. However, additional strong aerodynamic damping may be present in some cases, making the response broad-banded. One example could be the flap-wise vibrations of a wind turbine blade.

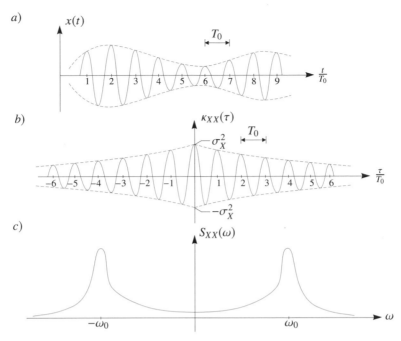

Fig. 3–7 Narrow-banded process. a) Realization. b) Auto-covariance function. c) Auto-spectral density function.

In equally vague terms, a *broad-banded stochastic process* is characterized as a process where the variance contributions are distributed in a way that no single dominating angular frequency can be identified in the auto-spectral density function. In this respect, a broad-banded process has some resemblance to white noise. However, the crucial difference is that any physical

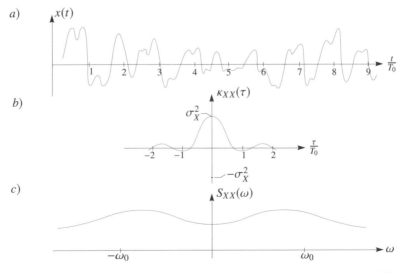

Fig. 3–8 Broad-banded process. a) Realization. b) Auto-covariance function. c) Auto-spectral density function.

broad-banded process will have finite variance, as well as continuous (and even differentiable) realizations.

Fig. 3-8a shows a typical realization $x(t)$ of a broad-banded process. The realization is rather irregular, and each crossing of the t-axis is often succeeded by more than one local maximum or local minimum, contrary to the narrow-banded case. Again, the t-axis has been normalized with respect to T_0, now indicating the averaged interval between succeeding up- or down-crossings of the t-axis.

Fig. 3-8b shows the auto-covariance function with the index separation variable τ normalized with respect to T_0. As seen, significant correlations merely take place for $|\tau| \lesssim T_0$. Hence, a broad-banded process is characterized with a correlation length fulfilling $\tau_0 \simeq T_0$.

Finally, Fig. 3-8c shows the auto-spectral density function of a broad-banded process.

The dynamic response of a civil engineering structure dominated by two or more vibration modes should be characterized as broad-banded even in the case of low structural damping. Dynamic loadings from wind, wave and traffic are typically broad-banded phenomena. However, vortex induced loadings on chimneys and towers should be modelled as narrow-banded load processes.

Negative angular frequencies cannot be interpreted physically. Due to the symmetry of $S_{XX}(\omega)$, we occasionally prefer to operate with the so-called *one-sided auto-spectral density function*, defined as:

$$S_X(\omega) = 2S_{XX}(\omega), \quad \omega \in [0, \infty[\tag{3–42}$$

Only one index X is applied for the one-sided auto-spectral density function. Occasionally, $S_{XX}(\omega)$ will be referred to as the *double-sided auto-spectral density function* (Fig. 3-9).

Expressed in the one-sided auto-spectral density function, the Wiener-Khintchine relations in Eq. (3-28) attain the form:

$$\left.\begin{aligned}
\kappa_{XX}(\tau) &= \int_0^\infty \cos(\omega\tau)\, S_X(\omega)\, d\omega \\
S_X(\omega) &= \frac{2}{\pi} \int_0^\infty \cos(\omega\tau)\, \kappa_{XX}(\tau)\, d\tau
\end{aligned}\right\} \tag{3–43}$$

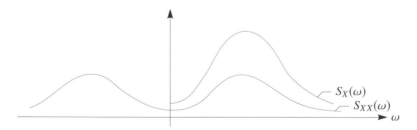

Fig. 3–9 Definition of one-sided auto-spectral density function.

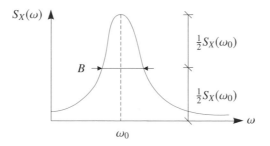

Fig. 3–10 Definition of the half bandwidth.

The variance becomes, cf. Eq. (3-29):

$$\sigma_X^2 = \int_0^\infty S_X(\omega)\,d\omega \tag{3–44}$$

As a measure of the narrow-bandedness, the *half bandwidth B* may be introduced, as defined in Fig. 3-10. Based on B, a non-dimensional *bandwidth parameter* ζ can be introduced:

$$\zeta = \frac{B}{2\omega_0} \tag{3–45}$$

If $\{X(t),\, t \in R\}$ is the response process of a slightly damped single-degree-of-freedom oscillator subjected to white noise excitation, ζ can be identified as the *damping ratio* of the oscillator. The use of the half bandwidth as a quantitative measure for narrow-bandedness only makes sense for $\zeta < 1$.

Let $\nu_0 = \frac{1}{T_0}$ denote the *expected number of up-crossings per unit time* of the t-axis (per unit length if t is a spatial parameter), and let μ_0 be the *expected number of local maxima per unit time*. Another bandwidth parameter γ can be defined as:

$$\gamma = \frac{\mu_0}{\nu_0} \geq 1 \tag{3–46}$$

where $\gamma > 1$ for a broad-banded process, $\gamma \gtrsim 1$ for a narrow-banded process and $\gamma = 1$ for a harmonic process. Based on γ, the following bandwidth parameter ε is introduced:

$$\varepsilon = \sqrt{1 - \frac{1}{\gamma^2}} \tag{3–47}$$

$\gamma = 1$ and $\gamma = \infty$, which represent the extremely narrow-banded and extremely broad-banded cases, are mapped into $\varepsilon = 0$ and $\varepsilon = 1$, respectively.

Based on the one-sided auto-spectral density function, the *spectral moments* are defined in the following way:

$$\lambda_j = \int_0^\infty \omega^j\, S_X(\omega)\, d\omega, \quad j = 0, 1, 2, \ldots \tag{3–48}$$

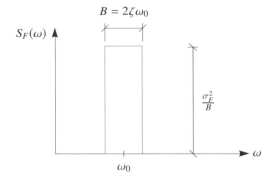

Fig. 3–11 One-sided auto-spectral density function of a band-limited white noise.

The fraction $S_X(\omega)/\lambda_0$ has the main characteristics of a probability density function (non-negativity and area=1). The "expected value" of this probability density function is $\frac{\lambda_1}{\lambda_0}$, and the "variance" is $\frac{\lambda_2}{\lambda_0} - (\frac{\lambda_1}{\lambda_0})^2$. Hence, the "variational coefficient" becomes, cf. Eq. (2-52):

$$\delta = \frac{(\lambda_0\lambda_2 - \lambda_1^2)^{\frac{1}{2}}}{\lambda_1} = \sqrt{\frac{\lambda_0\lambda_2}{\lambda_1^2} - 1} \tag{3–49}$$

δ was suggested as a bandwidth parameter by Vanmarcke.[1] It is left as an exercise to prove that $\lambda_0\lambda_2 \geq \lambda_1^2$. For a harmonic process, $\lambda_1 = \omega_0\lambda_0$ and $\lambda_2 = \omega_0^2\lambda_0$, implying $\delta = 0$. Therefore, small values of δ indicate narrow-bandedness.

Example 3-6: Band-limited white noise

A weakly homogeneous stochastic process is called a *band-limited white noise*, if the one-sided auto-spectral density function (as shown in Fig. 3-11) is written as:

$$S_F(\omega) = \begin{cases} \dfrac{\sigma_F^2}{B}, & \omega \in \left[\omega_0 - \dfrac{B}{2}, \omega_0 + \dfrac{B}{2}\right] \\[3mm] 0, & \omega \notin \left[\omega_0 - \dfrac{B}{2}, \omega_0 + \dfrac{B}{2}\right] \end{cases} \tag{3–50}$$

where ω_0 is termed the angular centre frequency, and B is the bandwidth of the band-limited white noise. These parameters are assumed to fulfil $B \leq 2\omega_0$.

The auto-covariance function follows from Eq. (3-43):

$$\kappa_{FF}(\tau) = \int\limits_{\omega_0 - \frac{B}{2}}^{\omega_0 + \frac{B}{2}} \cos(\omega\tau) \frac{\sigma_F^2}{B} d\omega = \frac{\sigma_F^2}{B\tau}\left(\sin\left(\left(\omega_0 + \frac{B}{2}\right)\tau\right) - \sin\left(\left(\omega_0 - \frac{B}{2}\right)\tau\right)\right) \tag{3–51}$$

[1]E. Vanmarcke. *Random fields: Analysis and synthesis*. The MIT Press, Cambridge, Massachusetts, 1983.

The spectral moments are obtained from Eq. (3-48):

$$\lambda_j = \frac{\sigma_F^2}{B} \int\limits_{\omega_0 - \frac{B}{2}}^{\omega_0 + \frac{B}{2}} \omega^j \, d\omega = \frac{\sigma_F^2}{(j+1)B} \left(\left(\omega_0 + \frac{B}{2} \right)^{j+1} - \left(\omega_0 - \frac{B}{2} \right)^{j+1} \right) \tag{3-52}$$

Vanmarcke's bandwidth parameter is obtained by inserting Eq. (3-52) into Eq. (3-49). Elimination of B in favour of the non-dimensional bandwidth parameter ζ provides:

$$\delta = \sqrt{\frac{1}{3}} \zeta \tag{3-53}$$

Assuming $H(z)$ to be a *rational function*, it needs to have the following form:

$$H(z) = \frac{P(z)}{Q(z)} \tag{3-54}$$

where the argument z may be complex, and:

$$P(z) = p_0 z^r + p_1 z^{r-1} + \cdots + p_{r-1} z + p_r \tag{3-55}$$

$$Q(z) = z^s + q_1 z^{s-1} + \cdots + q_{s-1} z + q_s = (z - z_1) \cdots (z - z_s) = \prod_{j=1}^{s} (z - z_j) \tag{3-56}$$

z_j indicate the roots to the polynomial equation $Q(z) = 0$, which are refer to as poles. We shall assume that $s > r$, i.e., the denominator polynomial is of higher degree than the numerator polynomial. Further, the parameters $p_0, \ldots, p_r$ and $q_1, \ldots, q_s$ are assumed to be real. Finally, it is assumed that the *real parts* of all poles are negative, i.e.:

$$\text{Re}(z_j) < 0, \qquad j = 1, \ldots, s \tag{3-57}$$

A weakly homogeneous stochastic process $\{X(t), t \in R\}$ is said to have *rational auto-spectral density function*, if $S_{XX}(\omega)$ can be written in the following form:

$$S_{XX}(\omega) = H(z)H(-z) S_0 = H(z)H^*(z) S_0 = |H(z)|^2 S_0, \quad z = i\omega \tag{3-58}$$

where S_0 is a positive real constant. The last statement of Eq. (3-58) follows, because for $z = i\omega$ we have $P^*(z) = P(-z)$, $Q^*(z) = Q(-z) \Rightarrow H^*(z) = H(-z)$.

By the method of *residue calculus* as explained in Appendix A, the auto-covariance function can be evaluated in the following expression:

$$\kappa_{XX}(\tau) = -\pi S_0 \sum_{j=1}^{s} \frac{e^{z_j |\tau|}}{z_j} \frac{P(z_j)P(-z_j)}{\prod\limits_{\substack{k=1 \\ k \neq j}}^{s} (z_k^2 - z_j^2)} \tag{3-59}$$

Because of this result and other nice mathematical properties, it is of great temptation to model homogeneous physical processes as processes with rational spectrum. However, some prudence should be considered in doing this if many frequencies in the spectrum affects the response. For example, turbulence and wind wave have non-rational spectra, which can hardly be calibrated to a rational function over a broad band of frequencies.

Example 3-7: Rational auto-spectral density function of the order (r,s)=(0,2)

Consider a homogeneous stochastic process $X(t)$ with the following auto-spectral density function:

$$S_{XX}(\omega) = \frac{S_0}{m^2\left((i\omega)^2 + 2\zeta\omega_0 i\omega + \omega_0^2\right)\left((-i\omega)^2 - 2\zeta\omega_0 i\omega + \omega_0^2\right)} \tag{3-60}$$

where $\zeta \in]0, 1[$, $\omega_0 > 0$, $S_0 > 0$.

Eq. (3-60) is seen to be a rational auto-spectral density function with:

$$\left.\begin{aligned} P(z) &= \frac{1}{m} \\[2mm] Q(z) &= z^2 + 2\zeta\omega_0 z + \omega_0^2 \end{aligned}\right\} \tag{3-61}$$

The poles of $Q(z)$ become:

$$\left.\begin{aligned} z_1 \\ z_2 \end{aligned}\right\} = \omega_0\left(-\zeta \pm i\sqrt{1-\zeta^2}\right) \tag{3-62}$$

It follows from Eq. (3-62) that $z_2 = z_1^*$ and $z_2^2 - z_1^2 = 4\omega_0^2\,\zeta\sqrt{1-\zeta^2}\,i$. Then, the use of Eq. (3-59) provides the auto-covariance function of $X(t)$:

$$\kappa_{XX}(\tau) = -\pi\frac{S_0}{m^2}\left(\frac{e^{z_1|\tau|}}{z_1} - \frac{e^{z_1^*|\tau|}}{z_1^*}\right)\frac{1}{4\omega_0^2\,\zeta\sqrt{1-\zeta^2}\,i} = -\frac{\pi S_0}{2\zeta\omega_0^2\sqrt{1-\zeta^2}\,m^2}\,\mathrm{Im}\left(\frac{e^{z_1|\tau|}}{z_1}\right)$$

$$= \sigma_X^2\,\rho_{XX}(\tau) \tag{3-63}$$

$$\sigma_X^2 = \frac{\pi S_0}{2\zeta\,\omega_0^3 m^2} \tag{3-64}$$

$$\rho_{XX}(\tau) = e^{-\zeta\omega_0|\tau|}\left(\cos\left(\omega_d\tau\right) + \frac{\zeta}{\sqrt{1-\zeta^2}}\sin\left(\omega_d|\tau|\right)\right) \tag{3-65}$$

$$\omega_d = \omega_0\sqrt{1-\zeta^2} \tag{3-66}$$

where $\rho_{XX}(\tau)$ is the auto-correlation coefficient function defined in Eq. (3-8). $\mathrm{Im}(\cdot)$ signifies the *imaginary part* of a complex quantity. If $\zeta \ll 1$, the parameter ζ can be identified as the non-dimensional bandwidth in Eq. (3-45).

3.3 Stochastic integration

Consider a stochastic process $\{F(t),\, t \in [t_0, t_1]\}$ defined on a certain sample space Ω. For a given $\omega \in \Omega$, the realization $f(t, \omega)$ is achieved, for which the following Riemann integral is assumed to exist:

$$x(t, \omega) = \int_{t_0}^{t} h(t, \tau) f(\tau, \omega)\, d\tau \tag{3-67}$$

In the following, the weighting function $h(t, \tau)$ will be termed the impulse response function. If the Riemann integral in Eq. (3-67) exists for all $\omega \in \Omega$, a stochastic variable $X(t)$ on Ω is defined from the sample values $x(t, \omega_1)$, $x(t, \omega_2)$, $\cdots$. Furthermore, if Eq. (3-67) exists for any $t \in [t_0, t_1]$, a set of stochastic variables $\{X(t),\, t \in [t_0, t_1]\}$ is defined. This stochastic process $\{X(t),\, t \in [t_0, t_1]\}$ is termed the *stochastic integral* or the *integrated process* of $\{F(t),\, t \in [t_0, t_1]\}$ with the impulse response function $h(t, \tau)$. Symbolically, the process is written as:

$$X(t) = \int_{t_0}^{t} h(t, \tau) F(\tau)\, d\tau \tag{3-68}$$

$X(t)$ as given by Eq. (3-68) should be considered as the limit as $m \to \infty$ of a sequence of random variables $X_m(t)$, which is defined as:

$$X_m(t) = \sum_{j=0}^{m-1} h(t, \tau_j) F(\tau_j) \Delta \tau_m \tag{3-69}$$

where $\Delta \tau_m = \frac{t - t_0}{m}$.

As shown in Fig. 3-12, the random variables $F(\tau_j)$ in the sum have been chosen at the start of the sub-intervals of the length $\Delta \tau$. As long as almost all the realizations of $\{F(t),\, t \in [t_0, t_1]\}$ are Riemann integrable, this is not a constraint.

The convergence of a sequence of random variables $X_m(t)$ to a stochastic variable $X(t)$ only makes sense, if a distance measure between $X_m(t)$ and $X(t)$ is defined. Here, we shall consider convergence in mean square, which means that $X_m(t)$ converges to $Y(t)$ if:

$$\lim_{m \to \infty} E\left[\left(X_m(t) - X(t)\right)^2\right] = 0 \tag{3-70}$$

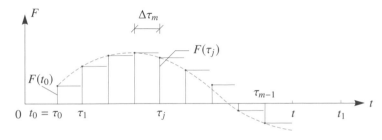

Fig. 3–12 Stochastic integration.

The integrated process $\{X(t), t \in [t_0, t_1]\}$ is characterized through its statistical moments. If Eq. (3-70) is fulfilled, it can be shown that the mean value function and the auto-covariance function of $X(t)$ become the mean value function and auto-covariance function of $X_m(t)$ as $m \to \infty$.[1] From Eq. (3-69), it follows that:

$$\mu_X(t) = \lim_{m \to \infty} E\left[X_m(t)\right] = \lim_{m \to \infty} \sum_{j=0}^{m-1} h(t, \tau_j) E\left[F(\tau_j)\right] \Delta \tau_m$$

$$= \int_{t_0}^{t} h(t, \tau) \mu_F(\tau) d\tau \tag{3-71}$$

$$\kappa_{XX}(t_1, t_2) = \lim_{m \to \infty} \lim_{n \to \infty} \sum_{j=0}^{m-1} \sum_{k=0}^{n-1} h(t_1, \tau_j) h(t_2, \tau_k) \kappa_{FF}(\tau_j, \tau_k) \Delta \tau_m \Delta \tau_n$$

$$= \int_{t_0}^{t_1} \int_{t_0}^{t_2} h(t_1, \tau_1) h(t_2, \tau_2) \kappa_{FF}(\tau_1, \tau_2) d\tau_1 d\tau_2 \tag{3-72}$$

It can be shown that the necessary and sufficient condition for the existence of the integrated process in mean square is that the planar integral in Eq. (3-72) exists.[1]

Higher order moments of the integrated process can be determined in a similar fashion. Thus, the joint nth order moment becomes:

$$E\left[X(t_1) \cdots X(t_n)\right] = \int_{t_0}^{t_1} \cdots \int_{t_0}^{t_n} h(t_1, \tau_1) \cdots h(t_n, \tau_n) E\left[F(\tau_1) \cdots F(\tau_n)\right] d\tau_1 \cdots d\tau_n \tag{3-73}$$

Therefore, the joint statistical moments of the integrated process $\{X(t), t \in [t_0, t_1]\}$ can be determined if the corresponding moments are known for the underlying process $\{F(t), t \in [t_0, t_1]\}$.

Loosely formulated, Eqs. (3-71), (3-72) and (3-73) state that the operations of expectation and stochastic integration commute.

Let $\{F(t), t \in [t_0, t_1]\}$ be a Gaussian process. Being the limit of a sum of normal stochastic variables, the stochastic integral $X(t)$ becomes normal itself. Next, assume that Eq. (3-69) is applied at n arbitrary instants of time $t = t_1$, $t = t_2$, $\cdots t = t_n$. The n-dimensional stochastic vector $\left[X(t_1), \cdots, X(t_n)\right]^T$ thus becomes an n-dimensional normal vector. Passing to the limit, it is concluded that the integrated process becomes a Gaussian process if $\{F(t), t \in [t_0, t_1]\}$ is Gaussian. Hence, Gaussianity is preserved under the linear operation of stochastic integration.

Example 3-8: Wiener process

Let $\{F(t), t \in R\}$ be a white noise process, with $\mu_F(t) = 0$, $\kappa_{FF}(\tau) = 2\pi S_0 \delta(\tau)$, cf. Eq. (3-40). A *Wiener process* or *Brownian motion* may formally be defined by the stochastic integral:

$$W(t) = \int_0^t F(\tau) d\tau \tag{3-74}$$

---

[1]A.H. Jazwinski. *Stochastic Processes and Filtering Theory*. Dover Publications, Inc., New York, 2007.

Due to the extreme behaviour of the realizations of $\{F(t), t \in R\}$, the indicated stochastic integral cannot be given any mathematical basis. Nevertheless, the following statements and results derived from the formal representation can all be proved to be correct.

It follows from Eq. (3-74) that $W(0) = 0$ for all realization of the stochastic integral. Further, since $\{F(t), t \in R\}$ is a Gaussian process, the Wiener process $\{W(t), t \in [0, \infty[\}$ becomes Gaussian as well. The mean value function of the Wiener process becomes, cf. Eq. (3-71):

$$\mu_W(t) = \int_0^t \mu_F(\tau)\, d\tau = 0 \tag{3--75}$$

The auto-covariance function becomes, cf. Eq. (3-72):

$$\kappa_{WW}(t_1, t_2) = \int_0^{t_1} \int_0^{t_2} \kappa_{FF}(\tau_1, \tau_2)\, d\tau_1 d\tau_2 = 2\pi S_0 \int_0^{t_1} \int_0^{t_2} \delta(\tau_2 - \tau_1)\, d\tau_2 d\tau_1 \tag{3--76}$$

It is assumed that $t_1 < t_2$ as shown in Fig. 3-13, so the variable $\tau_1 \in\,]0, t_2[$ and the following equation holds:

$$\int_0^{t_2} \delta(\tau_2 - \tau_1)\, d\tau_2 = 1 \quad \Rightarrow$$

$$\kappa_{WW}(t_1, t_2) = 2\pi S_0 \int_0^{t_1} d\tau_1 = 2\pi S_0\, t_1 \tag{3--77}$$

In the same way, it is shown that $t_2 < t_1$ implies $\kappa_{WW}(t_1, t_2) = 2\pi S_0\, t_2$. In combination, these two results can be written as:

$$\kappa_{WW}(t_1, t_2) = 2\pi S_0 \min(t_1, t_2) \tag{3--78}$$

Especially, the variance function is obtained from Eq. (3-78) using $t_1 = t_2 = t$:

$$\sigma_W^2(t) = \sigma_0^2\, t, \quad \sigma_0^2 = 2\pi S_0 \tag{3--79}$$

It should be noticed that $\sigma_0 = 1$ for a unit white noise.

Since the Wiener process is Gaussian, it follows from Eqs. (3-75) and (3-79) that $W(t) \sim N(0, \sigma_0^2 t)$. Hence, the probability density function of the first order is given by:

$$f_W(w, t) = \frac{1}{\sqrt{2\pi t}\, \sigma_0} \exp\left(-\frac{1}{2} \frac{w^2}{\sigma_0^2\, t}\right) \tag{3--80}$$

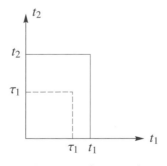

Fig. 3–13 Integration domain with $t_1 < t_2$.

By insertion, it can be seen that $f_W(w,t)$ fulfills the following parabolic partial differential equation:

$$\frac{\partial f_W(w,t)}{\partial t} = \frac{1}{2}\sigma_0^2 \frac{\partial^2 f_W(w,t)}{\partial w^2} \tag{3-81}$$

Eq. (3-81) is merely a one-dimensional heat equation with the diffusion constant $\frac{1}{2}\sigma_0^2 = \pi S_0$, which determines the evolution of the probability density function $f_W(w,t)$ with time. A similar parabolic partial differential equation, known as the Fokker-Planck equation, exists for all linear and non-linear dynamic systems driven by a white noise process.

Assuming $t_1 < t_2 < t_3 < t_4$, the following increments of the Wiener process are defined:

$$\Delta W(t_1, t_2) = W(t_2) - W(t_1) = \int_{t_1}^{t_2} F(\tau)\, d\tau \tag{3-82}$$

$$\Delta W(t_3, t_4) = W(t_4) - W(t_3) = \int_{t_3}^{t_4} F(\tau)\, d\tau \tag{3-83}$$

$\Delta W(t_1, t_2)$ and $\Delta W(t_3, t_4)$ are normally distributed with zero mean values. Hence, they will be stochastically independent if they are uncorrelated. The covariance between $\Delta W(t_1, t_2)$ and $\Delta W(t_3, t_4)$ is:

$$E\left[\Delta W(t_1, t_2)\Delta W(t_3, t_4)\right] = E\left[\int_{t_1}^{t_2} F(\tau_1)\, d\tau_1 \int_{t_3}^{t_4} F(\tau_2)\, d\tau_2\right]$$

$$= \int_{t_1}^{t_2}\int_{t_3}^{t_4} E\left[F(\tau_1)F(\tau_2)\right] d\tau_1 d\tau_2 = 2\pi S_0 \int_{t_1}^{t_2}\int_{t_3}^{t_4} \delta(\tau_2 - \tau_1)\, d\tau_1 d\tau_2 \tag{3-84}$$

As τ_1 and τ_2 belong to disjoint intervals, we always have $\tau_1 \neq \tau_2$, so $\delta(\tau_2 - \tau_1) = 0$, leading to the following result:

$$E\left[\Delta W(t_1, t_2)\Delta W(t_3, t_4)\right] = 0 \tag{3-85}$$

Since the auto-covariance function $\kappa_W(t_1, t_2) = 2\pi S_0 \min(t_1, t_2)$ is continuous at the diagonal $t_1 = t_2 = t$, it can be shown that the realizations of $\{W(t),\, t \in [0,\infty[\}$ are continuous.[1] A Wiener process is the only process with independent increments, which has continuous realizations. On the other hand, the realizations are not differentiable. As $\sigma_W(t) = \sigma_0\sqrt{t}$ increases with time, the oscillations of the realizations about $\mu_W(t) = 0$ will also increase. These observations are displayed in the realization shown in Fig. 3-14. The "fuzzy" appearance of the graph aims at illustrating the non-differentiability of the realization.

Let $dW(t) = W(t + dt) - W(t)$ indicate the differential increment of $\{W(t),\, t \in [0,\infty[\}$ in the interval $[t, t + dt[$. Then, $dW(t)$ fulfills the following property:

$$E\left[dW(t_1)dW(t_2)\right] = \begin{cases} 0, & t_1 \neq t_2 \\ 2\pi S_0\, dt, & t_1 = t_2 = t \end{cases} \tag{3-86}$$

It should be noticed that $dW(t)$ is of the magnitude $\sqrt{dt}$.

---

[1] A.H. Jazwinski. *Stochastic Processes and Filtering Theory*. Dover Publications, Inc., New York, 2007.

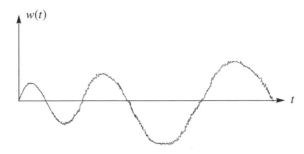

Fig. 3–14 Realization of a Wiener process.

Next, consider an N-dimensional stochastic vector process $\{\mathbf{F}(t),\, t \in [t_0, t_1]\}$, where $\mathbf{F}(t) = \left[F_1(t), \cdots, F_N(t)\right]^T$. The M-dimensional *integrated vector process* $\{\mathbf{X}(t),\, t \in [t_0, t_1]\},\, \mathbf{X}(t) = \left[X_1(t), \cdots, X_M(t)\right]^T$ can be symbolically written as:

$$X_j(t) = \sum_{k=1}^{N} \int_{t_0}^{t} h_{jk}(t, \tau) F_k(\tau)\, d\tau, \quad j = 1, \cdots, M \tag{3–87}$$

All addends in Eq. (3-87) are scalar stochastic integrals as in Eq. (3-68). $h_{jk}(t, \tau)$ are components of an $M \times N$ dimensional matrix $\mathbf{h}(t, \tau)$, termed the *impulse response matrix*.

The mean value function $\mu_{X_j}(t)$, the cross-covariance function $\kappa_{X_j X_k}(t_1, t_2)$ and the joint statistical moment of the nth order of the integrated vector process can immediately be written down, generalizing the corresponding results for the scalar case:

$$\mu_{X_j}(t) = \sum_{k=1}^{N} \int_{t_0}^{t} h_{jk}(t, \tau) \mu_{F_k}(\tau)\, d\tau \tag{3–88}$$

$$\kappa_{X_{j_1} X_{j_2}}(t_1, t_2) = \sum_{k_1=1}^{N} \sum_{k_2=1}^{N} \int_{t_0}^{t_1} \int_{t_0}^{t_2} h_{j_1 k_1}(t_1, \tau_1) h_{j_2 k_2}(t_2, \tau_2)\, \kappa_{F_{k1} F_{k2}}(\tau_1, \tau_2)\, d\tau_1 d\tau_2 \tag{3–89}$$

$$E\left[X_{j_1}(t_1) \cdots X_{j_n}(t_n)\right] = \sum_{k_1=1}^{N} \cdots \sum_{k_n=1}^{N} \int_{t_0}^{t_1} \cdots \int_{t_0}^{t_n} h_{j_1 k_1}(t_1, \tau_1) \cdots h_{j_n k_n}(t_n, \tau_n) \cdot$$
$$E\left[F_{k_1}(\tau_1) \cdots F_{k_n}(\tau_n)\right] d\tau_1 \cdots d\tau_n \tag{3–90}$$

The impulse response function is called *causal* if:

$$h(t, \tau) = 0, \quad \tau > t \tag{3–91}$$

In this case, the upper limits in Eqs. (3-67) and (3-68) may be replaced by ∞ without affecting the integrated process. $X(t)$ may be interpreted as the stochastic dynamic response to the stochastic load $F(t)$. Hence, causality means that the dynamic response $X(t)$ is unaffected by future loads $\{F(\tau),\, \tau > t\}$.

The system is called *stable* if the impulse response function fulfills:

$$\int_{t_0}^{t} |h(t, \tau)| \, d\tau \; < \; \infty \tag{3–92}$$

Eq. (3-92) insures that any limited realization of the load process produces a limited realization of the response. Actually, if $|f(\tau)| \leq K$ for all τ with K being a positive constant, we have:

$$x(t) \; = \; \int_{t_0}^{t} h(t, \tau) \, f(\tau) \, d\tau \; \leq \; \int_{t_0}^{t} |h(t, \tau)| \, |f(\tau)| \, d\tau \; \leq \; K \int_{t_0}^{t} |h(t, \tau)| \, d\tau \; < \; \infty \tag{3–93}$$

For a structural dynamic system, it can be shown that any eigenvibrations due to nonzero initial conditions applied at $t = 0$ will be gradually dissipated if and only if the stability condition in Eq. (3-92) is fulfilled .

The system is called *time-invariant* if the impulse response function fulfills $h(t, \tau) = h(t - \tau)$. For such a system, any realization $f(\tau)$ which produces the response $x(t)$, will produce exactly the same response if it is initiated at any previous or subsequent time with the same initial conditions.

These definitions may be directly generalized to similar conditions on the impulse response matrix $\mathbf{h}(t, \tau)$.

Example 3-9: Relation between the impulse response matrix h(t) and the frequency response matrix H(ω) for a linear time-invariant system

Fig. 3-15 shows a time-invariant linear structure, where a time-varying load $f_k(t)$ is applied at point k in a given direction. $x_j(t)$ indicates the response in a given direction at another point j. Further, we shall assume the system to be causal, so the impulse response matrix vanishes for negative argument.

A harmonically varying load with unit amplitude and the angular frequency ω, $f_k(t) = e^{i\omega t}$, is first applied at point k. Only the real part of the indicated complex quantity is related with physical meaning. The explicit indication of this has been omitted for ease in the rest of this textbook.

In the stationary state when response due to the initial conditions has been dissipated, the response at point j can be written as:

$$x_j(t) \; = \; H_{jk}(\omega) \, e^{i\omega t} \tag{3–94}$$

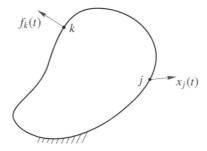

Fig. 3–15 Response $x_j(t)$ at point j of a linear time-invariant structure due to a dynamic load $f_k(t)$ at point k.

Per definition, $H_{jk}(\omega)$ is the *frequency response function* for the indicated experiment.

Next, a unit impulse load $f_k(t) = \delta(t)$, formally represented by the Dirach delta function, is applied at point k at time $t = 0$ when the structure is at rest. Then, the response $x_j(t)$ per definition is given by:

$$x_j(t) = h_{jk}(t) \tag{3-95}$$

where $h_{jk}(t)$ is the impulse response function for the indicated experiment.

On the other hand, we may construct the harmonic response in the first experiment as an infinite sum of responses from all past differential impulses $e^{i\omega\tau} d\tau$:

$$x_j(t) = \int_{-\infty}^{t} h_{jk}(t - \tau) e^{i\omega\tau} \, d\tau \tag{3-96}$$

where $h_{jk}(t - \tau)$ indicates the response at point j at the time t from a unit impulse load applied at point k at a previous time τ. $h_{jk}(t - \tau)e^{i\omega\tau} d\tau$ indicates the corresponding response from an impulse load of the magnitude $e^{i\omega\tau} d\tau$. Due to linearity of the structure, the total response may be summed up from these infinitesimal contributions.

Change of integration variable to $u = t - \tau$ in Eq. (3-96) yields:

$$x_j(t) = e^{i\omega t} \int_{0}^{\infty} e^{-i\omega u} h_{jk}(u) \, du \tag{3-97}$$

Comparing Eq. (3-94) with Eq. (3-97), we have:

$$H_{jk}(\omega) = \int_{0}^{\infty} e^{-i\omega u} h_{jk}(u) \, d\tau \tag{3-98}$$

Hence, $h_{jk}(t)$ and $H_{jk}(\omega)$ are mutual Fourier transforms. The inverse relation is written as:

$$h_{jk}(u) = \frac{1}{2\pi} \int_{-\infty}^{\infty} e^{i\omega u} H_{jk}(\omega) \, d\omega \tag{3-99}$$

Due to the relations in Eqs. (3-98) and (3-99), $h_{jk}(t)$ and $H_{jk}(\omega)$ contain exactly the same information of the dynamic property of the system.

In the following, it is assumed that the time-invariant impulse response function $h(t - \tau)$ or the component impulse response functions $h_{ij}(t - \tau)$ are causal and stable, i.e., they are absolute integrable with respect to the argument. Further, it is assumed that the excitation processes $\{F(t), t \in R\}$ and $\{\mathbf{F}(t), t \in R\}$ are weakly homogeneous, and the excitation has been acting for infinite long time so the lower limit of the stochastic integral in Eq. (3-68) is $t_0 = -\infty$. Then, it can be shown that the related integrated processes $\{X(t), t \in R\}$ and $\{\mathbf{X}(t), t \in R\}$ also become weakly homogeneous.

The mean value function in Eq. (3-71) thus becomes:

$$\mu_X(t) = \int_{-\infty}^{t} h(t - \tau)\mu_F \, d\tau = \mu_F \int_{-\infty}^{t} h(t - \tau) \, d\tau = \mu_F \int_{0}^{\infty} h(u) \, du = H(0) \, \mu_F \tag{3-100}$$

where $H(\omega)$ is the frequency response function related to the impulse response function $h(t)$, cf. Eq. (3-98):

$$H(\omega) = \int_0^\infty e^{-i\omega u} h(u) \, du \tag{3–101}$$

In Section 4.1, it will be shown that $H(0) = \frac{1}{k}$ for a *single-degree-of-freedom system*, where k is the *spring constant*. Hence, μ_X in Eq. (3-100) may be interpreted as the static displacement of the oscillator due to the static load μ_F.

Using substitutions $u_1 = t_1 - \tau_1$ and $u_2 = t_2 - \tau_2$ in Eq. (3-72), we have:

$$\kappa_{XX}(t_1, t_2) = \int_{-\infty}^{t_1} \int_{-\infty}^{t_2} h(t_1 - \tau_1) h(t_2 - \tau_2) \, \kappa_{FF}(\tau_2 - \tau_1) \, d\tau_1 d\tau_2$$

$$= \int_0^\infty \int_0^\infty h(u_1) h(u_2) \, \kappa_{FF}(t_2 - t_1 + u_1 - u_2) \, du_1 du_2 = \kappa_{XX}(t_2 - t_1) \tag{3–102}$$

The auto-spectral density function of $\{X(t), \, t \in R\}$ follows from Eqs. (3-23) and (3-102):

$$S_{XX}(\omega) = \frac{1}{2\pi} \int_{-\infty}^\infty e^{-i\omega\tau} \, \kappa_{XX}(\tau) \, d\tau$$

$$= \frac{1}{2\pi} \int_{-\infty}^\infty e^{-i\omega\tau} \int_0^\infty \int_0^\infty h(u_1) h(u_2) \, \kappa_{FF}(\tau + u_1 - u_2) \, du_1 du_2 \, d\tau$$

$$= \int_0^\infty e^{i\omega u_1} h(u_1) du_1 \int_0^\infty e^{-i\omega u_2} h(u_2) du_2 \, \frac{1}{2\pi} \int_{-\infty}^\infty e^{-i\omega(\tau + u_1 - u_2)} \kappa_{FF}(\tau + u_1 - u_2) d\tau$$

$$\tag{3–103}$$

It should be noted that u_1 and u_2 are constants in the innermost integral of Eq. (3-103). Hence, this integral becomes $S_{FF}(\omega)$, and Eq. (3-103) can then be written as:

$$S_{XX}(\omega) = H^*(\omega) H(\omega) \, S_{FF}(\omega) = |H(\omega)|^2 \, S_{FF}(\omega) \tag{3–104}$$

Eq. (3-104) has been illustrated in Fig. 3-16. If $S_{FF}(\omega)$ is broad-banded and $|H(\omega)|^2$ is narrow-banded as is the case in the figure, the integrated process becomes narrow-banded as well.

Upon comparison of Eqs. (3-58) and (3-104), the former equation may be interpreted as the auto-spectral density of an integrated process (with the *rational frequency response function* $H(z) = \frac{P(z)}{Q(z)}$, $z = i\omega$) of a white noise process $\{F(t), \, t \in R\}$ (with the auto-spectral density $S_{FF} = S_0$). This observation forms the basis of modelling zero-mean stationary Gaussian processes. The idea is to calibrate the parameters $p_0, \cdots, p_r$ and $q_1, \cdots, q_s$ of the numerator and the denominator polynomials, so that a given target auto-spectral density function for the integrated process is obtained. For this reason, the indicated rational frequency response function is occasionally known as a shaping filter, and the stochastic integration is known as a linear filtration. This procedure is illustrated in Fig. 3-17.

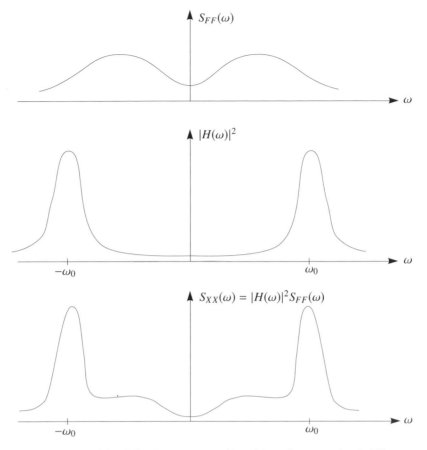

Fig. 3–16 Broad-banded excitation process filtered through a narrow-banded filter.

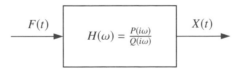

Fig. 3–17 Gaussian white noise filtered through a rational shaping filter.

The impulse response function related to a rational frequency response function is determined by inserting Eq. (3-54) into Eq. (3-99). Using the method of residues as shown in Appendix B, the resulting $h(t)$ can be expressed as:

$$h(t) = \begin{cases} 0 & , \quad t \leq 0 \\[2ex] \displaystyle\sum_{j=1}^{s} \mathrm{e}^{z_j t} \frac{P(z_j)}{\displaystyle\prod_{\substack{k=1 \\ k \neq j}}^{s}(z_j - z_k)}, & t > 0 \end{cases} \tag{3–105}$$

where z_j indicate the poles of the denominator polynomial, i.e., $Q(z_j) = 0$.

Example 3-10: Impulse response function of a rational filter of the order (r,s) = (0,2)

Consider a rational frequency response function with $P(z)$ and $Q(z)$ given by Eq. (3-61). The poles z_1 and z_2 are given by Eq. (3-62). For $t > 0$, the impulse response function follows from Eq. (3-105):

$$h(t) = \frac{1}{m} \left(e^{z_1 t} - e^{z_2 t} \right) \frac{1}{z_1 - z_2} = \frac{1}{m} \left(e^{z_1 t} - e^{z_1^* t} \right) \frac{1}{2\omega_0 i \sqrt{1 - \zeta^2}} = \frac{\mathrm{Im} \left(e^{z_1 t} \right)}{m \omega_0 \sqrt{1 - \zeta^2}}$$

$$= \frac{1}{m \omega_0 \sqrt{1 - \zeta^2}} \mathrm{Im} \left(e^{\omega_0 \left(-\zeta + i \sqrt{1 - \zeta^2} \right) t} \right) = \frac{e^{-\zeta \omega_0 t}}{m \omega_d} \sin \left(\omega_d t \right) \tag{3–106}$$

where ω_d is given by Eq. (3-66).

Example 3-11: Hilbert transform

Let $\{X(t), t \in R\}$ be a weakly homogeneous stochastic process. The *Hilbert transform* $\{\hat{X}(t), t \in R\}$ of $\{X(t), t \in R\}$ is defined by the stochastic integral:

$$\hat{X}(t) = \frac{1}{\pi} \fint_{-\infty}^{\infty} \frac{X(\tau)}{t - \tau} d\tau \tag{3–107}$$

The impulse response function $h(t - \tau) = \frac{1}{\pi} \frac{1}{t - \tau}$ is time-invariant, but it is neither causal nor stable. Due to the singularity of the integrand at $\tau = t$, the integral does not exists in the ordinary Riemann sense. Instead, as indicated by the symbol $\fint$, the integral needs to be calculated as a Cauchy principal value:

$$\hat{X}(t) = \lim_{\varepsilon \to 0} \left(\frac{1}{\pi} \int_{-\infty}^{t - \varepsilon} \frac{X(\tau)}{t - \tau} d\tau + \frac{1}{\pi} \int_{t + \varepsilon}^{\infty} \frac{X(\tau)}{t - \tau} d\tau \right) \tag{3–108}$$

Although the impulse response function $h(t) = \frac{1}{\pi t}$ is not absolute integrable, the frequency response function nevertheless exists as a Cauchy principal value:

$$H(\omega) = \frac{1}{\pi} \fint_{-\infty}^{\infty} e^{-i\omega t} \frac{1}{t} dt = \frac{1}{\pi} \fint_{-\infty}^{\infty} \left(\cos(\omega t) - i \sin(\omega t) \right) \frac{1}{t} dt$$

$$= -\frac{2i}{\pi} \int_{0}^{\infty} \frac{1}{t} \sin(\omega t) dt = \begin{cases} -i, & \omega > 0 \\ 0, & \omega = 0 \\ i, & \omega < 0 \end{cases} \tag{3–109}$$

where it has been used that $\cos(\omega t)$ is an even function of t, and $\sin(\omega t)$ and $\frac{1}{t}$ are odd functions of t. Further, the following semi-infinite integral has been applied:[1]

$$\int_{0}^{\infty} \frac{1}{t} \sin(\omega t) dt = \begin{cases} \dfrac{\pi}{2}, & \omega > 0 \\ 0, & \omega = 0 \\ -\dfrac{\pi}{2}, & \omega < 0 \end{cases} \tag{3–110}$$

Eq. (3-109) may be written on the form $H(\omega) = e^{-i \frac{\pi}{2} \mathrm{sign}(\omega)}$.

---

[1] I.S. Gradshteyn and I.M. Ryzhik. *Table of Integrals, Series and Products, 7th Edition*. Academic press, London, 2007.

Next, consider a harmonic excitation $X(t) = \text{Re}(X_0 e^{i\omega t})$. The Hilbert transform of the harmonic excitation becomes:

$$\hat{X}(t) = \text{Re}(H(\omega)X_0 e^{i\omega t}) = \text{Re}\left(X_0 e^{i\left(\omega t - \frac{\pi}{2}\text{sign}(\omega)\right)}\right) \qquad (3\text{–}111)$$

Hence, the Hilbert transform merely implies a phase lag of the magnitude $\frac{\pi}{2}\text{sign}(\omega)$ of the harmonic component $\text{Re}(X_0 e^{i\omega t})$, leaving the amplitude $|X_0|$ unchanged. This means that a realization of the harmonic process $x(t) = r\cos(\omega t + \varphi)$ is transformed into $\hat{x}(t) = r\cos\left(\omega t + \varphi - \frac{\pi}{2}\right) = -r\sin(\omega t + \varphi)$. As seen, in this case $\hat{x}(t) = \frac{1}{\omega}\dot{x}(t)$.

From Eq. (3-107), it follows that the mean value function of the Hilbert transform becomes:

$$\mu_{\hat{X}}(t) = \frac{1}{\pi}\int_{-\infty}^{\infty}\frac{1}{t-\tau}\mu_X \, d\tau = \frac{\mu_X}{\pi}\int_{-\infty}^{\infty}\frac{1}{u}\, du \equiv 0 \qquad (3\text{–}112)$$

Eqs. (3-104) and (3-109) imply that:

$$S_{\hat{X}\hat{X}}(\omega) = |H(\omega)|^2 S_{XX}(\omega) = S_{XX}(\omega) \qquad (3\text{–}113)$$

Then, from the Wiener-Khintchine relation, we have:

$$\kappa_{\hat{X}\hat{X}}(\tau) = \kappa_{XX}(\tau) \qquad (3\text{–}114)$$

Hence, the auto-spectral density function and the auto-covariance function of the Hilbert transform are identical to the corresponding quantities of the underlying process $\{X(t), t \in R\}$. This is just another statement of the previous finding that the Hilbert transform does not change the amplitudes of any harmonic component of the underlying process.

Because $E[\hat{X}(t)] = 0$, the covariance of $\hat{X}(t)$ and $X(t)$ becomes:

$$\begin{aligned}
\kappa_{\hat{X}(t)X(t)} &= E\left[\hat{X}(t)X(t)\right] = \frac{1}{\pi}\int_{-\infty}^{\infty}\frac{1}{t-\tau}E\left[X(t)X(\tau)\right]d\tau \\
&= \frac{1}{\pi}\int_{-\infty}^{\infty}\frac{1}{t-\tau}E\left[X(t)X(\tau) - \mu_X^2\right]d\tau = \frac{1}{\pi}\int_{-\infty}^{\infty}\frac{1}{t-\tau}\kappa_{XX}(t-\tau)\,d\tau \\
&= \frac{1}{\pi}\int_{-\infty}^{\infty}\frac{1}{u}\kappa_{XX}(u)\,du = 0
\end{aligned} \qquad (3\text{–}115)$$

Therefore, $X(t)$ and $\hat{X}(t)$ are uncorrelated for a weakly homogeneous process. This is a consequence of the phase shift of $\pm\frac{\pi}{2}$ of all harmonic components by the Hilbert transform. Especially, if $\{X(t), t \in R\}$ is Gaussian, $\{\hat{X}(t), t \in R\}$ becomes a zero-mean Gaussian process as well, and $\hat{X}(t)$ and $X(t)$ become stochastically independent.

Based on Eqs. (3-88) and (3-89), results similar to Eqs. (3-100), (3-102) and (3-104) may be derived by a similar procedure for a weakly homogeneous M-dimensional stochastic integral vector process $\{\mathbf{X}(t), t \in R\}$. These results can be written as:

$$\mu_{X_j} = \sum_{k=1}^{N}\mu_{F_k}\int_0^{\infty}h_{jk}(u)\,du = \sum_{k=1}^{N}H_{jk}(0)\,\mu_{F_k} \qquad (3\text{–}116)$$

$$\kappa_{X_{j_1} X_{j_2}}(\tau) = \sum_{k_1=1}^{N} \sum_{k_2=1}^{N} \int_0^\infty \int_0^\infty h_{j_1 k_1}(u_1) \, \kappa_{F_{k_1} F_{k_2}}(\tau + u_1 - u_2) \, h_{j_2 k_2}(u_2) \, du_1 du_2 \tag{3-117}$$

$$S_{X_{j_1} X_{j_2}}(\omega) = \sum_{k_1=1}^{N} \sum_{k_2=1}^{N} H^*_{j_1 k_1}(\omega) \, S_{F_{k_1} F_{k_2}}(\omega) \, H_{j_2 k_2}(\omega) \tag{3-118}$$

where $\tau = t_2 - t_1$, and:

$$H_{jk}(\omega) = \int_0^\infty e^{-i\omega u} \, h_{jk}(u) \, du \tag{3-119}$$

Equations (3-116), (3-117) and (3-118) may be reformulated in the following matrix form:

$$\boldsymbol{\mu_X} = \mathbf{H}(0) \, \boldsymbol{\mu_F} \tag{3-120}$$

$$\boldsymbol{\kappa_{XX}}(\tau) = \int_0^\infty \int_0^\infty \mathbf{h}(u_1) \, \boldsymbol{\kappa_{FF}}(\tau + u_1 - u_2) \, \mathbf{h}^T(u_2) \, du_1 du_2 \tag{3-121}$$

$$\mathbf{S_{XX}}(\omega) = \mathbf{H}^*(\omega) \, \mathbf{S_{FF}}(\omega) \, \mathbf{H}^T(\omega) \tag{3-122}$$

where $\boldsymbol{\kappa_{XX}}(\tau)$ and $\mathbf{S_{XX}}(\omega)$ are termed the cross-covariance matrix function (at stationarity) and the cross-spectral density matrix function, respectively.

3.4   Stochastic differentiation

Consider a stochastic process $\{X(t), \ t \in [t_0, t_1]\}$ defined on a certain sample space Ω. For a given $\omega \in \Omega$, the realization $x(t, \omega)$ is achieved, for which the following derivative is assumed to exist for any $t \in [t_0, t_1]$:

$$\dot{x}(t, \omega) = \frac{d}{dt} x(t, \omega) \tag{3-123}$$

If Eq. (3-123) exists for all realizations $\omega \in \Omega$, a stochastic variable on Ω symbolically denoted as $\dot{X}(t)$ can be defined with probability one. Furthermore, if $\dot{X}(t)$ exists for all $t \in [t_0, t_1]$, a set of stochastic variables $\{\dot{X}(t), \ t \in [t_0, t_1]\}$ is defined, termed the *derivative process*.

$\dot{X}(t)$ should be considered as the limit as $\Delta t_m \to 0$ of a sequence of random variables defined as:

$$\dot{X}_m(t) = \frac{1}{\Delta t_m}\left(X(t + \Delta t_m) - X(t) \right) \tag{3-124}$$

where it has been assumed that $\Delta t_m \to 0$ as $m \to \infty$.

The convergence of the random variables $\dot{X}_m(t)$ to $\dot{X}(t)$ for $\Delta t_m \to 0$ is considered in mean square, which means:

$$\lim_{m \to \infty} E\left[\left(\dot{X}_m(t) - \dot{X}(t) \right)^2 \right] = 0 \tag{3–125}$$

The derivative process $\{\dot{X}(t),\ t \in [t_0, t_1]\}$ is characterized through its statistical moments. If Eq. (3-125) is fulfilled, it can be shown that the mean value function and the auto-covariance function of $\dot{X}(t)$ become the mean value function and auto-covariance function of $\dot{X}_m(t)$ as $m \to \infty$.[1]

Then, the mean value function $\mu_{\dot{X}}(t)$ of $\{\dot{X}(t),\ t \in [t_0, t_1]\}$ follows from Eq. (3-124):

$$\mu_{\dot{X}}(t) = \lim_{m \to \infty} E\left[\dot{X}_m(t) \right] = \lim_{\Delta t_m \to 0} E\left[\frac{1}{\Delta t_m} \left(X(t + \Delta t_m) - X(t) \right) \right]$$

$$= \lim_{\Delta t_m \to 0} \frac{1}{\Delta t_m} \left(\mu_X(t + \Delta t_m) - \mu_X(t) \right) = \frac{d}{dt} \mu_X(t) \tag{3–126}$$

Similarly, the 2nd order moment $E\left[\dot{X}(t_1)\dot{X}(t_2) \right]$ can be written as:

$$E\left[\dot{X}(t_1)\dot{X}(t_2) \right] = \lim_{\substack{m \to \infty \\ n \to \infty}} E\left[\dot{X}_m(t_1)\dot{X}_n(t_2) \right]$$

$$= \lim_{\substack{\Delta t_m \to 0 \\ \Delta t_n \to 0}} E\left[\frac{1}{\Delta t_m} \left(X(t_1 + \Delta t_m) - X(t_1) \right) \frac{1}{\Delta t_n} \left(X(t_2 + \Delta t_n) - X(t_2) \right) \right]$$

$$= \lim_{\Delta t_n \to 0} \frac{1}{\Delta t_n} \left(\lim_{\Delta t_m \to 0} \frac{1}{\Delta t_m} E\left[X(t_1 + \Delta t_m)X(t_2 + \Delta t_n) - X(t_1)X(t_2 + \Delta t_n) \right] \right.$$

$$\left. \lim_{\Delta t_m \to 0} \frac{1}{\Delta t_m} E\left[X(t_1 + \Delta t_m)X(t_2) - X(t_1)X(t_2) \right] \right)$$

$$= \lim_{\Delta t_n \to 0} \frac{1}{\Delta t_n} \left(\frac{\partial}{\partial t_1} E\left[X(t_1)X(t_2 + \Delta t_n) \right] - \frac{\partial}{\partial t_1} E\left[X(t_1)X(t_2) \right] \right)$$

$$= \frac{\partial^2}{\partial t_1\, \partial t_2} E\left[X(t_1)X(t_2) \right] \tag{3–127}$$

It can be shown that the derivative process exists in mean square if and only if the mixed partial derivative in Eq. (3-127) exists at the main diagonal $t_1 = t_2 = t$.[1] Since Eq. (3-78) is not differentiable at the main diagonal, the Wiener process is not differentiable in means square.

In the same way, the *derivative process of the nth order* $\{X^{(n)}(t),\ t \in [t_0, t_1]\}$ can be defined with probability one, if the nth derivative $\frac{d^n}{dt^n} x(t, \omega)$ of the realization $x(t, \omega)$ exists for any $\omega \in \Omega$ and any $t \in R$. Similarly, the *derivative vector process of the nth order* $\{\mathbf{X}^{(n)}(t),\ t \in [t_0, t_1]\}$, based on the N-dimensional vector process $\{\mathbf{X}(t),\ t \in [t_0, t_1]\}$, is defined as the vector process with nth order derivative component processes $\{X_j^{(n)}(t),\ t \in [t_0, t_1]\}$. If t signifies time

---

[1]A.H. Jazwinski. *Stochastic Processes and Filtering Theory*. Dover Publications, Inc., New York, 2007.

and $\{X(t),\, t \in [t_0, t_1]\}$ denotes a *displacement process*, the derivative process of 1st order $\{\dot{X}(t),\, t \in [t_0, t_1]\}$ is termed the *velocity process*, and the derivative process of 2nd order $\{\ddot{X}(t),\, t \in [t_0, t_1]\}$ is called the *acceleration process*.

The auto-covariance function of the derivative process follows from Eqs. (3-126) and (3-127):

$$\kappa_{\dot{X}\dot{X}}(t_1, t_2) = E\big[\dot{X}(t_1)\dot{X}(t_2)\big] - \mu_{\dot{X}}(t_1)\mu_{\dot{X}}(t_2)$$

$$= \frac{\partial^2}{\partial t_1\, \partial t_2}\Big(E\big[X(t_1)X(t_2)\big] - \mu_X(t_1)\mu_X(t_2)\Big) = \frac{\partial^2}{\partial t_1\, \partial t_2}\kappa_{XX}(t_1, t_2) \qquad (3\text{–}128)$$

Higher order moments of the derivative process can be determined in the same way. The joint nth order moment becomes:

$$E\big[\dot{X}(t_1)\cdots\dot{X}(t_n)\big] = \frac{\partial^n}{\partial t_1\cdots\partial t_n}E\big[X(t_1)\cdots X(t_n)\big] \qquad (3\text{–}129)$$

The derivative process of 2nd order $\{\ddot{X}(t),\, t \in [t_0, t_1]\}$ can be considered as the derivative process of 1st order derivative process $\{\dot{X}(t),\, t \in [t_0, t_1]\}$. The mean value function $\mu_{\ddot{X}}(t)$ and the auto-covariance function $\kappa_{\ddot{X}\ddot{X}}(t_1, t_2)$ can be determined by repeated application of Eqs. (3-126) and (3-128):

$$\mu_{\ddot{X}}(t) = \frac{d}{dt}\mu_{\dot{X}}(t) = \frac{d^2}{dt^2}\mu_X(t) \qquad (3\text{–}130)$$

$$\kappa_{\ddot{X}\ddot{X}}(t_1, t_2) = \frac{\partial^2}{\partial t_1\, \partial t_2}\kappa_{\dot{X}\dot{X}}(t_1, t_2) = \frac{\partial^4}{\partial t_1^2\, \partial t_2^2}\kappa_{XX}(t_1, t_2) \qquad (3\text{–}131)$$

More generally, the mean value function $\mu_{X^{(n)}}(t)$ and the auto-covariance function $\kappa_{X^{(n)}X^{(n)}}(t_1, t_2)$ of the derivative process of the nth order $\{X^{(n)}(t),\, t \in [t_1, t_2]\}$ can be written as:

$$\mu_{X^{(n)}}(t) = \frac{d^n}{dt^n}\mu_X(t) \qquad (3\text{–}132)$$

$$\kappa_{X^{(n)}X^{(n)}}(t_1, t_2) = \frac{\partial^{2n}}{\partial t_1^n\, \partial t_2^n}\kappa_{XX}(t_1, t_2) \qquad (3\text{–}133)$$

Finally, the cross-covariance function $\kappa_{X^{(m)}X^{(n)}}(t_1, t_2)$ of the derivative processes of the mth and nth order becomes:

$$\kappa_{X^{(m)}X^{(n)}}(t_1, t_2) = \frac{\partial^{m+n}}{\partial t_1^m\, \partial t_2^n}\kappa_{XX}(t_1, t_2) \qquad (3\text{–}134)$$

Let $\{X(t),\, t \in [t_1, t_2]\}$ be a Gaussian process. The right-hand side of Eq. (3-124) determines the $\dot{X}_m(t)$ as a difference of two normally distributed stochastic variables. Then, $\dot{X}_m(t)$ is normally distributed throughout the limit passing, for which reason $\dot{X}(t)$ becomes normally distributed as well. Similarly, it is seen that $\dot{X}_m(t_1), \ldots, \dot{X}_m(t_n)$ are jointly normally distributed throughout the limit passing, so the limit $\dot{X}(t_1), \ldots, \dot{X}(t_n)$ will be jointly normally distributed as well. Hence,

$\{\dot{X}(t),\ t \in [t_0, t_1]\}$ becomes a Gaussian process if $\{X(t),\ t \in [t_0, t_1]\}$ is Gaussian. In other words, Gaussianity is preserved under the linear operation of stochastic differentiation. Due to the Gaussianity, the derivative process is completely described by Eqs. (3-126) and (3-128).

Further, it can be shown that if $\{X(t),\ t \in R\}$ is homogeneous in the strict or weak sense, $\{\dot{X}(t),\ t \in R\}$ also becomes strictly or weakly homogeneous. By induction, this statement holds for the derivative process of arbitrary order.

The mean value function and the auto-covariance function of $\{\dot{X}(t),\ t \in R\}$ in the homogeneous case follow from Eqs. (3-126) and (3-128):

$$\mu_{\dot{X}}(t) = \frac{d}{dt}\mu_X \equiv 0 \tag{3–135}$$

$$\kappa_{\dot{X}\dot{X}}(t_1, t_2) = \frac{\partial^2}{\partial t_1\, \partial t_2}\kappa_{XX}(t_2 - t_1) = -\frac{d^2}{d\tau^2}\kappa_{XX}(\tau), \quad \tau = t_2 - t_1 \tag{3–136}$$

In the latter statement of Eq. (3-136), we have applied $\partial t_2 = d\tau$ (with t_1 kept constant) and $\partial t_1 = -d\tau$ (with t_2 kept constant).

Similarly, the cross-covariance function of $\{X^{(m)}(t),\ t \in R\}$ and $\{X^{(n)}(t),\ t \in R\}$ becomes:

$$\kappa_{X^{(m)}X^{(n)}}(t_1, t_2) = \frac{\partial^{m+n}}{\partial t_1^m\, \partial t_2^n}\kappa_{XX}(t_2 - t_1)$$

$$= (-1)^m \frac{d^{m+n}}{d\tau^{m+n}}\kappa_{XX}(\tau), \quad \tau = t_2 - t_1 \tag{3–137}$$

Especially, for $m = 0$ and $n = 1$, the cross-covariance function of $\{X(t),\ t \in R\}$ and the derivative process $\{\dot{X}(t),\ \in R\}$ is obtained:

$$\kappa_{X\dot{X}}(\tau) = \frac{d}{d\tau}\kappa_{XX}(\tau), \quad \tau = t_2 - t_1 \tag{3–138}$$

Since $\kappa_{XX}(\tau)$ is a symmetric function of τ, its derivative must be zero at $\tau = 0$ as shown in Fig. 3-18. Then, $\kappa_{X\dot{X}}(0) = \frac{d}{d\tau}\kappa_{XX}(0) = 0$. This shows that for a weakly homogeneous process, the stochastic variables $X(t)$ and $\dot{X}(t)$ are uncorrelated. Moreover, if $\{X(t),\ t \in R\}$ is Gaussian, it can be stated that $X(t)$ and $\dot{X}(t)$ are stochastically independent.

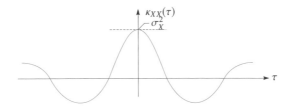

Fig. 3–18 Zero slope of the auto-covariance function at $\tau = 0$.

Using the Wiener-Khintchine relation Eq. (3-23) on the right-hand side of Eqs. (3-136) and (3-137), we have:

$$\kappa_{\dot{X}\dot{X}}(\tau) = -\frac{d^2}{d\tau^2} \int_{-\infty}^{\infty} e^{i\omega\tau} S_{XX}(\omega)\, d\omega = \int_{-\infty}^{\infty} e^{i\omega\tau} \omega^2 S_{XX}(\omega)\, d\omega \tag{3–139}$$

$$\kappa_{X^{(m)}X^{(n)}}(\tau) = (-1)^m \frac{d^{m+n}}{d\tau^{m+n}} \int_{-\infty}^{\infty} e^{i\omega\tau} S_{XX}(\omega)\, d\omega$$

$$= \int_{-\infty}^{\infty} e^{i\omega\tau} (-1)^m (i\omega)^{m+n} S_{XX}(\omega)\, d\omega \tag{3–140}$$

It follows from Eq. (3-23) that the auto-spectral density function $S_{\dot{X}\dot{X}}(\omega)$ of $\{\dot{X}(t),\, t \in R\}$, and the cross-spectral density function $S_{X^{(m)}X^{(n)}}(\omega)$ of $\{X^{(m)}(t),\, t \in R\}$ and $\{X^{(n)}(t),\, t \in R\}$ can be expressed as:

$$S_{\dot{X}\dot{X}}(\omega) = \omega^2 S_{XX}(\omega) \tag{3–141}$$

$$S_{X^{(m)}X^{(n)}}(\omega) = (-1)^m (i\omega)^{m+n} S_{XX}(\omega) \tag{3–142}$$

Then, the variance of $\{X^{(n)}(t),\, t \in R\}$ becomes:

$$\sigma^2_{X^{(n)}} = \int_{-\infty}^{\infty} \omega^{2n} S_{XX}(\omega)\, d\omega = \int_0^{\infty} \omega^{2n} S_X(\omega)\, d\omega = \lambda_{2n}, \quad n = 0, 1, \ldots \tag{3–143}$$

where λ_{2n} are the spectral moments defined in Eq. (3-48).

Time-series of physically realizable processes is continuous, and the variance function $\sigma^2_{X^{(n)}}(t)$ is finite. None of these requirements are fulfilled for a white noise process. Continuity of sample curves implies that the stochastic variables $X^{(n)}(t - \Delta t)$, $X^{(n)}(t)$ and $X^{(n)}(t + \Delta t)$ tend to become increasingly correlated as $\Delta t \to 0$. This means that the auto-covariance function $\kappa_{X^{(n)}X^{(n)}}(t,\, t + \Delta t)$ is continuous as $\Delta t \to 0$. For a homogeneous process, it follows from the first Wiener-Khintchine relation in Eq. (3-23) that continuity of $\kappa_{X^{(n)}X^{(n)}}(\tau)$ at $\tau = 0$ is achieved if the auto-spectral density function $S_{X^{(n)}X^{(n)}}(\omega) = \omega^{2n} S_{XX}(\omega)$ is integrable over the interval $]-\infty, \infty[$, i.e., the variance $\sigma^2_{X^{(n)}} < \infty$. Then, the requirements for existence of $\{X^{(n)}(t),\, t \in R\}$ can be stated in the following alternative ways:

1. $\kappa_{X^{(n)}X^{(n)}}(t_1, t_2)$ is continuous at the diagonal $t_1 = t_2 = t$.

2. $\kappa_{X^{(n)}X^{(n)}}(t, t) = \sigma^2_{X^{(n)}}(t) < \infty$.

For a stochastic process with rational auto-spectral density function, the variance function of the nth derivative process becomes, cf. Eqs. (3-54), (3-58) and (3-143):

$$\sigma^2_{X^{(n)}} = \int_{-\infty}^{\infty} \omega^{2n} \frac{P(i\omega)}{Q(i\omega)} \frac{P(-i\omega)}{Q(-i\omega)} S_0\, d\omega \tag{3–144}$$

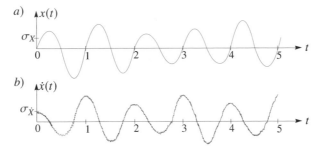

Fig. 3–19 Realizations of the displacement process and the velocity process of a process with rational spectrum of the order (0,2). a) Realization of $\{X(t),\ t \in R\}$. b) Realization of $\{\dot{X}(t),\ t \in R\}$.

$\{X^{(n)}(t),\ t \in R\}$ exists if the integral on the right-hand side of Eq. (3-144) exists. This will be the case if $n + r < s$, where r and s are the degrees of the polynomials $P(z)$ and $Q(z)$ as given by Eqs. (3-55) and (3-56). If $n + r = s - 1$, the realizations of $\{X^{(n)}(t),\ t \in R\}$ resemble the one shown in Fig. 3-19b.

Example 3-12: Velocity process and acceleration process of a process with rational spectrum of the order (r,s) = (0,2)

For a weakly stationary displacement process $\{X(t),\ t \in R\}$ with a rational auto-spectral density function given by Eq. (3-60), the auto-spectral density function of the velocity process $\{\dot{X}(t),\ t \in R\}$ and the acceleration process $\{\ddot{X}(t),\ t \in R\}$ can be calculated from Eqs. (3-141) and (3-142), respectively:

$$S_{\dot{X}\dot{X}}(\omega) = \frac{\omega^2 S_0}{m^2 \left((\omega_0^2 - \omega^2)^2 + 4\zeta^2 \omega_0^2 \omega^2 \right)} \tag{3–145}$$

$$S_{\ddot{X}\ddot{X}}(\omega) = \frac{\omega^4 S_0}{m^2 \left((\omega_0^2 - \omega^2)^2 + 4\zeta^2 \omega_0^2 \omega^2 \right)} \tag{3–146}$$

As $\omega \to \pm\infty$, $S_{\dot{X}\dot{X}}(\omega) \propto \frac{S_0}{m^2 \omega^2}$ and $S_{\ddot{X}\ddot{X}}(\omega) = \frac{S_0}{m^2}$. Therefore, the velocity process has a finite variance, which is not the case for the acceleration process. Correspondingly, $\{\dot{X}(t),\ t \in R\}$ is continuous but not differentiable, and the acceleration process $\{\ddot{X}(t),\ t \in R\}$ does not exists.

Typical realizations of $\{X(t),\ t \in R\}$ and $\{\dot{X}(t),\ t \in R\}$ are shown in Fig. 3-19. Similar to the realizations of the Wiener process in Fig. 3-14, the indicated ripples on top of $\dot{x}(t)$ are of infinitely small amplitude and infinitely small wave length, insuring that the tangent of $\dot{x}(t)$ is vertical everywhere, corresponding to infinite values of $\ddot{x}(t)$. Hence, the acceleration process has realizations that are discontinuous everywhere and with infinite variance, similar to a white noise process.

The auto-covariance function of $\{\dot{X}(t),\ t \in R\}$ follows from insertion of Eqs. (3-63), (3-64) and (3-65) into Eq. (3-128):

$$\kappa_{\dot{X}\dot{X}}(\tau) = \sigma_{\dot{X}}^2 \rho_{\dot{X}\dot{X}}(\tau) \tag{3–147}$$

$$\sigma_{\dot{X}}^2 = \frac{\pi S_0}{2\zeta \omega_0 m^2} = \omega_0^2 \sigma_X^2 \tag{3–148}$$

$$\rho_{\dot{X}\dot{X}}(\tau) = -\frac{\rho_{XX}''(\tau)}{\omega_0^2} = e^{-\zeta\omega_0|\tau|}\left(\cos\left(\omega_d\tau\right) - \frac{\zeta}{\sqrt{1-\zeta^2}}\sin\left(\omega_d|\tau|\right)\right) \tag{3–149}$$

where the primes indicate differentiation with respect to τ, and ω_d is given by Eq. (3-66). From Eq. (3-149), it follows that $\rho_{XX}''(0) = -\omega_0^2$.

Comparing Eq. (3-65) with Eq. (3-149), it is seen that in the narrow-banded case ($\zeta \ll 1$), we have:

$$\rho_{XX}(\tau) \simeq -\frac{\rho_{XX}''(\tau)}{\omega_0^2} \tag{3–150}$$

As follows from Eq. (3-149), $\frac{d^2}{d\tau^2}\kappa_{\dot{X}\dot{X}}(\tau)$ has different limits as $\tau \downarrow 0$ or $\tau \uparrow 0$. This is another indication of the non-existence of the acceleration process.

Example 3-13: Zero time-lag covariance matrix of $X(t)$, $\dot{X}(t)$ and $\ddot{X}(t)$ for a weakly homogeneous stochastic process

$\{X(t), t \in R\}$ is a weakly homogeneous stochastic process with the mean value function $\mu_X(t) = \mu_X$ and the auto-covariance function $\kappa_{XX}(\tau) = \sigma_X^2\,\rho(\tau)$. $\rho(\tau) = \rho_{XX}(\tau)$ signifies the auto-correlation coefficient function, cf. Eq. (3-8). The weakly homogeneous vector process $\{\mathbf{Y}(t), t \in R\}$ is considered:

$$\mathbf{Y}(t) = \begin{bmatrix} X(t) \\ \dot{X}(t) \\ \ddot{X}(t) \end{bmatrix} \tag{3–151}$$

At each time t, a covariance matrix $\mathbf{C_{YY}}(t)$ of $\mathbf{Y}(t)$ can be defined as $\mathbf{C_{YY}}(t) = E\left[\left(\mathbf{Y}(t) - \boldsymbol{\mu_Y}\right)\left(\mathbf{Y}(t) - \boldsymbol{\mu_Y}\right)^T\right]$, which is usually termed the zero time-lag covariance matrix. Due to the indicated homogeneity, $\mathbf{C_{YY}}(t)$ becomes independent of t, i.e., $\mathbf{C_{YY}}(t) = \mathbf{C_{YY}}(0) = \mathbf{C_{YY}}$.

From Eqs. (3-136), (3-137) and (3-138), the following auto- and cross-covariance functions at zero time-lag can be obtained:

$$\left. \begin{aligned} \kappa_{X\dot{X}}(0) &= \frac{d}{d\tau}\kappa_{XX}(0) = 0 \\[4pt] \kappa_{X\ddot{X}}(0) &= \frac{d^2}{d\tau^2}\kappa_{XX}(0) = \sigma_X^2\,\rho''(0) \\[4pt] \kappa_{\dot{X}\dot{X}}(0) &= -\frac{d^2}{d\tau^2}\kappa_{XX}(0) = -\sigma_X^2\,\rho''(0) \\[4pt] \kappa_{\dot{X}\ddot{X}}(0) &= -\frac{d}{d\tau}\kappa_{\dot{X}\dot{X}}(0) = 0 \\[4pt] \kappa_{\ddot{X}\ddot{X}}(0) &= \frac{d^4}{d\tau^4}\kappa_{XX}(0) = \sigma_X^2\,\rho^{(4)}(0) \end{aligned} \right\} \tag{3–152}$$

where $\rho''(0) = \frac{d^2}{d\tau^2}\rho(0)$ and $\rho^{(4)}(0) = \frac{d^4}{d\tau^4}\rho(0)$.

Hence, the mean value vector and the zero time-lag covariance matrix of $\{\mathbf{Y}(t),\ t \in R\}$ become:

$$\boldsymbol{\mu_Y} = \begin{bmatrix} \mu_X \\ 0 \\ 0 \end{bmatrix}, \quad \mathbf{C_{YY}} = \sigma_X^2 \begin{bmatrix} 1 & 0 & \rho''(0) \\ 0 & -\rho''(0) & 0 \\ \rho''(0) & 0 & \rho^{(4)}(0) \end{bmatrix} \tag{3-153}$$

Example 3-14: Joint probability density functions of $(X(t), \dot{X}(t))$ and of $(\dot{X}(t), \ddot{X}(t))$ for a weakly homogeneous Gaussian process

$\{X(t),\ t \in R\}$ is a weakly homogeneous Gaussian process. It follows from Eq. (3-152) that $X(t)$ and $\dot{X}(t)$ is uncorrelated, and $\dot{X}(t)$ and $\ddot{X}(t)$ is also uncorrelated. Since the indicated stochastic variables are all normally distributed, they are also stochastically independent. Then, the joint probability density functions become:

$$f_{X\dot{X}}(x, \dot{x}) = f_X(x) f_{\dot{X}}(\dot{x}) = \frac{1}{\sigma_X} \varphi\left(\frac{x - \mu_X}{\sigma_X}\right) \frac{1}{\sigma_{\dot{X}}} \varphi\left(\frac{\dot{x}}{\sigma_{\dot{X}}}\right) \tag{3-154}$$

$$f_{\dot{X}\ddot{X}}(\dot{x}, \ddot{x}) = f_{\dot{X}}(\dot{x}) f_{\ddot{X}}(\ddot{x}) = \frac{1}{\sigma_{\dot{X}}} \varphi\left(\frac{\dot{x}}{\sigma_{\dot{X}}}\right) \frac{1}{\sigma_{\ddot{X}}} \varphi\left(\frac{\ddot{x}}{\sigma_{\ddot{X}}}\right) \tag{3-155}$$

where $\varphi(x)$ is defined in Eq. (2-24), and $\sigma_{\dot{X}}$ and $\sigma_{\ddot{X}}$ are given by, cf. Eq. (3-152):

$$\left. \begin{aligned} \sigma_{\dot{X}} &= \sigma_X \sqrt{-\rho_{XX}''(0)} \\ \sigma_{\ddot{X}} &= \sigma_X \sqrt{\rho_{XX}^{(4)}(0)} \end{aligned} \right\} \tag{3-156}$$

3.5 Stochastic differential equations

A linear time-invariant differential system of order (r, s) has the form:

$$\left. \begin{aligned} R(t) &= p_0 Y^{(r)} + p_1 Y^{(r-1)} + \cdots + p_{r-1}\dot{Y} + p_r Y \\ Y^{(s)} &+ q_1 Y^{(s-1)} + q_2 Y^{(s-2)} + \cdots + q_{s-1}\dot{Y} + q_s Y = F(t) \end{aligned} \right\} \tag{3-157}$$

where $s > r$, and $p_0, \cdots, p_r, q_1, \cdots, q_s$ are deterministic real constants. In the case of a linear time-variant system, these are deterministic functions of time t.

The first equation is termed the *output differential equation*, and the stochastic process $\{R(t),\ t \in [t_0, \infty[\}$ is denoted the *output process* or the response process of the system.

The second equation is termed the *filter differential equation*, and the stochastic process $\{F(t),\ t \in [t_0, \infty[\}$ is denoted the *input process* or the load process of the system.

$\{Y(t),\ t \in [t_0, \infty[\}$ is an *auxiliary process* without any physical meaning, which is determined from the filter differential equation in combination with a set of prescribed initial values at the time $t = t_0$:

$$Y(t_0) = Y_0, \quad \dot{Y}(t_0) = \dot{Y}_0, \ \ldots, \quad Y^{(s-1)}(t_0) = Y_0^{(s-1)} \tag{3-158}$$

In what follows, it is assumed that the system is stable in the sense that any eigenvibration of the system due to the initial values in Eq. (3-158) will eventually be dissipated. On this assumption, a purely harmonic input with the complex amplitude F_0 and the angular frequency ω is considered:

$$F(t) = \mathrm{Re}\left(F_0\, e^{i\omega t}\right) \tag{3–159}$$

Due to the linearity of the system, after some time following the dissipation of initial values, the auxiliary process and the output process become harmonically varying with the same angular frequency as well:

$$Y(t) = \mathrm{Re}\left(Y_0\, e^{i\omega t}\right) \tag{3–160}$$

$$R(t) = \mathrm{Re}\left(R_0\, e^{i\omega t}\right) \tag{3–161}$$

Insertion of Eqs. (3-159), (3-160) and (3-161) into Eq. (3-157) provides the following relation between the complex amplitudes F_0, Y_0 and R_0 :

$$R_0\, e^{i\omega t} = \left(p_0(i\omega)^r + p_1(i\omega)^{r-1} + \ldots + p_{r-1}i\omega + +p_r\right) Y_0\, e^{i\omega t} \qquad \Rightarrow$$

$$R_0 = P(i\omega)\, Y_0 \tag{3–162}$$

$$\left((i\omega)^s + q_1(i\omega)^{s-1} + \ldots + q_{s-1}i\omega + q_s\right) Y_0\, e^{i\omega t} = F_0\, e^{i\omega t} \qquad \Rightarrow$$

$$Q(i\omega)\, Y_0 = F_0 \tag{3–163}$$

where the polynomials $P(z)$ and $Q(z)$ are given by Eqs. (3-55) and (3-56). Eliminating Y_0 from Eq. (3-162) and (3-163) yields:

$$R_0 = H(z)\, F_0, \quad z = i\omega \tag{3–164}$$

$$H(z) = \frac{P(z)}{Q(z)} \tag{3–165}$$

The frequency response function of the differential system, $H(i\omega)$, is seen to be identical to Eq. (3-54). Then, the solution to Eqs. (3-157) and (3-158) can be written in the following integral form:

$$R(t) = R^{(0)}(t) + \int_{t_0}^{t} h(t-\tau)F(\tau)\, d\tau \tag{3–166}$$

where $h(t)$ is the impulse response function given by Eq. (3-105).

$R^{(0)}(t)$ indicates the response from the initial values Y_0, $\dot{Y}_0$, $\ldots$, $Y_0^{(s-1)}$ of the auxiliary process.

Further, because of the linearity, $R^{(0)}(t)$ can be written as a linear combination of the initial values $Y_0, \dot{Y}_0, \ldots, Y_0^{(s-1)}$, i.e.:

$$R^{(0)}(t) = a_0(t)Y_0 + \ldots + a_{s-1}(t)Y_0^{(s-1)} \tag{3-167}$$

The deterministic functions $a_0(t), \cdots, a_{s-1}(t)$ can in turn be expressed as linear combinations of the impulse response function and its derivatives up to and including the order $s - 1$. As an example, these functions will be given for a single-degree-of-freedom system in Section 4.1. Hence, the complete solution of the problem is obtained if the impulse response function $h(t)$ of the differential system is known.

Since the linear time-invariant differential system of order (r, s) in Eq. (3-157) is equivalent to the stochastic integral equation in Eq. (3-166) with the impulse response function given by Eq. (3-105), the mean value function, the auto-covariance function, etc. can be evaluated from the corresponding results on stochastic integration presented in Section 3.2.

The filter differential equation in Eq. (3-157) may be reformulated in terms of the following *state vector* differential equation made up of s *coupled first order differential equations*:

$$\left. \begin{aligned} \frac{d}{dt}\mathbf{Y}(t) &= \mathbf{A}\,\mathbf{Y}(t) + \mathbf{b}\,F(t), \quad t > t_0 \\ \mathbf{Y}(t_0) &= \mathbf{Y}_0 \end{aligned} \right\} \tag{3-168}$$

$$\mathbf{Y}(t) = \begin{bmatrix} Y(t) \\ \dot{Y}(t) \\ \vdots \\ Y^{(s-2)}(t) \\ Y^{(s-1)}(t) \end{bmatrix}, \quad \mathbf{Y}_0 = \begin{bmatrix} Y_0 \\ \dot{Y}_0 \\ \vdots \\ Y_0^{(s-2)} \\ Y_0^{(s-1)} \end{bmatrix}, \quad \mathbf{b} = \begin{bmatrix} 0 \\ 0 \\ \vdots \\ 0 \\ 1 \end{bmatrix} \tag{3-169}$$

$$\mathbf{A} = \begin{bmatrix} 0 & 1 & 0 & \cdots & 0 \\ 0 & 0 & 1 & \cdots & 0 \\ \vdots & \vdots & \vdots & \ddots & \vdots \\ 0 & 0 & 0 & \cdots & 1 \\ -q_s & -q_{s-1} & -q_{s-2} & \cdots & -q_1 \end{bmatrix} \tag{3-170}$$

The output differential equation can be reformulated as:

$$R(t) = \mathbf{p}^T \mathbf{Y}(t) \tag{3-171}$$

$$\mathbf{p}^T = [p_r, p_{r-1}, \ldots, p_0, 0, \ldots, 0] \tag{3-172}$$

The first $s - 1$ of the differential equations in Eq. (3-168) simply expresses the identity $\frac{d}{dt}Y^{(i-1)}(t) = Y^{(i)}(t), i = 1, \ldots, s - 1$. The last equation is identical to the filter differential equation in Eq. (3-157). The state vector formulation in Eq. (3-168) is the standard form required in most explicit numerical integration schemes, such as the 4th order Runge-Kutta scheme.

It should be noticed that the differential system in Eq. (3-157) can be used to generate realizations of stochastic load processes, where the input process is a white noise process with the auto-spectral density S_0, and the parameters $p_0, \cdots, p_r$ and $q_1, \cdots, q_r$ of the filter are calibrated such the related auto-spectral density function $S_{RR}(\omega) = |H(\omega)|^2 S_0$ fits a given target auto-spectral density function, cf. Eq. (3-58). This approach will be considered in Section 6.1. Further, a certain response process $\{R(t), t \in [t_0, \infty[\}$ in a *multi-degree-of-freedom system* due to a single load process $\{F(t), t \in [t_0, \infty[\}$ applied at the response position or at another position in the structure, can also be analyzed by Eqs. (3-168) and (3-171). If s specifies the number of degrees of freedom of the system, the denominator polynomial will be of the degree $2s$, and the numerator polynomial will be of the degree $r \leq s - 1$. This approach will be demonstrated in Example 3-15 and in Section 4.1.

Example 3-15: Kanai-Tajimi filter

Fig. 3-20a shows a linear elastic one-storey frame based on a flexible sediment layer which in turn is settled upon bedrock. The columns and the beams are assumed to be massless and inextensible in the axial direction. The shear stiffness of the column is k, and the energy dissipation in the column is modelled by a linear viscous damper with the *damping coefficient c*. The storey beam with the mass m is assumed to have infinite bending and axial stiffness, and is rigidly connected to the columns. Hence, the horizontal displacement of the storey beam is described by only single degree of freedom, which is selected as the displacement $R(t)$ relative to the ground surface. Notice that the indicated structure should be considered as a symbolic representation of any continuous structure modelled by a single-degree-of-freedom system with the modal mass m, the modal stiffness k and the modal damping constant c.

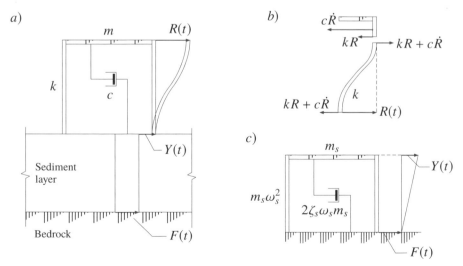

Fig. 3–20 Earthquake excitation with sediment layer modelled as a Kanai-Tajimi filter. a) Linear elastic one-storey shear frame. b) Internal forces applied to the free storey beam. c) Shear model for sediment layer.

The horizontal motion of the surface of the sediment layer relative to the bedrock is denoted $Y(t)$, and the horizontal motion of the surface of the bedrock is denoted $F(t)$, as shown in Fig. 3-20a.

The shear motion of the sediment layer is also modelled by a linear single-degree-of-freedom model, which may be represented by the shear frame shown in Fig. 3-20c. m_s is an equivalent mass, ω_s is the angular eigenfrequency and ζ_s is the damping ratio related to the considered mode of vibration. Since $m_s \gg m$, the reaction force from the structural shear frame onto the sediment layer is negligible compared to the inertia, damping and elastic forces of the sediment layer, and thus may be ignored.

The indicated model for the sediment layer is known as a *Kanai-Tajimi filter*,[1] which has wide applications in earthquake engineering. The main problem with the model is the selected linear viscous damping as the dissipation mechanism, which presumes a kind of mechanical friction between particles in the soil. In reality, the energy is removed by wave-propagation which is a much more effective way of removing mechanical energy. For this reason, relatively large values of ζ_s need to be selected in the Kanai-Tajimi model, typically $\zeta_s \simeq 0.4 - 0.6$.

The storey beam is cut free from the columns, and the internal elastic force $-kR(t)$ and the internal damping force $-c\dot{R}(t)$ are applied as equivalent external forces, as shown in Fig. 3-20b. Further, it is obvious that the total horizontal displacement of the storey beam is $F(t) + Y(t) + R(t)$. Then, use of Newton's 2nd law of motion for the free storey mass provides the equation of motion:

$$m\left(\ddot{F}(t) + \ddot{Y}(t) + \ddot{R}(t)\right) = -kR(t) - c\dot{R}(t) \qquad \Rightarrow$$

$$\ddot{R}(t) + 2\zeta\omega_0 \dot{R}(t) + \omega_0^2 R(t) = -\ddot{F}(t) - \ddot{Y}(t) \tag{3-173}$$

$$\omega_0 = \sqrt{\frac{k}{m}}, \quad \zeta = \frac{c}{2\sqrt{km}} \tag{3-174}$$

Next, Newton's 2nd law of motion is applied to the free storey beam of the equivalent frame representing the dynamic properties of the soil. The total horizontal displacement of the storey beam is $F(t) + Y(t)$, equal to the ground surface displacement. The equations of motion of the equivalent frame becomes:

$$m_s\left((\ddot{F}(t) + \ddot{Y}(t)\right) = -m_s\omega_s^2 Y(t) - 2\zeta_s\omega_s m_s \dot{Y}(t) \qquad \Rightarrow$$

$$\ddot{Y}(t) + 2\zeta_s\omega_s \dot{Y}(t) + \omega_s^2 Y(t) = -\ddot{F}(t) \tag{3-175}$$

Eq. (3-175) is solved for $\ddot{F}(t) + \ddot{Y}(t)$, and the result is inserted on the right-hand side of Eq. (3-173), yielding:

$$\ddot{R}(t) + 2\zeta\omega_0 \dot{R}(t) + \omega_0^2 R(t) = 2\zeta_s\omega_s \dot{Y}(t) + \omega_s^2 Y(t) \tag{3-176}$$

As seen from Eq. (3-175), the input stochastic process to the system is the earthquake acceleration process $\{\ddot{F}(t), t \in R\}$. To find the frequency response function for the output process $\{R(t), t \in R\}$, a harmonically varying input is considered:

$$\ddot{F}(t) = \ddot{F}_0 e^{i\omega t} \tag{3-177}$$

[1]H. Tajimi. A statistical method of determining the maximum response of a building structure during an earthquake. *Proc. 2nd World Conference on Earthquake Engineering*, Tokyo, 1960.

Due to linearity of the system, the stationary response of $Y(t)$ and $R(t)$ become harmonic as well:

$$Y(t) = Y_0 e^{i\omega t} \tag{3-178}$$

$$R(t) = R_0 e^{i\omega t} \tag{3-179}$$

Insertion of Eqs. (3-177), (3-178) and (3-179) into Eqs. (3-175) and (3-176), and removing the common factor $e^{i\omega t}$, provides the following relation between the complex amplitudes $\ddot{F}_0$, Y_0 and R_0:

$$\left.\begin{aligned}
\left((i\omega)^2 + 2\zeta\omega_0 (i\omega) + \omega_0^2\right) R_0 &= \left(2\zeta_s\omega_s (i\omega) + \omega_s^2\right) Y_0 \\
\left((i\omega)^2 + 2\zeta_s\omega_s (i\omega) + \omega_s^2\right) Y_0 &= -\ddot{F}_0
\end{aligned}\right\} \tag{3-180}$$

Elimination of Y_0 in Eq. (3-180) provides the following frequency response function:

$$H(z) = \frac{R_0}{\ddot{F}_0} = \frac{P(z)}{Q(z)}, \quad z = i\omega \tag{3-181}$$

$$P(z) = -2\zeta_s\omega_s z - \omega_s^2 \tag{3-182}$$

$$\begin{aligned}
Q(z) &= \left(z^2 + 2\zeta\omega_0 z + \omega_0^2\right)\left(z^2 + 2\zeta_s\omega_s z + \omega_s^2\right) \\
&= z^4 + \left(2\zeta\omega_0 + 2\zeta_s\omega_s\right) z^3 + \left(\omega_0^2 + \omega_s^2 + 4\zeta\zeta_s\omega_0\omega_s\right) z^2 + \left(2\zeta\omega_s + 2\zeta_s\omega_0\right)\omega_0\omega_s z + \omega_0^2\omega_s^2
\end{aligned} \tag{3-183}$$

Hence, the relative displacement process $\{R(t), t \in [0, \infty[\}$ can be calculated by the stochastic differential equations Eq. (3-157), with the input process given by $\{\ddot{F}(t), t \in [0, \infty[\}$. The system is of the order $(r, s) = (1, 4)$, and the parameters are given by:

$$\left.\begin{aligned}
p_0 &= -2\zeta_s\omega_s & , \quad p_1 &= -\omega_s^2 \\
q_1 &= 2\zeta\omega_0 + 2\zeta_s\omega_s & , \quad q_2 &= \omega_0^2 + \omega_s^2 + 4\zeta\zeta_s\omega_0\omega_s \\
q_3 &= \left(2\zeta\omega_s + 2\zeta_s\omega_0\right)\omega_0\omega_s, & q_4 &= \omega_0^2\omega_s^2
\end{aligned}\right\} \tag{3-184}$$

Next, state vector formulation of Eqs. (3-175) and (3-176) is to be established, which directly provides the displacement and velocity processes $\{R(t), t \in [0, \infty[\}$ and $\{\dot{R}(t), t \in [0, \infty[\}$. Applying Eqs. (3-168), (3-169) and (3-170), we have the state vector formulation for the present example:

$$\left.\begin{aligned}
\frac{d}{dt}\mathbf{Y}(t) &= \mathbf{A}\,\mathbf{Y}(t) + \mathbf{b}\,\ddot{F}(t), \quad t > 0 \\
\mathbf{Y}(0) &= \mathbf{0}
\end{aligned}\right\} \tag{3-185}$$

$$\mathbf{Y}(t) = \begin{bmatrix} R(t) \\ \dot{R}(t) \\ Y(t) \\ \dot{Y}(t) \end{bmatrix}, \quad \mathbf{b} = \begin{bmatrix} 0 \\ 0 \\ 0 \\ -1 \end{bmatrix}, \quad \mathbf{A} = \begin{bmatrix} 0 & 1 & 0 & 0 \\ -\omega_0^2 & -2\zeta\omega_0 & \omega_s^2 & 2\zeta_s\omega_s \\ 0 & 0 & 0 & 1 \\ 0 & 0 & -\omega_s^2 & -2\zeta_s\omega_s \end{bmatrix} \tag{3-186}$$

Even though Eq. (3-168) has been derived from the differential system in Eq. (3-157), this equation indicates the general state vector format for a linear multi-degree-of-freedom system exposed to a single input $F(t)$. Of course, the physical meanings of the state vector $\mathbf{Y}(t)$, the system matrix $\mathbf{A}$ and the load distribution vector $\mathbf{b}$ can be different in different cases, and the Kanai-Tajimi filter is just one example. Therefore, the following analysis based on Eq. (3-168) can be used to cover a much broader range of applications.

Let $\{F(t), t \in R\}$ be a white noise process with the auto-spectral density S_0. Further, assume that the initial value vector $\mathbf{Y}_0$ is stochastic and stochastically independent of the input process. Upon taking the expectations on both sides of Eq. (3-168) and using $E[F(t)] = 0$, the differential equation for the mean value vector function $\boldsymbol{\mu}_{\mathbf{Y}}(t)$ is obtained:

$$\frac{d}{dt} \boldsymbol{\mu}_{\mathbf{Y}}(t) = \mathbf{A}\, \boldsymbol{\mu}_{\mathbf{Y}}(t), \quad \boldsymbol{\mu}_{\mathbf{Y}}(t_0) = E[\mathbf{Y}_0] \tag{3-187}$$

A similar differential equation can be derived for the zero time-lag 2nd order joint moment matrix $\mathbf{R}_{\mathbf{YY}}(t) = E[\mathbf{Y}(t)\mathbf{Y}^T(t)]$. Using Eq. (3-168), we have:

$$\frac{d}{dt} E\left[\mathbf{Y}(t)\mathbf{Y}^T(t)\right] = E\left[\dot{\mathbf{Y}}(t)\mathbf{Y}^T(t)\right] + E\left[\mathbf{Y}(t)\dot{\mathbf{Y}}^T(t)\right]$$

$$= \mathbf{A}\, E\left[\mathbf{Y}(t)\mathbf{Y}^T(t)\right] + E\left[\mathbf{Y}(t)\mathbf{Y}^T(t)\right]\mathbf{A}^T + \mathbf{b}\, E\left[F(t)\mathbf{Y}^T(t)\right] + E\left[\mathbf{Y}(t)F(t)\right]\mathbf{b}^T$$

$$\Rightarrow \quad \frac{d}{dt} \mathbf{R}_{\mathbf{YY}}(t) = \mathbf{A}\,\mathbf{R}_{\mathbf{YY}}(t) + \mathbf{R}_{\mathbf{YY}}(t)\,\mathbf{A}^T + \mathbf{b}\, E\left[F(t)\mathbf{Y}^T(t)\right] + E\left[\mathbf{Y}(t)F(t)\right]\mathbf{b}^T \tag{3-188}$$

Further, it follows from Eq. (3-168) that:

$$\mathbf{Y}(t) = \mathbf{Y}_0 + \int_{t_0}^{t} \left(\mathbf{A}\,\mathbf{Y}(\tau) + \mathbf{b}\, F(\tau)\right) d\tau \quad \Rightarrow$$

$$E\left[\mathbf{Y}(t)F(t)\right] = E\left[\mathbf{Y}_0 F(t)\right] + \int_{t_0}^{t} \left(\mathbf{A}\, E\left[\mathbf{Y}(\tau)F(t)\right] + \mathbf{b}\, E\left[F(\tau)F(t)\right]\right) d\tau \tag{3-189}$$

In the last statement of Eq. (3-189), it has been applied that the integration is with respect to τ. Hence, $F(t)$ is considered as a constant for fixed t and can be entered into the integral. Since $\{F(t), t \in R\}$ has been assumed to be stochastically independent of the initial value vector $\mathbf{Y}_0$, the following expression holds:

$$E\left[\mathbf{Y}_0 F(t)\right] = E\left[\mathbf{Y}_0\right] E\left[F(t)\right] = E\left[\mathbf{Y}_0\right] \cdot 0 = \mathbf{0} \tag{3-190}$$

Moreover, $\mathbf{Y}(\tau)$ depends on the initial value vector $\mathbf{Y}_0$ and the excitations in the interval $]t_0, \tau[$, i.e., on the set of stochastic variables $\{F(u), u \in]t_0, \tau[\}$. Since $\{F(t), t \in R\}$ is a Gaussian white noise process independent of $\mathbf{Y}_0$, $F(t)$ for $t > \tau$ is stochastically independent of $\{F(u), u \in]t_0, \tau[\}$ and $\mathbf{Y}_0$. Hence, $F(t)$ and $\mathbf{Y}(\tau)$ are stochastically independent, leading to:

$$E\left[\mathbf{Y}(\tau)F(t)\right] = E\left[\mathbf{Y}(\tau)\right] E\left[F(t)\right] = \boldsymbol{\mu}_{\mathbf{Y}}(\tau) \cdot 0 = \mathbf{0}, \quad t > \tau \tag{3-191}$$

Recall from Eq. (3-40) that:

$$E\left[F(\tau)F(t)\right] = 2\pi S_0\, \delta(t - \tau) \tag{3-192}$$

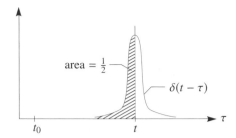

Fig. 3–21 Interpretation of the Dirach delta function.

Inserting Eqs. (3-190), (3-191) and (3-192) into Eq. (3-189) provides:

$$E\left[\mathbf{Y}(t)F(t)\right] = \int_{t_0}^{t} \mathbf{b}\, 2\pi S_0\, \delta(t-\tau)\, d\tau = \frac{1}{2}\,\mathbf{b}\, 2\pi S_0 \tag{3–193}$$

The factor $\frac{1}{2}$ on the right-hand side of Eq. (3-193) is brought forward because the integration is only performed over half a delta-spike, as shown in Fig. 3-21. Finally, insertion of Eq. (3-193) into Eq. (3-188) provides the following differential equation:

$$\frac{d}{dt}\mathbf{R_{YY}}(t) = \mathbf{A}\,\mathbf{R_{YY}}(t) + \mathbf{R_{YY}}(t)\,\mathbf{A}^T + 2\pi S_0\,\mathbf{b}\mathbf{b}^T, \quad \mathbf{R_{YY}}(t_0) = E\left[\mathbf{Y}_0\mathbf{Y}_0^T\right] \tag{3–194}$$

Because of the symmetry condition $\mathbf{R_{YY}}(t) = \mathbf{R_{YY}}^T(t)$, Eq. (3-194) merely contains $\frac{1}{2}s(s+1)$ nontrivial coupled linear differential equations. Moreover, if $\mathbf{Y}_0$ is assumed to be normally distributed, the vector process $\{\mathbf{Y}(t),\ t \in [t_0,\infty[\}$ also becomes a Gaussian (normal) process due to the linearity. Then, the stochastic state vector $\mathbf{Y}(t)$ is completely described by the solutions of Eq. (3-187) and (3-194).

As stated in Example 3-13, at each time t, a covariance matrix of $\mathbf{Y}(t)$ can be defined as $\mathbf{C_{YY}}(t){=}E\left[(\mathbf{Y}(t)-\boldsymbol{\mu_Y})(\mathbf{Y}(t)-\boldsymbol{\mu_Y})^T\right]$, which is termed the zero time-lag covariance matrix. This can also be written as $\mathbf{C_{YY}}(t) = \mathbf{R_{YY}}(t) - \boldsymbol{\mu_Y}(t)\boldsymbol{\mu_Y}^T(t)$. From Eq. (3-187) and (3-194), it follows that:

$$\frac{d}{dt}\mathbf{C_{YY}}(t) = \frac{d}{dt}\mathbf{R_{YY}}(t) - \dot{\boldsymbol{\mu}}_\mathbf{Y}(t)\boldsymbol{\mu_Y}^T(t) - \boldsymbol{\mu_Y}(t)\dot{\boldsymbol{\mu}}_\mathbf{Y}^T(t)$$

$$= \mathbf{A}\left(\mathbf{R_{YY}}(t) - \boldsymbol{\mu_Y}(t)\boldsymbol{\mu_Y}^T(t)\right) + \left(\mathbf{R_{YY}}(t) - \boldsymbol{\mu_Y}(t)\boldsymbol{\mu_Y}^T(t)\right)\mathbf{A}^T + 2\pi S_0\,\mathbf{b}\mathbf{b}^T$$

$$= \mathbf{A}\,\mathbf{C_{YY}}(t) + \mathbf{C_{YY}}(t)\,\mathbf{A}^T + 2\pi S_0\,\mathbf{b}\mathbf{b}^T, \quad t > t_0 \tag{3–195}$$

where $\mathbf{C_{YY}}(t)$ fulfills the initial value condition $\mathbf{C_{YY}}(t_0) = E[\mathbf{Y}_0\mathbf{Y}_0^T] - E[\mathbf{Y}_0]E[\mathbf{Y}_0^T]$.

As time evolves, the initial value response is dissipated, and $\{\mathbf{Y}(t),\ t \in [t_0,\infty[\}$ eventually approaches to a stationary vector process. In this state, $\boldsymbol{\mu_Y}(t) = \boldsymbol{\mu_Y}$ and $\mathbf{C_{YY}}(t) = \mathbf{C_{YY}}$ become time-independent, implying $\frac{d}{dt}\boldsymbol{\mu_Y} = \mathbf{0}$, $\frac{d}{dt}\mathbf{C_{YY}} = \mathbf{0}$. Hence, the mean value vector and the zero time-lag covariance matrix at stationarity fulfill:

$$\boldsymbol{\mu_Y} = \mathbf{0} \tag{3–196}$$

$$\mathbf{A} \, \mathbf{C_{YY}} + \mathbf{C_{YY}} \, \mathbf{A}^T + 2\pi S_0 \, \mathbf{b} \mathbf{b}^T = \mathbf{0} \tag{3–197}$$

Eq. (3-197), which is termed the *Lyapunov equation*, can be recast into a system of $\frac{1}{2}s(s+1)$ linear equations for determining the stationary covariances (or variances) of different components of $\mathbf{Y}(t)$.

If the mean value vector function $\boldsymbol{\mu_Y}(t)$ and the zero time-lag covariance matrix $\mathbf{C_{YY}}(t)$ of $\{\mathbf{Y}(t), \, t \in [t_0, \infty[\}$ have been calculated, the mean value function and the variance function of the output process $\{R(t), \, t \in [t_0, \infty[\}$ in Eq. (3-157) follow from Eq. (3-171):

$$\mu_R(t) = \mathbf{p}^T \boldsymbol{\mu_Y}(t) \tag{3–198}$$

$$\sigma_R^2(t) = \mathbf{p}^T \mathbf{C_{YY}}(t) \, \mathbf{p} \tag{3–199}$$

Example 3-16: Ornstein-Uhlenbeck process

The *Ornstein-Uhlenbeck process* is the output process of the following first order stochastic differential equation:

$$\left. \begin{array}{l} \dot{Y}(t) + \alpha \, Y(t) = \alpha \, F(t) \\ Y(0) = y_0 \end{array} \right\} \tag{3–200}$$

where $\{F(t), \, t \in [0, \infty[\}$ is a white noise process with the auto-spectral density function S_0, and $\alpha > 0$.

The frequency response function of the indicated system becomes:

$$H(\omega) = \frac{\alpha}{z + \alpha}, \quad z = i\omega \tag{3–201}$$

Hence, the frequency response function is rational of the order $(r, s) = (0, 1)$, with $P(z) = \alpha$ and $Q(z) = z + \alpha$. The pole is $z_1 = -\alpha$.

Then, the auto-covariance function and the auto-spectral density function of $Y(t)$ in the stationary state, where the response from the initial value y_0 has been dissipated, follow from Eqs. (3-59) and (3-104):

$$\kappa_{YY}(\tau) = \pi S_0 \frac{e^{-\alpha|\tau|}}{\alpha} \alpha^2 = \pi S_0 \, \alpha \, e^{-\alpha|\tau|} \tag{3–202}$$

$$S_{YY}(\tau) = |H(\omega)|^2 \, S_0 = \frac{\alpha^2}{\alpha^2 + \omega^2} \, S_0 \tag{3–203}$$

The frequency response function of the Ornstein-Uhlenbeck process is a so-called low-pass filter. Harmonic components in a signal with angular frequencies $\omega \ll \alpha$ are unaffected by the filtration, since $H(\omega) \simeq 1$ at these frequencies. The Ornstein-Uhlenbeck process has continuous but non-differentiable realizations, quite similar to those of the Wiener process, although the stationary variance attains a constant value $\pi S_0 \, \alpha$.

Example 3-17: Rational filter of the order $(r, s)=(0,2)$ exposed to a white noise process

The polynomials $P(z)$ and $Q(z)$ of the frequency response function are given by Eq. (3-61). Then, the differential system in Eq. (3-157) attains the form:

$$\left.\begin{aligned} X(t) &= \frac{1}{m} Y(t) \\ \ddot{Y}(t) + 2\zeta\omega_0 \dot{Y}(t) + \omega_0^2 Y(t) &= F(t) \end{aligned}\right\} \tag{3–204}$$

where the input process $\{F(t),\ t \in [0, \infty[\}$ is a white noise with the auto-spectral density function S_0.

In the present case, Eqs. (3-169) and (3-170) attain the form:

$$\mathbf{A} = \begin{bmatrix} 0 & 1 \\ -\omega_0^2 & -2\zeta\omega_0 \end{bmatrix}, \quad \mathbf{b} = \begin{bmatrix} 0 \\ 1 \end{bmatrix} \tag{3–205}$$

The stationary (zero time-lag) covariance matrix is written as:

$$\mathbf{C_{YY}} = \begin{bmatrix} \sigma_Y^2 & \kappa_{Y\dot{Y}} \\ \kappa_{Y\dot{Y}} & \sigma_{\dot{Y}}^2 \end{bmatrix} \tag{3–206}$$

Then, the Lyapunov equation Eq. (3-197) attains the following form for the present example:

$$\begin{bmatrix} \sigma_Y^2 & \kappa_{Y\dot{Y}} \\ \kappa_{Y\dot{Y}} & \sigma_{\dot{Y}}^2 \end{bmatrix} \begin{bmatrix} 0 & -\omega_0^2 \\ 1 & -2\zeta\omega_0 \end{bmatrix} + \begin{bmatrix} 0 & 1 \\ -\omega_0^2 & -2\zeta\omega_0 \end{bmatrix} \begin{bmatrix} \sigma_Y^2 & \kappa_{Y\dot{Y}} \\ \kappa_{Y\dot{Y}} & \sigma_{\dot{Y}}^2 \end{bmatrix} + 2\pi S_0 \begin{bmatrix} 0 & 0 \\ 0 & 1 \end{bmatrix} = \begin{bmatrix} 0 & 0 \\ 0 & 0 \end{bmatrix} \Rightarrow$$

$$\begin{bmatrix} 2\kappa_{Y\dot{Y}} & -\omega_0^2\sigma_Y^2 - 2\zeta\omega_0\kappa_{Y\dot{Y}} + \sigma_{\dot{Y}}^2 \\ -\omega_0^2\sigma_Y^2 - 2\zeta\omega_0\kappa_{Y\dot{Y}} + \sigma_{\dot{Y}}^2 & -2\omega_0^2\kappa_{Y\dot{Y}} - 4\zeta\omega_0\sigma_{\dot{Y}}^2 \end{bmatrix} + 2\pi S_0 \begin{bmatrix} 0 & 0 \\ 0 & 1 \end{bmatrix} = \begin{bmatrix} 0 & 0 \\ 0 & 0 \end{bmatrix} \tag{3–207}$$

From the three non-trivial relations in Eq. (3-207), the following solutions for $\kappa_{Y\dot{Y}}$, σ_Y^2 and $\sigma_{\dot{Y}}^2$ can be obtained:

$$\begin{bmatrix} \kappa_{Y\dot{Y}} \\ \sigma_Y^2 \\ \sigma_{\dot{Y}}^2 \end{bmatrix} = \begin{bmatrix} 0 \\ \frac{\pi S_0}{2\zeta\,\omega_0^3} \\ \frac{\pi S_0}{2\zeta\,\omega_0} \end{bmatrix} \tag{3–208}$$

Since in the present example $m = 1$ in the filter equation, the results in Eq. (3-208) are actually in perfect agreement with previous derivations of these results in Eqs. (3-64), (3-148) and (3-152), based on a totally different approach.

Next, Eq. (3-168) is generalized to the following multi-dimensional input case:

$$\frac{d}{dt}\mathbf{Y}(t) = \mathbf{A}\,\mathbf{Y}(t) + \mathbf{B}(t)\,\mathbf{F}(t), \quad t > t_0, \quad \mathbf{Y}(t_0) = \mathbf{Y}_0 \tag{3–209}$$

$\{\mathbf{F}(t),\ t \in R\}$ is an m-dimensional input stochastic vector process, of which the component processes $\{F_j(t),\ t \in R\},\ j = 1, \ldots, m$ are mutually independent unit white noise processes, cf. Eq. (3-40):

$$\left.\begin{aligned} E[F_j(t)] &= 0 \\ E[F_j(t_1)F_k(t_2)] &= \delta_{jk}\,\delta(t_2 - t_1) \end{aligned}\right\} \tag{3–210}$$

where δ_{jk} signifies *Kronecker's delta*. $\mathbf{B}(t)$ is a time-varying, deterministic $n \times m$-dimensional matrix, specifying the intensity and possible correlations among the loads on the system. The differential equation for the mean value vector function is unchanged as given by Eq. (3-187), whereas the differential equation for the zero time-lag covariance matrix becomes:

$$\frac{d}{dt}\mathbf{C_{YY}}(t) = \mathbf{A}\,\mathbf{C_{YY}}(t) + \mathbf{C_{YY}}(t)\,\mathbf{A}^T + \mathbf{B}(t)\mathbf{B}^T(t), \quad t > t_0 \tag{3-211}$$

Example 3-18: Two linear 1st order stochastic differential equations with two white noise input processes

It is assumed that the state vector $\mathbf{Y}(t)$ and the vector of the mutually independent unit white noise processes $\mathbf{F}(t)$ in Eq. (3-209) are both 2-dimensional, i.e.:

$$\mathbf{Y}(t) = \begin{bmatrix} Y_1(t) \\ Y_2(t) \end{bmatrix}, \quad \mathbf{F}(t) = \begin{bmatrix} F_1(t) \\ F_2(t) \end{bmatrix} \tag{3-212}$$

As a numerical example, let the system matrices $\mathbf{A}$ and $\mathbf{B}$ and the deterministic initial value vector $\mathbf{y}_0$ be written as:

$$\mathbf{A} = \begin{bmatrix} 2 & -8 \\ 5 & -10 \end{bmatrix}, \quad \mathbf{B} = \begin{bmatrix} 2 & -1 \\ 1 & -1 \end{bmatrix}, \quad \mathbf{y}_0 = \begin{bmatrix} 1 \\ 1 \end{bmatrix} \tag{3-213}$$

The differential equations, Eq. (3-187) for the mean value functions $\mu_{Y_j}(t)$, and Eq. (3-211) for the zero time-lag auto- and cross-covariance functions $\kappa_{Y_iY_j}(t,t)$, thus attain the form:

$$\frac{d}{dt}\begin{bmatrix} \mu_{Y_1} \\ \mu_{Y_2} \end{bmatrix} = \begin{bmatrix} 2 & -8 \\ 5 & -10 \end{bmatrix}\begin{bmatrix} \mu_{Y_1} \\ \mu_{Y_2} \end{bmatrix}, \quad \begin{bmatrix} \mu_{Y_1}(0) \\ \mu_{Y_2}(0) \end{bmatrix} = \begin{bmatrix} 1 \\ 1 \end{bmatrix} \tag{3-214}$$

$$\frac{d}{dt}\begin{bmatrix} \sigma_{Y_1}^2 & \kappa_{Y_1Y_2} \\ \kappa_{Y_1Y_2} & \sigma_{Y_2}^2 \end{bmatrix} = \begin{bmatrix} 2 & -8 \\ 5 & -10 \end{bmatrix}\begin{bmatrix} \sigma_{Y_1}^2 & \kappa_{Y_1Y_2} \\ \kappa_{Y_1Y_2} & \sigma_{Y_2}^2 \end{bmatrix} + \begin{bmatrix} \sigma_{Y_1}^2 & \kappa_{Y_1Y_2} \\ \kappa_{Y_1Y_2} & \sigma_{Y_2}^2 \end{bmatrix}\begin{bmatrix} 2 & 5 \\ -8 & -10 \end{bmatrix} + \begin{bmatrix} 2 & -1 \\ 1 & -1 \end{bmatrix}\begin{bmatrix} 2 & 1 \\ -1 & -1 \end{bmatrix}$$

$$= \begin{bmatrix} 4\sigma_{Y_1}^2 - 16\kappa_{Y_1Y_2} & 5\sigma_{Y_1}^2 - 8\kappa_{Y_1Y_2} - 8\sigma_{Y_2}^2 \\ 5\sigma_{Y_1}^2 - 8\kappa_{Y_1Y_2} - 8\sigma_{Y_2}^2 & 10\kappa_{Y_1Y_2} - 20\sigma_{Y_2}^2 \end{bmatrix} + \begin{bmatrix} 5 & 3 \\ 3 & 2 \end{bmatrix} \tag{3-215}$$

The three nontrivial differential equations for $\sigma_{Y_1}^2(t)$, $\kappa_{Y_1Y_2}(t)$ and $\sigma_{Y_2}^2(t)$ may be reformulated in terms of the following vector differential equation:

$$\frac{d}{dt}\begin{bmatrix} \sigma_{Y_1}^2 \\ \kappa_{Y_1Y_2} \\ \sigma_{Y_2}^2 \end{bmatrix} = \begin{bmatrix} 4 & -16 & 0 \\ 5 & -8 & -8 \\ 0 & 10 & -20 \end{bmatrix}\begin{bmatrix} \sigma_{Y_1}^2 \\ \kappa_{Y_1Y_2} \\ \sigma_{Y_2}^2 \end{bmatrix} + \begin{bmatrix} 5 \\ 3 \\ 2 \end{bmatrix}, \quad \begin{bmatrix} \sigma_{Y_1}^2(0) \\ \kappa_{Y_1Y_2}(0) \\ \sigma_{Y_2}^2(0) \end{bmatrix} = \begin{bmatrix} 0 \\ 0 \\ 0 \end{bmatrix} \tag{3-216}$$

Eqs. (3-214) and (3-216) are linear 1st order differential equations with constant coefficient matrices of the dimension 2 and 3, respectively. Solutions can be obtained from the standard theory for ordinary differential equations: [1]

$$\begin{bmatrix} \mu_{Y_1}(t) \\ \mu_{Y_2}(t) \end{bmatrix} = \begin{bmatrix} 1 \\ 1 \end{bmatrix} e^{-4t}\cos(2t) - \begin{bmatrix} 1 \\ \frac{1}{2} \end{bmatrix} e^{-4t}\sin(2t) \tag{3-217}$$

---

[1]D.G. Zill and M.R. Cullen. *Differential Equations with Boundary-Value Problems, 7th Edition.* Brooks/Cole Publishing Company, Belmont, California, 2009.

$$\begin{bmatrix} \sigma_{Y_1}^2(t) \\ \kappa_{Y_1 Y_2}(t) \\ \sigma_{Y_2}^2(t) \end{bmatrix} = \frac{1}{320}\left(\begin{bmatrix} 248 \\ 162 \\ 113 \end{bmatrix} - \begin{bmatrix} 200 \\ 150 \\ 125 \end{bmatrix} e^{-8t} - \begin{bmatrix} 48 \\ 12 \\ -12 \end{bmatrix} e^{-8t}\cos(4t) - \begin{bmatrix} 96 \\ 84 \\ 66 \end{bmatrix} e^{-8t}\sin(4t)\right) \tag{3-218}$$

As seen from Eqs. (3-217) and (3-218), the mean value functions and the zero time-lag covariance functions approach the following stationary values as $t \to \infty$:

$$\begin{bmatrix} \mu_{Y_1}(\infty) \\ \mu_{Y_2}(\infty) \end{bmatrix} = \begin{bmatrix} 0 \\ 0 \end{bmatrix}, \qquad \begin{bmatrix} \sigma_{Y_1}^2(\infty) \\ \kappa_{Y_1 Y_2}(\infty) \\ \sigma_{Y_2}^2(\infty) \end{bmatrix} = \frac{1}{320}\begin{bmatrix} 248 \\ 162 \\ 113 \end{bmatrix} \tag{3-219}$$

It should be noticed that the convergence towards stationarity of the mean value functions and the covariance functions is controlled by the factors $\exp(-4t)$ and $\exp(-8t)$, respectively. Hence, the covariances approach the stationary state significantly faster than the mean values do.

3.6 Ergodic sampling and ergodic processes

If a physical phenomenon is to be modelled by a stochastic process $\{X(t),\ t \in [0, \infty[\}$, the statistical properties of the process such as the mean value function and the auto-covariance function need to be determined from the available measurements. Assuming that N independent realizations $x_1(t),\ x_2(t),\ \ldots,\ x_N(t)$ are available, the mean value function and the auto-covariance function can be estimated from the ensemble averages:

$$\mu_X(t) \simeq \frac{1}{N}\sum_{n=1}^{N} x_n(t) \tag{3-220}$$

$$\kappa_{XX}(t_1, t_2) \simeq \frac{1}{N}\sum_{n=1}^{N} x_n(t_1)x_n(t_2) - \mu_X(t_1)\mu_X(t_2) \tag{3-221}$$

where the estimate in Eq. (3-220) for the mean value function is applied on the right-hand side of Eq. (3-221). The accuracy of the estimates depends on the number N of the available time-series.

Quite often, only a single long time-series $x(t)$ of the length T is available. Hence, the probabilistic structure of the underlying stochastic process must be estimated based on this time-series. This is possible, if from the physical nature of the underlying problem, and based on a preliminary statistical examination of the available time-series, that no systematic changes in the mean value function, the variance function and the spectral contents with time are revealed.

It is assumed that the time-series covers the interval $[0, T]$ of a realization of a stationary process $\{X(t),\ t \in R\}$. This time-series is divided into N sub time-series of the length $T_0 = \frac{T}{N}$, as shown

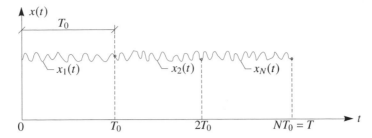

Fig. 3–22 Time-series of a stationary process $\{X(t), t \in R\}$ in the interval $[0, T]$.

in Fig. 3-22:

$$\left.\begin{aligned} x_1(t) &= x(t) \\ x_2(t) &= x(t + T_0) \\ &\vdots \\ x_N(t) &= x\big(t + (N-1)T_0\big) \end{aligned}\right\} , \quad t \in [0, T_0] \tag{3--222}$$

The length T_0 must be sufficiently large, so that the stochastic variables $X(t)$, $X(t + T_0)$, $\ldots$, $X\big(t+(N-1)T_0\big)$ may be assumed to be stochastically independent for the same $t \in [0, T_0]$. It does not mean that $x_1(t)$, $x_2(t)$, $\ldots$, $x_N(t)$ are realizations of mutually independent stochastic processes $\{X_j(t), t \in [0, T_0]\}$. Actually, $X_1(T_0) = X_2(0)$, $\ldots$, $X_{N-1}(T_0) = X_N(0)$. The requirement may be assumed to be valid if T_0 is comparable to the correlation length τ_0 of the process, which can only be verified after the auto-correlation coefficient function $\rho_{XX}(\tau)$ has been estimated from the time-series, cf. Eqs. (3-19) and (3-20). All these heuristic arguments suggest that the statistical properties of the process may be estimated based on such sub time-series.

Then, the mean value function may be estimated from, cf. Eq. (3-220):

$$\mu_X \simeq \frac{1}{N} \sum_{n=1}^{N} x_n(t) \tag{3--223}$$

Due to the assumed stationarity, Eq. (3-223) is valid for any $t \in [0, T_0]$. Especially, Eq. (3-223) may be calculated for the index values $t = 0$, $t = \Delta t$, $\ldots$, $t = (M-1)\Delta t$, where $\Delta t = \frac{T_0}{M}$. The best estimate is obtained as the mean value of these local M estimates:

$$\mu_X \simeq \frac{1}{M} \sum_{m=1}^{M} \frac{1}{N} \sum_{n=1}^{N} x_n\big((m-1)\Delta t\big)$$

$$= \frac{1}{MN\Delta t} \sum_{m=1}^{M} \sum_{n=1}^{N} x_n\big((m-1)\Delta t\big) \Delta t \xrightarrow[\Delta t \to 0]{} \frac{1}{T} \int_0^T x(t)\, dt \tag{3--224}$$

where it has been used that $T = NT_0 = MN\Delta t$, and $x(t)$ is the given realization of the length T.

The auto-covariance function may be estimated from, cf. Eq. (3-221):

$$\kappa_{XX}(t, t + \tau) = \kappa_{XX}(\tau) \simeq \frac{1}{N} \sum_{n=1}^{N} x_n(t) x_n(t + \tau) - \mu_X^2 \tag{3–225}$$

where μ_X is estimated from Eq. (3-224).

Using a similar averaging technique for local estimates of $\kappa_{XX}(t, t + \tau)$ as given by Eq. (3-225) at the times $t = 0$, $t = \Delta t$, ..., $t = (M - 1)\Delta t$, and letting $\Delta t \to 0$, it may be shown that the best estimate of the auto-covariance function is obtained from the following *time average*:

$$\kappa_{XX}(\tau) \simeq \frac{1}{T - \tau} \int_0^{T-\tau} x(t) x(t + \tau) \, dt - \mu_X^2, \quad \tau > 0 \tag{3–226}$$

By Eqs. (3-224) and (3-226), the problem of estimating the mean value function and the auto-covariance function of the stochastic process model has been reduced to the evaluation of certain time averages of the given time-series $x(t)$. This is the so-called *ergodic sampling*. The length T of the time-series must be sufficiently large, so that an extra increase of T does not alter the estimates significantly. N should be larger than 50 to ensure the stabilization of the averages in Eqs. (3-223) and (3-225). Hence, T must be at least of the magnitude of $50T_0$, or 50 times the correlation length of the process. Of course, this requires longer time-series for narrow-banded processes than for broad-banded processes.

An *ergodic process* $\{X(t), t \in R\}$ is a stationary process of which the same statistical moments (mean value function, auto-covariance function, etc.) are obtained from the related time-averages as $T \to \infty$, no matter which realization of the process is inserted. If only a single realization is available, it follows that we need to presume this ergodic property for estimating the statistical moments.

For processes with known stochastic properties, sufficient conditions for ergodicity of statistical moments can be formulated in terms of differentiability and integrability requirement to the auto-covariance function. If the requirement is fulfilled for the mean value time average in Eq. (3-224), the process is said to be *ergodic in the mean value*. If the requirement is fulfilled for the auto-covariance time average in Eq. (3-226), the process is said to be *ergodic in the auto-covariance function*, etc. Ergodicity in the higher order joint moments involves ergodicity in the lower order, whereas the reverse statement is not necessarily true. Hence, ergodicity in the auto-covariance function implies ergodicity in the mean value. These requirements can hardly be tested from the data, but have to be assumed.

In any case, the estimates in Eqs. (3-224) and (3-226) will be unbiased, i.e., on the average these time averages will give the correct values for any sample length T. Actually, if $\{X(t), t \in R\}$ is weakly stationary (but not necessarily ergodic), we have:

$$\mu_X = E\left[\frac{1}{T} \int_0^T X(t) \, dt\right] = \frac{1}{T} \int_0^T E\left[X(t)\right] dt = \frac{1}{T} \int_0^T \mu_X \, dt = \mu_X \tag{3–227}$$

$$\kappa_{XX}(\tau) = E\left[\frac{1}{T-\tau}\int_0^{T-\tau} X(t)X(t+\tau)\,dt - \mu_X^2\right]$$

$$= \frac{1}{T-\tau}\int_0^{T-\tau}\left(E\left[X(t)X(t+\tau)\right] - \mu_X^2\right)dt$$

$$= \frac{1}{T-\tau}\int_0^{T-\tau}\kappa_{XX}(\tau)\,dt = \kappa_{XX}(\tau) \tag{3-228}$$

The probability density function of the process can also be estimated by *ergodic sampling*. The procedure for doing this has been illustrated in Fig. 3-23. The sample interval T is divided into N equidistant sub-intervals of the length $\Delta t = \frac{T}{N}$, where N is a large number. The sample space (or ordinate) of $x(t)$ is in a similar way divided into disjoint sub-intervals $[x, x + \Delta x[$. At each sub-interval $\left[i\Delta t, (i+1)\Delta t\right[$, $i = 0, \ldots, N-1$, the value of the realization $x_i = x(i\Delta t)$ is registered in one of the sample sub-intervals. Denoting $\Delta N(x)$ the number of samples in $[x, x + \Delta x[$, the 1st order probability density function of the process may be estimated as:

$$f_X(x) \simeq \frac{1}{\Delta x}\frac{\Delta N(x)}{N} = \frac{1}{\Delta x}\frac{\Delta N(x)\Delta t}{N\Delta t} = \frac{1}{\Delta x}\frac{\Delta N(x)\Delta t}{T} \tag{3-229}$$

Eq. (3-229) is supposed to give the correct answer as $N \to \infty \Leftrightarrow \Delta t \to 0$, so the time sub-interval $\left[i\Delta t, (i+1)\Delta t\right[$ is mapped entirely into a given sample sub-interval $[x, x + \Delta x[$. Then, $\Delta N(x)\Delta t$ indicates the part of the time axis that is mapped into $[x, x + \Delta x[$. Eq. (3-229) may formally be written as:

$$f_X(x) \simeq \frac{1}{\Delta x}\frac{\Delta N(x)\,\Delta t}{T} \xrightarrow[\Delta \to 0]{} \frac{1}{\Delta x}\frac{1}{T}\int_0^T \mathbf{1}_{[x,x+\Delta x[}\left(x(t)\right)dt \tag{3-230}$$

where $\mathbf{1}_{[x,x+\Delta x[}(y)$ is an indicator function defined as:

$$\mathbf{1}_{[x,x+\Delta x[}(y) = \begin{cases} 1, & y \in [x, x + \Delta x[\\ 0, & y \notin [x, x + \Delta x[\end{cases} \tag{3-231}$$

Hence, $f_X(x)\Delta x$ is estimated as the time average of the non-linear function $\mathbf{1}_{[x,x+\Delta x[}\left(x(t)\right)$ of the time-series $x(t)$. In Fig. 3-23, the part of the abscissa axis with $\mathbf{1}_{[x,x+\Delta x[}\left(x(t)\right) = 1$ has been

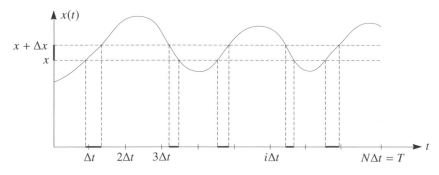

Fig. 3–23 Estimation of the probability density function by ergodic sampling.

marked with a bold red line.

As the probability density function is tantamount to the knowledge of all statistical moments of the considered process, it is only expected to be estimated correctly by the indicated procedure, if the process is ergodic in statistical moments of arbitrary high order.

The probability distribution function specifies the relative number of samples in $]-\infty, x]$, and it can be estimated from the following time average (ergodic sampling):

$$F_X(x) \simeq \frac{1}{T} \int_0^T \mathbf{1}_{]-\infty,x]}(x(t))\, dt \tag{3–232}$$

Example 3-19: First order probability density function of a harmonic process obtained by ergodic sampling

Consider the harmonic process defined as $X(t) = R\cos(\omega_0 t + \Phi)$, $X(t) \sim N(0, \sigma_X^2)$, cf. Eq. (3-9). Its first order probability density function is given by:

$$f_X(x) = \frac{1}{\sqrt{2\pi}\sigma_X} \exp\left(-\frac{1}{2}\frac{x^2}{\sigma_X^2}\right) \tag{3–233}$$

Let r and φ denote the amplitude and the phase of the realization, respectively. The corresponding realization $x(t) = r\cos(\omega_0 t + \varphi)$ has been illustrated in Fig. 3-24a. Applying Eqs. (3-224) and (3-226) to this time-series, the following results are obtained for the mean value function and the auto-covariance function as $T \to \infty$:

$$\mu_X = \lim_{T\to\infty} \frac{1}{T} \int_0^T r\cos(\omega_0 t + \varphi)\, dt = \lim_{T\to\infty} \frac{1}{T}\frac{1}{\omega_0}\left(r\sin\left(\omega_0 T + \varphi\right) - r\sin(\varphi)\right) = 0 \tag{3–234}$$

$$\kappa_{XX}(\tau) = \lim_{T\to\infty} \frac{1}{T-\tau} \int_0^{T-\tau} r\cos(\omega_0 t + \varphi)\, r\cos\left(\omega_0(t+\tau) + \varphi\right) dt$$

$$= \lim_{T\to\infty} \frac{1}{T-\tau}\frac{r^2}{2}\left[t\cos(\omega_0\tau) + \frac{1}{2\omega_0}\sin\left(\omega_0(2t+\tau) + 2\varphi\right)\right]_0^{T-\tau} = \frac{r^2}{2}\cos(\omega_0\tau) \tag{3–235}$$

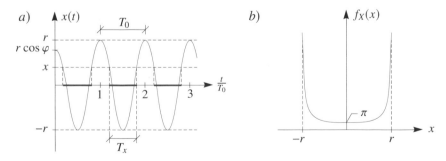

Fig. 3–24 a) Realization of a harmonic process. b) Probability density function obtained by ergodic sampling.

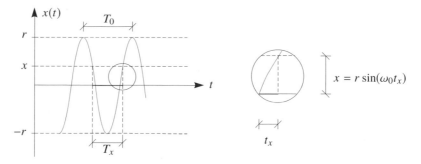

Fig. 3–25 Estimation of the first order probability distribution function of harmonic process by ergodic sampling.

Eq. (3-234) provides the correct mean value function of the harmonic process for all samples of R and Φ. Hence, the harmonic process is ergodic in the mean value.

In general, the variance $\frac{r^2}{2}$ obtained by ergodic sampling of the realization is different from the variance σ_X^2 of the harmonic process. By comparison of Eqs. (3-11) and (3-235), it is concluded that the harmonic process is not ergodic in the auto-covariance function.

Because of the periodicity of the time-series, the estimation of probability distribution function in Eq. (3-232) can be obtained by ergodic sampling over merely a single period $T_0 = \frac{2\pi}{\omega_0}$. The part T_x of the period T_0 that is mapped into the interval $[-r, x]$ has been bold-line marked on the abscissa in Fig. 3-24a. Hence, the probability distribution function can be written as:

$$F_X(x) = \frac{T_x}{T_0} = \frac{\frac{1}{2}T_0 + 2t_x}{T_0} = \frac{1}{2} + \frac{\omega_0}{\pi} t_x \qquad (3–236)$$

where t_x as shown in Fig. 3-25 is given by:

$$t_x = \frac{1}{\omega_0} \arcsin\left(\frac{x}{r}\right) \qquad (3–237)$$

Then, the probability distribution function in Eq. (3-236) becomes:

$$F_X(x) = \begin{cases} \frac{1}{2} + \frac{1}{\pi} \arcsin\left(\frac{x}{r}\right), & x \in [-r, r] \\ 0, & x \notin [-r, r] \end{cases} \qquad (3–238)$$

The first order probability density function is obtained by differentiation of $F_X(x)$ with respect to x:

$$f_X(x) = \begin{cases} \dfrac{1}{\pi\sqrt{r^2 - x^2}}, & x \in [-r, r] \\ 0, & x \notin [-r, r] \end{cases} \qquad (3–239)$$

Eq. (3-239), which has been shown in Fig. 3-24b, differs completely from the correct one in Eq. (3-233). This is due to the limited ergodicity property of the harmonic process.

Different from the harmonic process, a *Resemblant process* $\{X(t), t \in R\}$ is defined as $X(t) = r_0 \cos(\omega_0 t + \Phi)$, where $\Phi \sim U(0, 2\pi)$ and r_0 is a deterministic amplitude. This process is also known as the *random phase process*, and is generated by the single stochastic variable Φ. For this process, ergodicity

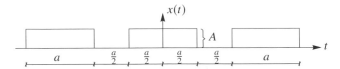

Fig. 3–26 One realization of an ergodic process. The amplitude A and the interval a are deterministic constants.

in the moments of any order holds. Therefore, the first order probability density function given by Eq. (3-239) with $r = r_0$ can be exactly (accurately) obtained by ergodic sampling.

Example 3-20: Ergodic sampling of a process with periodic realizations

A stochastic process $\{X(t), t \in R\}$ with periodic realizations is assumed to be ergodic in the mean value, in the auto-covariance function and in the probability distribution function. A typical realization $x(t)$ is shown in Fig. 3-26, based on which the mean value function μ_X, the auto-covariance function $\kappa_{XX}(\tau)$, the auto-spectral density function $S_{XX}(\omega)$, the probabilities $P(X(t) = x)$ of the various states of $X(t)$, as well as the probability distribution function of the 1st order $F_X(x)$ will be determined by ergodic sampling.

The periodicity of the time-series implies that:

$$x(t) = x(t + nT_0), \quad n = 0, \pm 1, \pm 2, \ldots \tag{3–240}$$

where the period is given by:

$$T_0 = \frac{3}{2} a \tag{3–241}$$

Because of the periodicity of the time-series, the required quantities can be determined by time-averaging over merely a single period T_0.

The mean value function and the auto-covariance function become, cf. Eqs. (3-224) and (3-226):

$$\mu_X = \frac{1}{\frac{3}{2}a} \int_{-\frac{3}{4}a}^{\frac{3}{4}a} x(t)\,dt = \frac{1}{\frac{3}{2}a} A\,a = \frac{2}{3} A \tag{3–242}$$

$$\kappa_{XX}(\tau) = \frac{1}{\frac{3}{2}a} \int_{-\frac{3}{4}a}^{\frac{3}{4}a} x(t)x(t+\tau)\,dt - \mu_X^2 \tag{3–243}$$

For $\tau > 0$, $x(t)x(t + \tau)$ is interpreted as the product of the realization $x(t)$ and the same realization translated the time interval τ to the left. This observation is used at the evaluation of Eq. (3-243). Three qualitatively different cases need to be considered, corresponding to $\tau \in \left[0, \frac{a}{2}\right[$, $\tau \in \left[\frac{a}{2}, a\right[$ and $\tau \in \left[a, \frac{3}{2}a\right[$, respectively. The corresponding positions of $x(t + \tau)$ have been shown in Figs. 3-27b, 3-27c and 3-27d. The red hatched areas represent the part of $x(t + \tau)$, which overlaps with $x(t)$ in the interval $\left[-\frac{T_0}{2}, \frac{T_0}{2}\right]$. Hence, the auto-covariance function in Eq. (3-243) can be further expressed as:

$$\kappa_{XX}(\tau) = \begin{cases} \dfrac{1}{\frac{3}{2}a}(a - \tau)A^2 - \dfrac{4}{9}A^2 = \dfrac{2}{9}\left(1 - 3\dfrac{\tau}{a}\right)A^2, & \tau \in \left[0, \dfrac{a}{2}\right[\\[3mm] \dfrac{1}{\frac{3}{2}a}\dfrac{a}{2}A^2 - \dfrac{4}{9}A^2 = -\dfrac{1}{9}A^2, & \tau \in \left[\dfrac{a}{2}, a\right[\\[3mm] \dfrac{1}{\frac{3}{2}a}\left(\tau - \dfrac{a}{2}\right)A^2 - \dfrac{4}{9}A^2 = \dfrac{1}{9}\left(6\dfrac{\tau}{a} - 7\right)A^2, & \tau \in \left[a, \dfrac{3}{2}a\right[\end{cases} \tag{3–244}$$

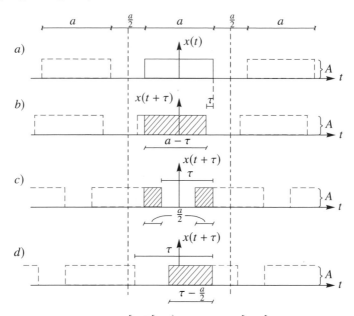

Fig. 3–27 a) $x(t)$. b) $x(t + \tau)$, $\tau \in \left[0, \frac{a}{2}\right[$. c) $x(t + \tau)$, $\tau \in \left[\frac{a}{2}, a\right[$. d) $x(t + \tau)$, $\tau \in \left[a, \frac{3}{2}a\right[$.

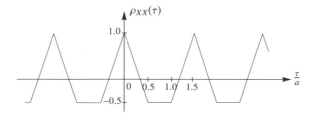

Fig. 3–28 Auto-correlation coefficient function of the ergodic process $\{X(t), t \in R\}$.

Values of $\kappa_{XX}(\tau)$ outside the interval $\left[0, \frac{3}{2}a\right[$ follow from the symmetry property Eq. (3-18) and the periodicity of the problem, i.e.:

$$\left.\begin{array}{ll} \kappa_{XX}(\tau) & = \kappa_{XX}(-\tau), \quad \tau \in [0, \infty[\\[2mm] \kappa_{XX}(\tau + nT_0) = \kappa_{XX}(\tau), \quad n = \pm 1, \pm 2, \dots \end{array}\right\} \tag{3-245}$$

The variance σ_X^2 of the process becomes, cf. Eq. (3-244):

$$\sigma_X^2 = \kappa_{XX}(0) = \frac{2}{9} A^2 \tag{3-246}$$

The auto-correlation coefficient function $\rho_{XX}(\tau) = \kappa_{XX}(\tau)/\sigma_X^2$ has been depicted in Fig. 3-28.

The auto-covariance function is periodic with the period T_0, and hence can be expanded into a Fourier series:

$$\kappa_{XX}(\tau) = \frac{a_0}{2} + \sum_{n=1}^{\infty} a_n \cos(\omega_n \tau) + b_n \sin(\omega_n \tau) \tag{3-247}$$

where:

$$\omega_n = n \frac{2\pi}{T_0}, \quad n = 1, 2, \ldots \tag{3–248}$$

$$\left. \begin{aligned} a_n &= \frac{2}{T_0} \int_{-\frac{T_0}{2}}^{\frac{T_0}{2}} \kappa_{XX}(\tau) \cos(\omega_n \tau) \, d\tau = \frac{4}{T_0} \int_{0}^{\frac{T_0}{2}} \kappa_{XX}(\tau) \cos(\omega_n \tau) \, d\tau \\ b_n &= \frac{2}{T_0} \int_{-\frac{T_0}{2}}^{\frac{T_0}{2}} \kappa_{XX}(\tau) \sin(\omega_n \tau) \, d\tau = 0 \end{aligned} \right\} \tag{3–249}$$

The last statements in Eq. (3-249) are due to the fact that $\kappa_{XX}(\tau) \cos(\omega_n \tau)$ is an even function of τ, and $\kappa_{XX}(\tau) \sin(\omega_n \tau)$ is an odd function of τ.

From Eqs. (3-244), (3-247) and (3-249), it follows that:

$$\kappa_{XX}(\tau) = \frac{a_0}{2} + \sum_{n=1}^{\infty} a_n \cos(\omega_n \tau) \tag{3–250}$$

$$a_n = \frac{4}{T_0} \frac{2}{9} A^2 \left(\int_{0}^{\frac{T_0}{3}} \left(1 - 3\frac{\tau}{a}\right) \cos(\omega_n \tau) \, d\tau - 0.5 \int_{\frac{T_0}{3}}^{\frac{T_0}{2}} \cos(\omega_n \tau) \, d\tau \right)$$

$$= \begin{cases} 0 & , \quad n = 0 \\ \frac{A^2}{n^2 \pi^2} \left(1 - \cos\left(\frac{2}{3} n \pi\right)\right), & n = 1, 2, \ldots \end{cases} \tag{3–251}$$

Since $a_0 = 0$, the auto-covariance function of an ergodic process with periodic realizations is of the same form as the auto-covariance function of a process made up of a sum of harmonic processes as given by Eq. (3-36). As will be shown in Chapter 6, it is also of the same form as that of of a process made up of a sum of random phase processes.

From the Wiener-Khintchine relation Eq. (3-23) and from Eq. (3-31), the auto-spectral density function can be written as:

$$S_{XX}(\omega) = \frac{1}{2\pi} \int_{-\infty}^{\infty} \left(\frac{a_0}{2} + \sum_{n=1}^{\infty} a_n \cos(\omega_n \tau) \right) e^{-i\omega\tau} \, d\tau$$

$$= \frac{1}{2} \sum_{n=1}^{\infty} \left(\delta(\omega - \omega_n) + \delta(\omega + \omega_n) \right) a_n \tag{3–252}$$

With a_n as given by Eq. (3-251), this auto-spectral density function has been demonstrated in Fig. 3-29.

$X(t)$ is a discrete stochastic variable, which can only attains the values $x_1 = 0$ and $x_2 = A$. Using the periodicity of the realization $x(t)$, the probability of the events, $\{X(t) = 0\}$ and $\{X(t) = A\}$, can be evaluated from the following time average, cf. Eq. (3-231):

$$P(X(t) = x_i) = \frac{1}{\frac{3}{2}a} \int_{-\frac{3}{4}a}^{\frac{3}{4}a} \mathbf{1}_{x(t)=x_i}(x(t)) \, dt = \begin{cases} \frac{\frac{1}{2}a}{\frac{3}{2}a} = \frac{1}{3}, & x = 0 \\[2mm] \frac{a}{\frac{3}{2}a} = \frac{2}{3}, & x = A \end{cases} \tag{3–253}$$

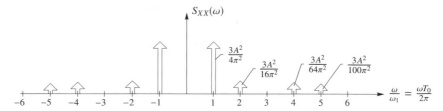

Fig. 3–29 Auto-spectral density function of the ergodic process $\{X(t),\, t \in R\}$.

Because the distribution function is continuous to the right, the probability distribution function of the 1st order becomes:

$$F_X(x) = \begin{cases} 0, & x \in]-\infty, 0[\\ \dfrac{1}{3}, & x \in [0, A[\\ 1, & x \in [A, \infty[\end{cases} \qquad (3\text{–}254)$$

This probability distribution function has been depicted in Fig. 3-30.

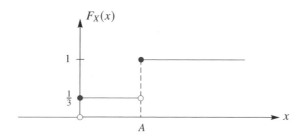

Fig. 3–30 Probability distribution function of the 1st order of the process $\{X(t),\, t \in R\}$.

Chapter 4
STOCHASTIC VIBRATION THEORY FOR LINEAR SYSTEMS

In the present chapter, some essential results of the stochastic vibration theory for linear structural systems are presented. First, *single-degree-of-freedom* (SDOF) systems are considered. Then, the theory is generalized to *multi-degrees-of-freedom* (MDOF) systems. The emphasis is laid on the methods available for obtaining the joint probability density functions and the statistical moments for discrete structural systems. Finally, some results for *continuous systems* have been indicated.

4.1 Single-degree-of-freedom systems

Fig. 4-1 shows a linear time-invariant SDOF system characterized by a linear elastic spring with the spring constant k, acting parallel to a viscous damper with the *damping coefficient c*, both in series with a *point mass m*. These system parameters are taken as deterministic constants. The system is excited by an external dynamic load, which is modelled as a stochastic process $\{F(t),\ t \in [0, \infty[\}$. The *displacement process* caused by the dynamic loading is modelled as a stochastic process $\{X(t),\ t \in [0, \infty[\}$. $X(t)$ is unidirectional to $F(t)$, and is measured from the *static equilibrium state* which the oscillator eventually attains when the dynamic load is removed.

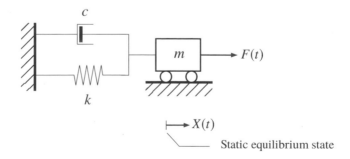

Fig. 4–1 Symbolic representation of a SDOF system.

As is the case for the shear frame in Fig. 3-20a, the system in Fig. 4-1 is merely a symbolic representation of the approximate dynamics of a real structure modelled by a so-called generalized single degree of freedom. The real structure is continuous and has infinite many degrees of freedom in the exact description.

The load process $\{F(t), t \in [0, \infty[\}$ is related to the displacement process $\{X(t), t \in [0, \infty[\}$, the velocity process $\{\dot{X}(t), t \in [0, \infty[\}$ and the acceleration process $\{\ddot{X}(t), t \in [0, \infty[\}$ by the stochastic differential equation of motion:

$$\left. \begin{array}{l} m\ddot{X}(t) + c\dot{X}(t) + kX(t) = F(t), \quad t > 0 \\ X(0) = X_0, \quad \dot{X}(0) = \dot{X}_0 \end{array} \right\} \tag{4–1}$$

where the initial values form a 2-dimensional stochastic vector $[X_0, \dot{X}_0]^T$.

In accordance with the integral representation Eq. (3-166) as the solution to the differential system in Eq. (3-157), the solution to Eq. (4-1) is given by the following stochastic integral known as Duhamel's integral representation:[1]

$$X(t) = X^{(0)}(t) + \int_0^t h(t - \tau)F(\tau)\,d\tau \tag{4–2}$$

where the impulse response function is written as:

$$h(t) = \begin{cases} 0 & , \quad t \le 0 \\ \dfrac{e^{-\zeta\omega_0 t}}{m\omega_d}\sin(\omega_d t), & \quad t > 0 \end{cases} \tag{4–3}$$

The *undamped angular eigenfrequency* ω_0, the *damping ratio* ζ and the *damped angular eigenfrequency* ω_d are given by:

$$\omega_0 = \sqrt{\frac{k}{m}} \tag{4–4}$$

$$\zeta = \frac{c}{2\sqrt{km}} \tag{4–5}$$

$$\omega_d = \omega_0\sqrt{1 - \zeta^2} \tag{4–6}$$

Eq. (4-3) has been indicated in Eq. (3-106) without any reference to single-degree-of-freedom dynamics.

$X^{(0)}(t)$ indicates the response from the initial values, cf. Eq. (3-167):

$$X^{(0)}(t) = a_0(t)X_0 + a_1(t)\dot{X}_0 \tag{4–7}$$

---

[1]S.R.K. Nielsen. *Vibration Theory, Vol. 1: Linear Vibration Theory.* Aalborg University Press, Aalborg, 2004.

$$a_0(t) = ch(t) + m\dot{h}(t) \tag{4–8}$$

$$a_1(t) = mh(t) \tag{4–9}$$

As seen from Eqs. (4-2), (4-8) and (4-9), both the responses from the external dynamic load and from the initial values are determined by the impulse response function $h(t)$.

All results indicated in Section 3.2 on stochastic integration can immediately be applied to Eq. (4-2). In doing this, it will be assumed that the initial value vector $\left[X_0, \dot{X}_0\right]^T$ is stochastically independent of the load process $\{F(t), t \in [0, \infty[\}$.

Based on Eqs. (3-71) and (3-72), it is straightforward to evaluate the mean value function $\mu_X(t)$ and the auto-covariance function $\kappa_{XX}(t_1, t_2)$ of the displacement process $\{X(t), t \in [0, \infty[\}$:

$$\mu_X(t) = \mu_{X^{(0)}}(t) + \int_0^t h(t - \tau)\mu_F(\tau)\,d\tau \tag{4–10}$$

$$\kappa_{XX}(t_1, t_2) = \kappa_{X^{(0)}X^{(0)}}(t_1, t_2) + \int_0^{t_1}\int_0^{t_2} h(t_1 - \tau_1)h(t_2 - \tau_2)\,\kappa_{FF}(\tau_1, \tau_2)\,d\tau_1 d\tau_2 \tag{4–11}$$

where $\mu_F(t)$ and $\kappa_{FF}(t_1, t_2)$ signify the mean value function and the auto-covariance function of the load process, respectively. $\mu_{X^{(0)}}(t)$ and $\kappa_{X^{(0)}X^{(0)}}(t_1, t_2)$ indicate the mean value function and the auto-covariance function of the *initial value process*, given by:

$$\mu_{X^{(0)}}(t) = a_0(t)\mu_{X_0} + a_1(t)\mu_{\dot{X}_0} \tag{4–12}$$

$$\kappa_{X^{(0)}X^{(0)}}(t_1, t_2) = a_0(t_1)a_0(t_2)\sigma_{X_0}^2 + \left(a_0(t_1)a_1(t_2) + a_0(t_2)a_1(t_1)\right)\kappa_{X_0\dot{X}_0} + a_1(t_1)a_1(t_2)\sigma_{\dot{X}_0}^2 \tag{4–13}$$

$\mu_{X_0}, \mu_{\dot{X}_0}, \sigma_{X_0}^2, \sigma_{\dot{X}_0}^2$ signify the mean values and variances of the initial values, and $\kappa_{X_0\dot{X}_0}$ indicates the covariance between X_0 and $\dot{X}_0$. If the initial values are deterministic, then $\mu_{X_0} = x_0, \mu_{\dot{X}_0} = \dot{x}_0$ and $\sigma_{X_0}^2 = \sigma_{\dot{X}_0}^2 = \kappa_{X_0\dot{X}_0} = 0$.

If $[X_0, \dot{X}_0]^T$ is a normally distributed stochastic vector and $\{F(t), t \in [0, \infty[\}$ is a Gaussian process, $\{X(t), t \in [0, \infty[\}$ becomes a Gaussian process as well. In this case, $\{X(t), t \in [0, \infty[\}$ can be completely determined by Eqs. (4-10) and (4-11).

If the damping ratio ζ is positive (this is not the case in some problems in aeroelasticity), the *initial value response* $X^{(0)}(t)$ will eventually be dissipated. Further, if the load process is weakly or strictly stationary, the displacement process $\{X(t), t \in [0, \infty[\}$ and the related derivative processes eventually approach stationarity in the same sense. The mean value function μ_X, the auto-spectral density function function $S_{XX}(\omega)$ and the auto-covariance function $\kappa_{XX}(\tau)$ in the stationary state follow from Eqs. (3-100), (3-104) and (3-23):

$$\mu_X = H(0)\,\mu_F = \frac{1}{k}\,\mu_F \tag{4–14}$$

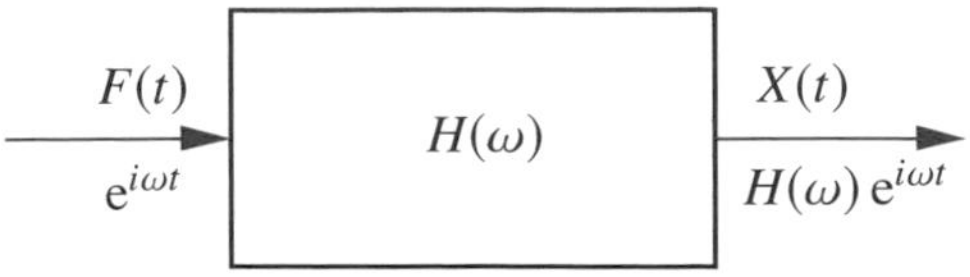

Fig. 4–2 Block diagram representation of input-output of a SDOF system in the stationary state.

$$S_{XX}(\omega) = |H(\omega)|^2 S_{FF}(\omega) \qquad (4\text{–}15)$$

$$\kappa_{XX}(\tau) = \int_{-\infty}^{\infty} e^{i\omega\tau} S_{XX}(\omega)\, d\omega = \int_{-\infty}^{\infty} e^{i\omega\tau} |H(\omega)|^2 S_{FF}(\omega)\, d\omega \qquad (4\text{–}16)$$

where $H(\omega)$ signifies the frequency response function of the system:

$$H(\omega) = \int_{0}^{\infty} e^{-i\omega t} h(t)\, dt = \frac{1}{m\left(z^2 + 2\zeta\omega_0 z + \omega_0^2\right)}, \qquad z = i\omega \qquad (4\text{–}17)$$

The relationship between the load process and the displacement process in the stationary state may be represented by the block diagram shown in Fig. 4-2.

Example 4-1: Joint probability density function of $X(t)$ and $\dot{X}(t)$ resulting from deterministic initial condition and stationary Gaussian load process

It is assumed that the displacement process has reached the stationary state, where the response from the initial values has been dissipated, and is given by the stationary Gaussian process $\{X(t),\ t \in R\}$ with the mean value μ_X given by Eq. (4-14), and the auto-covariance function $\kappa_{XX}(\tau) = \sigma_X^2\, \rho(\tau)$ given by Eq. (4-16). $\rho(\tau) = \rho_{XX}(\tau)$ signifies the auto-correlation coefficient function, cf. Eq. (3-19).

Based on this information, the joint probability density function of $[X(t), \dot{X}(t)]^T$ in the transient phase is to be determined, where the response from the deterministic initial values $X(0) = x_0$ and $\dot{X}(0) = \dot{x}_0$ is still felt.

First, consider the following 4-dimensional stochastic vector process defined from the related stationary stochastic process, assuming the initial values $X(0)$ and $\dot{X}(0)$ to be part of the stationary processes:

$$\mathbf{Z} = \begin{bmatrix} \mathbf{X} \\ \mathbf{Y} \end{bmatrix}, \quad \mathbf{X} = \begin{bmatrix} X(t) \\ \dot{X}(t) \end{bmatrix}, \quad \mathbf{Y} = \begin{bmatrix} X(0) \\ \dot{X}(0) \end{bmatrix} \qquad (4\text{–}18)$$

The mean value vector and the covariance matrix of $\mathbf{Z}$ become:

$$\mu_{\mathbf{Z}} = \begin{bmatrix} \mu_{\mathbf{X}} \\ \mu_{\mathbf{Y}} \end{bmatrix} = \begin{bmatrix} \mu_X \\ 0 \\ \mu_X \\ 0 \end{bmatrix} \qquad (4\text{–}19)$$

$$\mathbf{C_{ZZ}} = \begin{bmatrix} \mathbf{C_{XX}} & \mathbf{C_{XY}} \\ \\ \mathbf{C_{XY}^T} & \mathbf{C_{YY}} \end{bmatrix} = \sigma_X^2 \begin{bmatrix} 1 & 0 & \rho(t) & -\rho'(t) \\ 0 & -\rho''(0) & \rho'(t) & -\rho''(t) \\ \\ \rho(t) & \rho'(t) & 1 & 0 \\ -\rho'(t) & -\rho''(t) & 0 & -\rho''(0) \end{bmatrix} \tag{4-20}$$

where it has been used that $\kappa_{X(0)X(t)} = \sigma_X^2 \rho(t)$, $\kappa_{X(0)X(0)} = \kappa_{X(t)X(t)} = \sigma_X^2$, $\kappa_{X(0)\dot{X}(0)} = \kappa_{X(t)\dot{X}(t)} = 0$, $\kappa_{X(0)\dot{X}(t)} = -\kappa_{\dot{X}(0)X(t)} = \sigma_X^2 \rho'(t)$, $\kappa_{\dot{X}(0)\dot{X}(t)} = -\sigma_X^2 \rho''(t)$, cf. Eq. (3-137). $'$ denotes differentiation with respect to the argument.

Then, it's clear that the searched function is the probability density function of $\mathbf{X}$ on condition of deterministic initial values $\mathbf{Y} = \mathbf{y} = [x_0, \dot{x}_0]^T$. As shown in Eq. (2-78), this distribution is normal with the conditional mean value vector $\boldsymbol{\mu}_{\mathbf{X}|\mathbf{Y}}$ and the conditional covariance matrix $\mathbf{C}_{\mathbf{XX}|\mathbf{Y}}$. Inserting Eqs. (4-19) and (4-20) into Eqs. (2-75) and (2-76), the following results for $\boldsymbol{\mu}_{\mathbf{X}|\mathbf{Y}}$ and $\mathbf{C}_{\mathbf{XX}|\mathbf{Y}}$ are obtained:

$$\boldsymbol{\mu}_{\mathbf{X}|\mathbf{Y}} = \boldsymbol{\mu}_{\mathbf{X}} + \mathbf{C_{XY}}\mathbf{C_{YY}^{-1}}(\mathbf{y} - \boldsymbol{\mu}_{\mathbf{Y}}) = \begin{bmatrix} \mu_X \\ 0 \end{bmatrix} + \begin{bmatrix} \rho(t) & -\rho'(t) \\ \rho'(t) & -\rho''(t) \end{bmatrix} \begin{bmatrix} 1 & 0 \\ 0 & \omega_0^2 \end{bmatrix}^{-1} \begin{bmatrix} x_0 - \mu_X \\ \dot{x}_0 \end{bmatrix}$$

$$= \begin{bmatrix} (1 - \rho(t))\mu_X + \rho(t)x_0 - \rho'(t)\dot{x}_0/\omega_0^2 \\ -\rho'(t)\mu_X + \rho'(t)x_0 - \rho''(t)\dot{x}_0/\omega_0^2 \end{bmatrix} \tag{4-21}$$

$$\mathbf{C}_{\mathbf{XX}|\mathbf{Y}} = \mathbf{C_{XX}} - \mathbf{C_{XY}}\mathbf{C_{YY}^{-1}}\mathbf{C_{XY}^T}$$

$$= \sigma_X^2 \begin{bmatrix} 1 & 0 \\ 0 & \omega_0^2 \end{bmatrix} - \sigma_X^2 \begin{bmatrix} \rho(t) & -\rho'(t) \\ \rho'(t) & -\rho''(t) \end{bmatrix} \begin{bmatrix} 1 & 0 \\ 0 & \omega_0^2 \end{bmatrix}^{-1} \begin{bmatrix} \rho(t) & \rho'(t) \\ -\rho'(t) & -\rho''(t) \end{bmatrix}$$

$$= \sigma_X^2 \begin{bmatrix} 1 - \rho^2(t) - (\rho'(t))^2/\omega_0^2 & -\rho'(t)(\rho(t) + \rho''(t)/\omega_0^2) \\ -\rho'(t)(\rho(t) + \rho''(t)/\omega_0^2) & \omega_0^2 - (\rho'(t))^2 - (\rho''(t))^2/\omega_0^2 \end{bmatrix} \tag{4-22}$$

where $\omega_0^2 = -\rho''(0)$ as shown by Eq. (3-149) has been introduced.

Especially, for a SDOF oscillator exposed to a Gaussian white noise, $\rho(\tau) = \rho_{XX}(\tau)$ has been given by Eq. (3-65).

Example 4-2: Indirectly acted external dynamic load and damping force

Fig. 4-3 shows a *Bernoulli-Euler beam* with the constant bending stiffness EI. At point 1, a point mass of magnitude m is attached. The external dynamic load and a linear viscous damper with the damping constant c are acting indirectly on the mass m through the linear elastic massless beam (free of damping). The displacements from the static equilibrium state at locations of the point mass, of the damper and of the external load are termed as $X_1(t)$, $X_2(t)$, $X_3(t)$, respectively. Using *d'Alembert's principle*,[1] these

---

[1]S.R.K. Nielsen. *Vibration Theory, Vol. 1: Linear Vibration Theory*. Aalborg University Press, Aalborg, 2004.

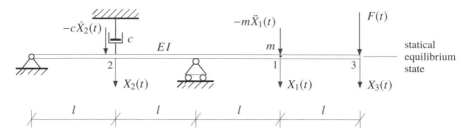

Fig. 4–3 Bernoulli-Euler beam with indirectly acting external dynamic load and damping force.

displacements can be obtained by adding the displacement contributions from the inertial force $-m\ddot{X}_1(t)$ at point 1, the damping force $-c\dot{X}_2(t)$ at point 2, and the dynamic load $F(t)$ at point 3:

$$\begin{bmatrix} X_1(t) \\ X_2(t) \\ X_3(t) \end{bmatrix} = \begin{bmatrix} \delta_{11} & \delta_{12} & \delta_{13} \\ \delta_{21} & \delta_{22} & \delta_{23} \\ \delta_{31} & \delta_{32} & \delta_{33} \end{bmatrix} \begin{bmatrix} -m\ddot{X}_1(t) \\ -c\dot{X}_2(t) \\ F(t) \end{bmatrix} \tag{4–23}$$

Eq. (4-23) is based on the *force method* of structural mechanics. The *flexibility coefficient* δ_{jk} indicates the displacement at point j due to a unit force at point k. Due to *Maxwell-Betti's reciprocal theorem* for linear elastic structures, the symmetry condition $\delta_{jk} = \delta_{kj}$ is valid. The flexibility coefficient may be calculated by the principle of complementary virtual work. In the present case, the following flexibility coefficients are obtained:

$$\mathbf{D} = \begin{bmatrix} \delta_{11} & \delta_{12} & \delta_{13} \\ \delta_{21} & \delta_{22} & \delta_{23} \\ \delta_{31} & \delta_{32} & \delta_{33} \end{bmatrix} = \frac{1}{12} \frac{l^3}{EI} \begin{bmatrix} 12 & -3 & 26 \\ -3 & 2 & -6 \\ 26 & -6 & 64 \end{bmatrix} \tag{4–24}$$

where $\mathbf{D}$ signifies the flexibility matrix.

Premultiplying Eq. (4-23) with the inverse flexibility matrix yields the matrix equation of motion:

$$\mathbf{M}\ddot{\mathbf{X}}(t) + \mathbf{C}\dot{\mathbf{X}}(t) + \mathbf{K}\mathbf{X}(t) = \mathbf{e}_3 F(t) \tag{4–25}$$

$$\left. \begin{aligned} \mathbf{X}(t) &= \begin{bmatrix} X_1(t) \\ X_2(t) \\ X_3(t) \end{bmatrix} \quad , \quad \mathbf{e}_j = \begin{bmatrix} 0 \\ 1 \\ 0 \end{bmatrix} \leftarrow j\text{th component} \\[2ex] \mathbf{M} &= \begin{bmatrix} m & 0 & 0 \\ 0 & 0 & 0 \\ 0 & 0 & 0 \end{bmatrix}, \quad \mathbf{C} = \begin{bmatrix} 0 & 0 & 0 \\ 0 & c & 0 \\ 0 & 0 & 0 \end{bmatrix} \\[2ex] \mathbf{K} &= \mathbf{D}^{-1} = \frac{1}{28} \frac{EI}{l^3} \begin{bmatrix} 276 & 108 & -102 \\ 108 & 276 & -18 \\ -102 & -18 & 45 \end{bmatrix} \end{aligned} \right\} \tag{4–26}$$

where $\mathbf{M}$ is the *mass matrix*, $\mathbf{C}$ is the *damping matrix* and $\mathbf{K}$ is the *stiffness matrix* of the system.

A harmonically varying load is considered:

$$F(t) = \text{Re}\left(F_0 e^{i\omega t}\right) \tag{4–27}$$

Then, the stationary harmonic response of $\mathbf{X}(t)$ becomes:

$$\mathbf{X}(t) = \mathrm{Re}\left(\mathbf{X}_0\, e^{i\omega t}\right) \tag{4-28}$$

$$\mathbf{X}_0 = \mathbf{H}(\omega)\, \mathbf{e}_3\, F_0 \tag{4-29}$$

$$\mathbf{H}(\omega) = \left(\mathbf{K} - \omega^2 \mathbf{M} + i\omega \mathbf{C}\right)^{-1} \tag{4-30}$$

It follows from Eq. (4-30) that the frequency response function $H_{13}(\omega)$, relating the displacement $X_1(t)$ at point 1 to the harmonic load at point 3, becomes:

$$H_{13}(\omega) = \mathbf{e}_1^T \mathbf{H}(\omega)\mathbf{e}_3 = \frac{P(z)}{Q(z)}, \quad z = i\omega \tag{4-31}$$

$$P(z) = p_0 z + p_1 \tag{4-32}$$

$$Q(z) = z^3 + q_1 z^2 + q_2 z + q_3 \tag{4-33}$$

$$\left.\begin{aligned}
p_0 &= -\frac{K_{13}}{mK_{33}} = \frac{34}{15}\frac{1}{m}, & p_1 &= \frac{K_{12}K_{23} - K_{13}K_{22}}{mcK_{33}} = \frac{104}{5}\frac{1}{mc} \\
q_1 &= \frac{K_{22}K_{33} - K_{23}^2}{cK_{33}} = \frac{48}{5}\frac{EI}{cl^3}, & q_2 &= \frac{K_{11}K_{33} - K_{13}^2}{mK_{33}} = \frac{8}{5}\frac{EI}{ml^3} \\
q_3 &= \frac{\det(\mathbf{K})}{mcK_{33}} = \frac{48}{5}\frac{EI}{mcl^3}
\end{aligned}\right\} \tag{4-34}$$

Eqs. (4-32) and (4-33) show that the frequency response function relating the displacement $X_1(t)$ to the load $F(t)$ becomes a rational function of the order $(r, s) = (1, 3)$. According to Eq. (3-157), the displacement process $\{X_1(t),\, t \in [0, \infty[\}$ is then related to the dynamic load process $\{F(t),\, t \in [0, \infty[\}$ by the following stochastic differential equations:

$$\left.\begin{aligned}
X_1(t) &= p_0 \dot{Y}(t)(t) + p_1 Y(t) \\
Y^{(3)}(t) &+ q_1 \ddot{Y}(t) + q_2 \dot{Y}(t) + q_3 Y(t) = F(t)
\end{aligned}\right\} \tag{4-35}$$

The filter differential equation becomes of the 3rd order because the damping force is acting indirectly at the point mass. Jocularly, such a system is sometimes referred to as a 1.5-degree-of-freedom oscillator.

If the load process $\{F(t),\, t \in R\}$ is weakly stationary with the mean value function μ_F and the auto-spectral density function $S_{FF}(\omega)$, the mean value function and the auto-spectral density function of the displacement process in the stationary state follow from Eqs. (3-100) and (3-104):

$$\mu_{X_1} = H(0)\,\mu_F = \frac{p_1}{q_3}\,\mu_F = \frac{13}{6}\frac{l^3}{EI}\,\mu_F = \delta_{13}\,\mu_F \tag{4-36}$$

$$S_{X_1 X_1}(\omega) = \left|\frac{P(i\omega)}{Q(i\omega)}\right|^2 S_{FF}(\omega) \tag{4-37}$$

Finally, if $\{F(t),\, t \in R\}$ is modelled as a Gaussian white noise with the auto-spectral density S_0, the auto-covariance function of $\{X(t),\, t \in R\}$ in the stationary state can be calculated from Eq. (3-59).

In Eq. (4-1), if the load process $\{F(t), \in R\}$ is idealized as a Gaussian white noise with the auto-spectral density S_0, the auto-spectral density functions of the displacement process $\{X(t), \in R\}$ and the velocity process $\{\dot{X}(t), \in R\}$ in the stationary state, where the response from the initial values has been dissipated, are obtained from, cf. Eqs. (4-15) and (3-141):

$$S_{XX}(\omega) = |H(\omega)|^2 S_0 = \frac{S_0}{m^2\left((\omega_0^2 - \omega^2)^2 + 4\zeta^2\omega_0^2\omega^2\right)} \tag{4–38}$$

$$S_{\dot{X}\dot{X}}(\omega) = \omega^2 |H(\omega)|^2 S_0 = \frac{\omega^2 S_0}{m^2\left((\omega_0^2 - \omega^2)^2 + 4\zeta^2\omega_0^2\omega^2\right)} \tag{4–39}$$

Stochastic processes with the indicated auto-spectral density functions have been analyzed in Example 3-12. Fig. 3-19 shows typical realizations of the displacement process $\{X(t), t \in R\}$ and the velocity process $\{\dot{X}(t), t \in R\}$, when the oscillator is exposed to white noise. In this case, the realizations of the displacement process are continuous and differentiable, and the velocity has continuous but non-differentiable realizations.

The auto-covariance functions of $\{X(t), t \in R\}$ and $\{\dot{X}(t), t \in R\}$ are given by Eqs. (3-63), (3-64), (3-65) and Eqs. (3-147), (3-148), (3-149), respectively, and are repeated in the following:

$$\left.\begin{aligned} &\kappa_{XX}(\tau) = \sigma_X^2 \rho_{XX}(\tau) \\ &\sigma_X^2 = \frac{\pi S_0}{2\zeta\omega_0^3 m^2} \\ &\rho_{XX}(\tau) = e^{-\zeta\omega_0|\tau|}\left(\cos\left(\omega_d\tau\right) + \frac{\zeta}{\sqrt{1-\zeta^2}}\sin\left(\omega_d|\tau|\right)\right) \end{aligned}\right\} \tag{4–40}$$

$$\left.\begin{aligned} &\kappa_{\dot{X}\dot{X}}(\tau) = \sigma_{\dot{X}}^2 \rho_{\dot{X}\dot{X}}(\tau) \\ &\sigma_{\dot{X}}^2 = \omega_0^2\sigma_X^2 = \frac{\pi S_0}{2\zeta\omega_0 m^2} \\ &\rho_{\dot{X}\dot{X}}(\tau) = e^{-\zeta\omega_0|\tau|}\left(\cos\left(\omega_d\tau\right) - \frac{\zeta}{\sqrt{1-\zeta^2}}\sin\left(\omega_d|\tau|\right)\right) \end{aligned}\right\} \tag{4–41}$$

In real applications, the dynamic load process $\{F(t), t \in R\}$ is often modelled as a stationary Gaussian process, obtained as the output process of a unit Gaussian white noise process $\{W(t), t \in R\}$ passing through a time-invariant rational filter $H_0(\omega)$ of the order (r, s), $r < s$. The frequency response function of the filter is given by, cf. Eqs. (3-54), (3-55) and (3-56):

$$H_0(\omega) = \frac{P(z)}{Q(z)} = \frac{p_0 z^r + \cdots + p_{r-1}z + p_r}{z^s + q_1 z + \cdots + q_{s-1}(z) + q_s}, \quad z = i\omega \tag{4–42}$$

Then, $\{F(t), t \in R\}$ can be obtained as the output process of the differential equations in Eq. (3-157). The idea is to combine these differential equations with the structural equation of motion in Eq. (4-1), in order to establish a extended state vector differential equation for the integrated system. As an example, this method is illustrated for a shaping filter of the order $(r, s) = (2, 3)$, in which case Eq. (3-157) attains the form:

$$F(t) = p_0\ddot{Y}(t) + p_1\dot{Y}(t) + p_2Y(t) \atop Y^{(3)} + q_1\ddot{Y}(t) + q_2\dot{Y}(t) + q_3Y(t) = W(t) \Bigg\}$$ (4-43)

Next, Eq. (4-43) is combined with the equation of motion in Eq. (4-1), leading to the following state vector differential equation:

$$\frac{d}{dt}\mathbf{Z}(t) = \mathbf{A}\,\mathbf{Z}(t) + \mathbf{b}\,W(t), \quad t > 0 \atop \mathbf{Z}(0) = \mathbf{z}_0 \Bigg\}$$ (4-44)

$$\mathbf{Z}(t) = \begin{bmatrix} X(t) \\ \dot{X}(t) \\ Y(t) \\ \dot{Y}(t) \\ \ddot{Y}(t) \end{bmatrix}, \quad \mathbf{b} = \begin{bmatrix} 0 \\ 0 \\ 0 \\ 0 \\ 1 \end{bmatrix}, \quad \mathbf{A} = \begin{bmatrix} 0 & 1 & 0 & 0 & 0 \\ -\omega_0^2 & -2\zeta\omega_0 & \frac{p_2}{m} & \frac{p_1}{m} & \frac{p_0}{m} \\ 0 & 0 & 0 & 1 & 0 \\ 0 & 0 & 0 & 0 & 1 \\ 0 & 0 & -q_3 & -q_2 & -q_1 \end{bmatrix}$$ (4-45)

Finally, the stationary variances σ_X^2 and $\sigma_{\dot{X}}^2$ may be calculated as entries in the stationary zero time-lag covariance matrix $\mathbf{C_{ZZ}}$ from the Lyapunov equation Eq. (3-197) with $S_0 = \frac{1}{2\pi} \Rightarrow 2\pi S_0 = 1$.

The auto-spectral density functions of the load process $\{F(t), t \in R\}$ and the displacement process $\{X(t), t \in R\}$ in the stationary state become:

$$S_{FF}(\omega) = |H_0(\omega)|^2 S_{WW} \atop S_{XX}(\omega) = |H(\omega)|^2 S_{FF}(\omega) \Bigg\} \quad \Rightarrow$$

$$S_{XX}(\omega) = |H_1(\omega)|^2 S_{WW}$$ (4-46)

where:

$$H_1(\omega) = H_0(\omega)\,H(\omega)$$ (4-47)

Hence, the frequency response function of the combined system is obtained as the product of the frequency response functions of the component systems, as demonstrated in Fig. 4-4.

Example 4-3: Load process obtained by a linear filtration of white noise through an Ornstein-Uhlenbeck filter

The load process $\{F(t), t \in R\}$ is assumed to be obtained by a linear filtration of a white noise process $\{W(t), t \in R\}$ with the auto-spectral density S_0 passing through an Ornstein-Uhlenbeck filter, corresponding to the stochastic differential equation, cf. Eq. (3-200):

$$\dot{F}(t) + \alpha F(t) = \alpha W(t)$$ (4-48)

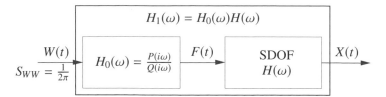

Fig. 4–4 The frequency response function of the combined system where the load process is obtained as the output of a unit white noise passing through a rational shaping filter.

The frequency response function $H_0(\omega)$ of this system is given by, cf. Eq. (3-201):

$$H_0(\omega) = \frac{\alpha}{z + \alpha}, \quad z = i\omega \tag{4-49}$$

From Eqs. (4-17), (4-47) and (4-49), it follows that the frequency response function of the displacement response due to the white noise input becomes:

$$H_1(\omega) = \frac{\alpha}{z + \alpha} \frac{1}{m\left(z^2 + 2\zeta\omega_0 z + \omega_0^2\right)} = \frac{\frac{\alpha}{m}}{\left(z^2 + 2\zeta\omega_0 z + \omega_0^2\right)(z + \alpha)}, \quad z = i\omega \tag{4-50}$$

As seen, $H_1(z)$ is a rational function of the order $(r, s) = (0, 3)$ with $P(z)$ and $Q(z)$ given by:

$$P = \frac{\alpha}{m} \tag{4-51}$$

$$Q(z) = \left(z^2 + 2\zeta\omega_0 z + \omega_0^2\right)\left(z + \alpha\right) = (z - z_1)(z - z_2)(z - z_3) \tag{4-52}$$

where the poles are:

$$\left.\begin{aligned}
z_1 &= \omega_0\left(-\zeta + i\sqrt{1 - \zeta^2}\right) \\
z_2 &= \omega_0\left(-\zeta - i\sqrt{1 - \zeta^2}\right) \\
z_3 &= -\alpha
\end{aligned}\right\} \tag{4-53}$$

Then, the auto-covariance function of the displacement process in the stationary state follows from Eq. (3-59):

$$\kappa_{XX}(\tau) = -\pi S_0 \frac{\alpha^2}{m^2}\left(\frac{e^{z_1|\tau|}}{z_1\left(z_1^2 - z_2^2\right)\left(z_1^2 - z_3^2\right)} + \frac{e^{z_2|\tau|}}{z_2\left(z_2^2 - z_1^2\right)\left(z_2^2 - z_3^2\right)} + \frac{e^{z_3|\tau|}}{z_3\left(z_3^2 - z_1^2\right)\left(z_3^2 - z_2^2\right)}\right) \tag{4-54}$$

Eq. (4-54) can be further reduced. However, for numerical calculations the indicated form is sufficiently appropriate.

If the system is lightly damped ($\zeta \ll 1$) and the weakly stationary load process $\{F(t), t \in R\}$ is broad-banded, the displacement process $\{X(t), t \in R\}$ in the stationary state becomes narrow-banded as shown in Fig. 4-5. Moreover, the variance of the displacement response σ_X^2 is mainly determined by the areas under the peaks at $\omega = \pm\omega_0$. Assuming that $S_{FF}(\omega)$ has a smooth

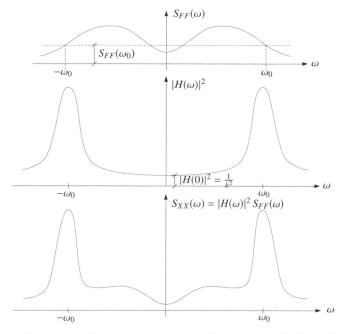

Fig. 4–5 Broad-banded load process applied to a lightly damped SDOF oscillator.

variation in the crucial intervals, so $S_{FF}(\omega) \simeq S_{FF}(\omega_0)$, and the displacement variance can consequently be approximated as:

$$\sigma_X^2 = \int_{-\infty}^{\infty} \left|H(\omega)\right|^2 S_{FF}(\omega)\, d\omega \simeq S_{FF}(\omega_0) \int_{-\infty}^{\infty} \left|H(\omega)\right|^2 d\omega = \frac{\pi S_{FF}(\omega_0)}{2\zeta \omega_0^3 m^2} \tag{4–55}$$

Eq. (4-55) is recognized as the displacement variance of a linear SDOF oscillator subjected to white noise with the auto-spectral density $S_{FF}(\omega_0)$, cf. Eq. (4-40). Hence, the considered approximate solution for the variance is equivalent to replacing the actual excitation process by a white noise with the auto-spectral density $S_{FF}(\omega_0)$. For this reason, the solution is referred to as the *equivalent white noise approximation*.

The variance $\sigma_{\ddot{X}}^2$ of the acceleration process cannot be determined from the equivalent white noise approximation even if the acceleration process does exist. The variance of the velocity process predicted by the equivalent white noise approximation becomes, cf. Eq. (4-41):

$$\sigma_{\dot{X}}^2 = \frac{\pi S_{FF}(\omega_0)}{2\zeta \omega_0 m^2} \tag{4–56}$$

Recalling that $S_{\dot{X}\dot{X}}(\omega) = \omega^2 S_{XX}(\omega) = \omega^2 |H(\omega)|^2 S_{FF}(\omega)$, Eq. (4-56) is seen to be tantamount to applying the equivalent white noise approximation when the oscillator is excited by a process with the auto-spectral density function $\omega^2 S_{FF}(\omega)$, which is still broad-banded in the vicinity of the angular eigenfrequency.

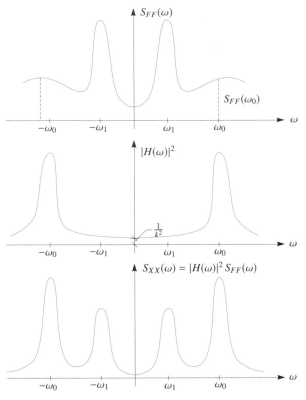

Fig. 4–6 Load process unsuitable for using equivalent white noise approximation.

Application of the equivalent white noise approximation Eq. (4-55) is only justified if the auto-spectral density function of the load process is sufficiently flat over the range of frequencies that are able to excite the oscillator. If the load process has a narrow-banded peak at the angular frequencies $\omega = \pm\omega_1$ outside the resonance interval, significant variance contribution at $\omega = \pm\omega_1$ may be present in $S_{XX}(\omega)$ as illustrated in Fig. 4-6. Then, Eq. (4-55) will underestimate the variance σ_X^2 significantly. However, if the peak of $S_{FF}(\omega)$ at $\omega = \omega_1$ is well separated from the peak at ω_0 ($\omega_1 < 0.5\omega_0$), a useful approximation may even be derived in this case. Since $|H(\omega)|^2$ is approximately flat at the static value $|H(0)|^2 = \frac{1}{k^2}$ for angular eigenfrequencies ω well below ω_1, we have:

$$
\begin{aligned}
\sigma_X^2 &= 2 \int_0^\infty |H(\omega)|^2 \, S_{FF}(\omega) \, d\omega \\
&= 2 \int_0^{\frac{1}{2}(\omega_0+\omega_1)} |H(\omega)|^2 \, S_{FF}(\omega) \, d\omega + 2 \int_{\frac{1}{2}(\omega_0+\omega_1)}^\infty |H(\omega)|^2 \, S_{FF}(\omega) \, d\omega \\
&\simeq 2 |H(0)|^2 \int_0^{\frac{1}{2}(\omega_0+\omega_1)} S_{FF}(\omega) \, d\omega \; + \; 2 S_{FF}(\omega_0) \int_{\frac{1}{2}(\omega_0+\omega_1)}^\infty |H(\omega)|^2 d\omega
\end{aligned}
\tag{4–57}
$$

The first integral considers that $|H(\omega)|^2$ is almost constant under the peak of $S_{FF}(\omega)$, and the last

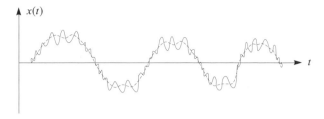

Fig. 4–7 Narrow-banded dynamic response at the top of a low-frequency quasi-static carrier signal.

integral follows because $S_{FF}(\omega)$ is almost constant under the peak of $|H(\omega)|^2$ (the equivalent white noise approximation). Further, the following approximations hold:

$$2 \int_0^{\frac{1}{2}(\omega_0+\omega_1)} S_{FF}(\omega)\, d\omega \simeq 2 \int_0^{\infty} S_{FF}(\omega)\, d\omega = \sigma_F^2 \tag{4-58}$$

$$2 \int_{\frac{1}{2}(\omega_0+\omega_1)}^{\infty} |H(\omega)|^2\, d\omega \simeq 2 \int_0^{\infty} |H(\omega)|^2\, d\omega = \frac{\pi}{2\zeta\omega_0^3 m^2} \tag{4-59}$$

Combining Eqs. (4-57), (4-58) and (4-59) provides:

$$\sigma_X^2 \simeq \frac{1}{k^2}\sigma_F^2 + \frac{\pi S_{FF}(\omega_0)}{2\zeta\omega_0^3 m^2} \tag{4-60}$$

where the first term on the right-hand side of Eq. (4-60) represents the *quasi-static variance* contribution from the peak region of the load spectrum, whereas the second term is the *dynamic variance* contribution. Eq. (4-60) has been implemented in the Danish code of practice for wind load. A possible realization of the displacement response for the considered load scenario has been shown in Fig. 4-7.

Example 4-4: SDOF system with deterministic initial values subjected to a white noise

The initial values in Eq. (4-1) are assumed to be deterministic, $\mathbf{x}_0 = [x_0, \dot{x}_0]^T$. Further, $\{F(t), t \in R\}$ is a white noise process with the auto-spectral density S_0. Then, the stationary displacement process becomes a Gaussian process with the mean value $\mu_X = 0$, and with the variance σ_X^2 and the auto-correlation coefficient function $\rho(\tau) = \rho_{XX}(\tau)$ given by Eq. (4-40).

Consider the stochastic state vector:

$$\mathbf{X}(t) = \begin{bmatrix} X(t) \\ \dot{X}(t) \end{bmatrix} \tag{4-61}$$

The mean value vector function $\boldsymbol{\mu}_{\mathbf{X}}(t) = E[X(t)]$ and the zero time-lag covariance matrix $\mathbf{C}_{\mathbf{XX}}(t) = E[(\mathbf{X}(t) - \boldsymbol{\mu}_{\mathbf{X}}(t))(\mathbf{X}(t) - \boldsymbol{\mu}_{\mathbf{X}}(t))^T]$ during the transient state, where the initial values are still felt in response, are given by Eqs. (4-21) and (4-22):

$$\boldsymbol{\mu}_{\mathbf{X}}(t) = \begin{bmatrix} \rho(t)x_0 - \rho'(t)\dot{x}_0/\omega_0^2 \\ \rho'(t)x_0 - \rho''(t)\dot{x}_0/\omega_0^2 \end{bmatrix} \tag{4-62}$$

$$\mathbf{C}_{\mathbf{XX}}(t) = \sigma_X^2 \begin{bmatrix} 1 - \rho^2(t) - (\rho'(t))^2/\omega_0^2 & -\rho'(t)(\rho(t) + \rho''(t)/\omega_0^2) \\ -\rho'(t)(\rho(t) + \rho''(t)/\omega_0^2) & \omega_0^2 - (\rho'(t))^2 - (\rho''(t))^2/\omega_0^2 \end{bmatrix} \tag{4–63}$$

where the auto-correlation coefficient function and its derivatives with respect to the argument have been given by Eq. (4-40):

$$\left. \begin{aligned} \rho(t) &= e^{-\zeta\omega_0|t|} \left(\cos(\omega_d t) + \frac{\zeta\omega_0}{\omega_d} \sin\left(\omega_d|t|\right) \right) \\ \rho'(t) &= -\frac{\omega_0^2}{\omega_d} e^{-\zeta\omega_0|t|} \sin\left(\omega_d t\right) \\ \rho''(t) &= -\omega_0^2 e^{-\zeta\omega_0|t|} \left(\cos(\omega_d t) - \frac{\zeta\omega_0}{\omega_d} \sin\left(\omega_d|t|\right) \right) \end{aligned} \right\} \tag{4–64}$$

From Eqs. (4-62), (4-63) and (4-64), it follows that the convergence towards the stationary value of the mean value functions and the covariance functions is controlled by the factors $\exp(-\zeta\omega_0 t)$ and $\exp(-2\zeta\omega_0 t)$, respectively. Similar to what was observed in Example 3-18, this implies that the covariance functions attain stationarity much faster than the mean value functions.

From Eq. (4-64), it can also be proved that:

$$\rho''(t) + 2\zeta\omega_0\rho'(t) + \omega_0^2\rho(t) = 0 \tag{4–65}$$

Hence, $\rho(t)$ fulfills the homogeneous differential equation (4-1), and can be characterized as a special example of an eigenvibration.

By differentiation of Eq. (4-65), it is seen that $\rho'(t)$ becomes an eigenvibration as well. Eliminating $\rho''(t)$ in Eq. (4-62) in favour of $\rho(t)$ and $\rho'(t)$ by means of Eq. (4-65), it follows that $\mu_X(t)$ and $\mu_{\dot{X}}(t)$ become linear functions of $\rho(t)$ and $\rho'(t)$, and hence become eigenvibrations themselves.

The solution for $\kappa_{X(t)\dot{X}(t)}$ in Eq. (4-63) may be further reduced by the use of Eq. (4-65), resulting in:

$$\kappa_{X(t)\dot{X}(t)} = 2\zeta\sigma_X^2 \left(\rho'(t)\right)^2/\omega_0 \tag{4–66}$$

Eq. (4-66) shows that $X(t)$ and $\dot{X}(t)$ are always non-negatively correlated in the transient phase, and will become uncorrelated as $t \to \infty$.

4.2 Multi-degree-of-freedom systems

A linear time-invariant n-degree-of-freedom system is excited by an external n-dimensional stochastic load vector process $\{\mathbf{F}(t), t \in [0,\infty[\}$. The load vector process is related to the *displacement vector process* $\{\mathbf{X}(t), t \in [0,\infty[\}$, the *velocity vector process* $\{\dot{\mathbf{X}}(t), t \in [0,\infty[\}$ and the *acceleration vector process* $\{\ddot{\mathbf{X}}(t), t \in [0,\infty[\}$ by the stochastic vector differential equation of motion:[1]

$$\left. \begin{aligned} \mathbf{M}\,\ddot{\mathbf{X}}(t) + \mathbf{C}\,\dot{\mathbf{X}}(t) + \mathbf{K}\,\mathbf{X}(t) &= \mathbf{F}(t), \quad t > 0 \\ \mathbf{X}(0) &= \mathbf{X}_0, \quad \dot{\mathbf{X}}(0) = \dot{\mathbf{X}}_0 \end{aligned} \right\} \tag{4–67}$$

---

[1]S.R.K. Nielsen. *Vibration Theory, Vol. 1: Linear Vibration Theory.* Aalborg University Press, Aalborg, 2004.

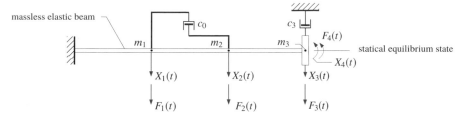

Fig. 4–8 Linear elastic, linear viscous damped multi-degree-of-freedom (MDOF) system.

where $\mathbf{M}$ is the mass matrix , $\mathbf{C}$ is the damping matrix and $\mathbf{K}$ is the stiffness matrix. Generally, the initial value vector $\left[\mathbf{X}_0, \dot{\mathbf{X}}_0\right]$ is considered to be stochastic.

The component $X_j(t)$ of the displacement vector $\mathbf{X}(t)$ is measured from the static equilibrium state which the system occupies at rest when the external loads are removed. $X_j(t)$ is considered positive in the same direction of the corresponding component $F_j(t)$ of the load vector. As a consequence of the choice of referential state, all static loads disappear from the equation of motion, and $\{\mathbf{F}(t), \; t \in [0, \infty[\}$ is a purely dynamic load vector process. Finally, the components $\dot{X}_j(t)$ and $\ddot{X}_j(t)$ of the velocity and acceleration vectors are considered positive, when acting in the same direction of $X_j(t)$.

The indicated definitions have been illustrated in Fig. 4-8, which shows a linear elastic system with two point masses m_1 and m_2, and a distributed rigid mass m_3 with the mass moment m_4. Since $X_4(t)$ is a rotation, the work-conjugated load $F_4(t)$ is an external dynamic moment.

The mass matrix $\mathbf{M}$, the stiffness matrix $\mathbf{K}$ and the damping matrix $\mathbf{C}$ fulfill the following symmetry and definite properties:

$$
\left.
\begin{aligned}
&\mathbf{a}^T \mathbf{M} \mathbf{a} > 0, \quad \mathbf{M} = \mathbf{M}^T \\
&\mathbf{a}^T \mathbf{C} \mathbf{a} > 0 \\
&\mathbf{a}^T \mathbf{K} \mathbf{a} \geq 0, \quad \mathbf{K} = \mathbf{K}^T
\end{aligned}
\right\}, \quad \forall \mathbf{a} \neq \mathbf{0}
\tag{4–68}
$$

$\mathbf{M}$ and $\mathbf{K}$ need to be symmetric in order that kinetic energy and strain energy (potential energy) can be defined for the structure. The symmetry property of $\mathbf{K}$ may also be explained from the Maxwell-Betti's reciprocal theorem in structural mechanics. The kinetic energy is always positive, which is the reason $\mathbf{M}$ is positive definite. For any rigid body motion of an unsupported structure (such as airplanes, ships and the rotor of a wind turbine), the strain energy is zero, resulting in the fact that $\mathbf{K}$ has been indicated as merely positive semi-definite in Eq. (4-68). However for supported structures, $\mathbf{K}$ is always positive definite. $\mathbf{C}$ needs to be positive definite, which is the necessary and sufficient condition that energy is dissipated per unit of time for any non-zero velocity vector of the structure. $\mathbf{C}$ is not required to be symmetric. However, only the symmetric part $\frac{1}{2}(\mathbf{C} + \mathbf{C}^T)$ implies energy dissipation.[1]

---

[1]S.R.K. Nielsen. *Vibration Theory, Vol. 1: Linear Vibration Theory*. Aalborg University Press, Aalborg, 2004.

Alternatively, the stochastic response processes can also be obtained from the following linear state vector differential equation of the dimension $2n$:

$$\dot{\mathbf{Y}}(t) = \mathbf{A}\,\mathbf{Y}(t) + \mathbf{B}\mathbf{F}(t), \quad t > 0 \\ \mathbf{Y}(0) = \mathbf{Y}_0 \Bigg\} \tag{4-69}$$

where:

$$\mathbf{Y}(t) = \begin{bmatrix} \mathbf{X}(t) \\ \dot{\mathbf{X}}(t) \end{bmatrix}, \quad \mathbf{Y}_0 = \begin{bmatrix} \mathbf{X}_0 \\ \dot{\mathbf{X}}_0 \end{bmatrix}, \quad \mathbf{A} = \begin{bmatrix} \mathbf{0} & \mathbf{I} \\ -\mathbf{M}^{-1}\mathbf{K} & -\mathbf{M}^{-1}\mathbf{C} \end{bmatrix}, \quad \mathbf{B} = \begin{bmatrix} \mathbf{0} \\ \mathbf{M}^{-1} \end{bmatrix} \tag{4-70}$$

Analogous to Eqs. (4-2), (4-7), (4-8) and (4-9), the solution to Eq. (4-67) can be written as:[1]

$$\mathbf{X}(t) = \mathbf{X}^{(0)}(t) + \int_0^t \mathbf{h}(t-\tau)\,\mathbf{F}(\tau)\,d\tau \tag{4-71}$$

$$\mathbf{X}^{(0)}(t) = \mathbf{a}_0(t)\,\mathbf{X}_0 + \mathbf{a}_1(t)\,\dot{\mathbf{X}}_0 \tag{4-72}$$

$$\mathbf{a}_0(t) = \mathbf{h}(t)\,\mathbf{C} + \dot{\mathbf{h}}(t)\,\mathbf{M} \tag{4-73}$$

$$\mathbf{a}_1(t) = \mathbf{h}(t)\,\mathbf{M} \tag{4-74}$$

where $\mathbf{h}(t)$ is the impulse response matrix of the system. $\mathbf{h}(t)$ is obtained as the inverse Fourier transform of the frequency response matrix, cf. Eq. (3-99):

$$\mathbf{h}(t) = \frac{1}{2\pi} \int_{-\infty}^{\infty} e^{i\omega t}\,\mathbf{H}(\omega)\,d\omega \tag{4-75}$$

$$\mathbf{H}(\omega) = \left(\mathbf{M}\,z^2 + \mathbf{C}\,z + \mathbf{K}\right)^{-1}, \quad z = i\omega \tag{4-76}$$

The jth column $\mathbf{h}_j(t)$ of $\mathbf{h}(t)$ is determined as the solution of Eq. (4-67) for the excitation $\mathbf{F}(t) = \mathbf{e}_j\,\delta(t)$ with the system at rest, and the jth column $\mathbf{H}_j(\omega)$ of $\mathbf{H}(\omega)$ is determined as the amplitude of the stationary solution of Eq. (4-67) from the excitation $\mathbf{F}(t) = \mathbf{e}_j\,e^{i\omega t}$, where:

$$\mathbf{e}_j = \begin{bmatrix} 0 \\ \vdots \\ 0 \\ 1 \\ 0 \\ \vdots \\ 0 \end{bmatrix} \leftarrow j\text{th component} \tag{4-77}$$

Undamped eigenvibrations of the system are obtained as solution to Eq. (4-67) for $\mathbf{C} = \mathbf{0}$ and $\mathbf{F}(t) \equiv \mathbf{0}$. They are given in the form:[1]

$$\mathbf{X}(t) = \boldsymbol{\Phi}\cos(\omega t - \Psi) \tag{4-78}$$

where Ψ is an arbitrary undeterminate phase. The amplitude vector Φ and the angular frequency ω are obtained as solutions to the *generalized eigenvalue problem*:

$$\left(\mathbf{K} - \omega^2\mathbf{M}\right)\Phi = \mathbf{0} \tag{4-79}$$

Non-trivial solutions $\Phi \neq \mathbf{0}$ exist only in case of a singular coefficient matrix, leading to the *frequency condition* or the *characteristic equation*:

$$\det\left(\mathbf{K} - \omega^2\mathbf{M}\right) = 0 \tag{4-80}$$

The solutions $0 \leq \omega_1 \leq \omega_2 \leq \ldots \leq \omega_n$ to Eq. (4-80) determine the *undamped angular eigenfrequencies* ω_j of the structure. It should be noticed that $\omega_1 = 0$ corresponds to a rigid body mode of an unsupported structure. For supported structures, we always have $\omega_1 > 0$.

For each angular eigenfrequency, a non-trivial solution Φ_j to Eq. (4-79) exists. These vectors are termed the *undamped eigenmodes*, which are determined within an arbitrary factor a, i.e., if Φ_j is an eigenmode, so is $a\Phi_j$.

The eigenmodes fulfill the following *orthogonality conditions*, which are a consequence of the symmetry properties $\mathbf{M} = \mathbf{M}^T$ and $\mathbf{K} = \mathbf{K}^T$:[1]

$$\Phi_j^T\mathbf{M}\,\Phi_k = \begin{cases} 0 \,, & j \neq k \\ m_j\,, & j = k \end{cases} \tag{4-81}$$

$$\Phi_j^T\mathbf{K}\,\Phi_k = \begin{cases} 0 \,, & j \neq k \\ k_j\,, & j = k \end{cases} \tag{4-82}$$

where m_j and k_j are indicated the *modal mass* and the *modal stiffness*, respectively. It follows from Eq. (4-79) that $k_j = \omega_j^2\, m_j$.

Further, we shall assume that the eigenmodes are also orthogonal with respect to the damping matrix:

$$\Phi_j^T\mathbf{C}\,\Phi_k = \begin{cases} 0 \,, & j \neq k \\ c_j\,, & j = k \end{cases} \tag{4-83}$$

where c_j is termed the *modal damping coefficient*.

A sufficient condition for Eq. (4-83) is that $\mathbf{C}$ is modelled by a *Rayleigh* or *Caughey damping model*.[1] In any case, modal decoupling can be assumed if the eigenfrequencies are well-separated and the eigenvibrations in all modes are lightly damped.[1] Eq. (4-83) is the underlying assumption in modal analysis, which leads to a decoupling of the vector equation of motion in Eq. (4-67) into n independent SDOF equations of motion.

---

[1]S.R.K. Nielsen. *Vibration Theory, Vol. 1: Linear Vibration Theory*. Aalborg University Press, Aalborg, 2004.

The orthogonality conditions in Eqs. (4-81), (4-82) and (4-83) may be assembled in the following matrix formulation:

$$\left.\begin{array}{l} \mathbf{m} = \mathbf{P}^T \mathbf{M} \mathbf{P} \\ \mathbf{c} = \mathbf{P}^T \mathbf{C} \mathbf{P} \\ \mathbf{k} = \mathbf{P}^T \mathbf{K} \mathbf{P} \end{array}\right\} \tag{4–84}$$

where $\mathbf{P}$ indicates the so-called *modal matrix* made up of the eigenmodes (stored column-wise):

$$\mathbf{P} = \begin{bmatrix} \mathbf{\Phi}_1 \, \mathbf{\Phi}_2 \cdots \mathbf{\Phi}_n \end{bmatrix} \tag{4–85}$$

$\mathbf{m}$, $\mathbf{c}$ and $\mathbf{k}$ indicate the *modal mass matrix*, the *modal damping matrix* and the *modal stiffness matrix*, respectively. These are diagonal matrices written as:

$$\mathbf{m} = \begin{bmatrix} m_1 & 0 & \cdots & 0 \\ 0 & m_2 & \cdots & 0 \\ \vdots & \vdots & \ddots & \vdots \\ 0 & 0 & \cdots & m_n \end{bmatrix}, \quad \mathbf{c} = \begin{bmatrix} c_1 & 0 & \cdots & 0 \\ 0 & c_2 & \cdots & 0 \\ \vdots & \vdots & \ddots & \vdots \\ 0 & 0 & \cdots & c_n \end{bmatrix}, \quad \mathbf{k} = \begin{bmatrix} k_1 & 0 & \cdots & 0 \\ 0 & k_2 & \cdots & 0 \\ \vdots & \vdots & \ddots & \vdots \\ 0 & 0 & \cdots & k_n \end{bmatrix} \tag{4–86}$$

It follows from Eqs. (4-74) and (4-84) that:

$$\mathbf{H}^{-1}(\omega) = \left(\mathbf{P}^T\right)^{-1} \left(\mathbf{m}\, z^2 + \mathbf{c}\, z + \mathbf{k}\right) \mathbf{P}^{-1} \quad \Rightarrow$$

$$\mathbf{H}(\omega) = \mathbf{P} \left(\mathbf{m}\, z^2 + \mathbf{c}\, z + \mathbf{k}\right)^{-1} \mathbf{P}^T = \sum_{j=1}^{n} H_j(\omega)\, \mathbf{\Phi}_j \mathbf{\Phi}_j^T, \quad z = i\omega \tag{4–87}$$

where $H_j(\omega)$ signifies the *modal frequency response function*, cf. Eq. (4-17):

$$H_j(\omega) = \frac{1}{m_j\, z^2 + c_j z + k_j} = \frac{1}{m_j \left(z^2 + 2\,\zeta_j\, \omega_j\, z + \omega_j^2\right)}, \quad z = i\omega \tag{4–88}$$

with ζ_j indicating the *modal damping ratio* in the jth mode:

$$\zeta_j = \frac{c_j}{2\sqrt{m_j k_j}} \tag{4–89}$$

Eq. (4-87) shows an expansion of the frequency response matrix in outer products (dyads) of the eigenmodes.

From Eqs. (4-75) and (4-87), it follows that the impulse response matrix may be given by a similar expansion:

$$\mathbf{h}(t) = \sum_{j=1}^{n} h_j(t)\, \mathbf{\Phi}_j \mathbf{\Phi}_j^T \tag{4–90}$$

where $h_j(t)$ signifies the *modal impulse response function* given by, cf. Eq. (4-3):

$$h_j(t) = \begin{cases} 0 & , \quad t \le 0 \\ \dfrac{e^{-\zeta\omega_j t}}{m_j\,\omega_{d,j}}\sin(\omega_{d,j}t), & t > 0 \end{cases} \tag{4–91}$$

with $\omega_{d,j}$ being the damped angular eigenfrequency in the jth mode:

$$\omega_{d,j} = \omega_j\sqrt{1 - \zeta_j^2} \tag{4–92}$$

Finally, insertion of Eq. (4-90) into Eqs. (4-73) and (4-74) provides the following expansions for the matrix functions $\mathbf{a}_0(t)$ and $\mathbf{a}_1(t)$:

$$\mathbf{a}_0(t) = \sum_{j=1}^{n} \boldsymbol{\Phi}_j\boldsymbol{\Phi}_j^T\Big(h_j(t)\,\mathbf{C} + \dot{h}_j(t)\,\mathbf{M}\Big) \tag{4–93}$$

$$\mathbf{a}_1(t) = \sum_{j=1}^{n} h_j(t)\,\boldsymbol{\Phi}_j\boldsymbol{\Phi}_j^T\,\mathbf{M} \tag{4–94}$$

Since the linear independent eigenmodes form a basis in R^n, the solution vector $\mathbf{X}(t)$ to Eq. (4-67) admits the *eigenmode expansion*:

$$\mathbf{X}(t) = \sum_{j=1}^{n} Q_j(t)\,\boldsymbol{\Phi}_j = \mathbf{P}\,\mathbf{Q}(t) \tag{4–95}$$

$$\mathbf{Q}(t) = \big[Q_1(t), Q_2(t), \ldots, Q_n(t)\big]^T \tag{4–96}$$

Eq. (4-95) is merely a coordinate transformation with the transformation matrix $\mathbf{P}$ between the physical coordinates $X_j(t)$ and the new coordinates $Q_j(t)$, which are termed the *modal coordinates*. They form an n-dimensional stochastic vector process $\{\mathbf{Q}(t),\ t \in [0,\infty[\}$. Insertion of Eq. (4-95) into Eq. (4-67) and application of the orthogonality properties in Eq. (4-84) provides:

$$\left.\begin{array}{l} \mathbf{m}\,\ddot{\mathbf{Q}}(t) + \mathbf{c}\,\dot{\mathbf{Q}}(t) + \mathbf{k}\,\mathbf{Q}(t) = \mathbf{f}(t), \quad t > 0 \\ \mathbf{Q}(0) = \mathbf{Q}_0, \quad \dot{\mathbf{Q}}(0) = \dot{\mathbf{Q}}_0 \end{array}\right\} \tag{4–97}$$

where $\mathbf{f}(t)$ is the *modal load vector* given by:

$$\mathbf{f}(t) = \mathbf{P}^T\mathbf{F}(t) = \begin{bmatrix} f_1(t) \\ f_2(t) \\ \vdots \\ f_n(t) \end{bmatrix} = \begin{bmatrix} \boldsymbol{\Phi}_1^T\mathbf{F}(t) \\ \boldsymbol{\Phi}_2^T\mathbf{F}(t) \\ \vdots \\ \boldsymbol{\Phi}_n^T\mathbf{F}(t) \end{bmatrix} \tag{4–98}$$

Further, the initial value vectors in Eq. (4-97) are given by:

$$\left.\begin{array}{l} \mathbf{Q}_0 = \mathbf{P}^{-1}\mathbf{X}_0 = \mathbf{m}^{-1}\mathbf{P}^T\mathbf{M}\mathbf{X}_0 \\ \dot{\mathbf{Q}}_0 = \mathbf{P}^{-1}\dot{\mathbf{X}}_0 = \mathbf{m}^{-1}\mathbf{P}^T\mathbf{M}\dot{\mathbf{X}}_0 \end{array}\right\} \tag{4–99}$$

From Eq. (4-97), the *modal equation of motion* for the jth mode becomes:

$$\left. \begin{array}{l} m_j\,\ddot{Q}_j(t) + c_j\,\dot{Q}_j(t) + k_j\,Q_j(t) = f_j(t), \quad t > 0 \\ Q_j(0) = Q_{j,0}, \quad \dot{Q}_j(0) = \dot{Q}_{j,0} \end{array} \right\} \tag{4-100}$$

Alternatively, the stochastic differential equation in Eq. (4-100) can also be reformulated as:

$$\ddot{Q}_j(t) + 2\zeta_j\omega_j\,\dot{Q}_j(t) + \omega_j^2\,Q_j(t) = \frac{1}{m_j}\,f_j(t), \quad t > 0 \tag{4-101}$$

According to Eq. (4-2), the solution to Eq. (4-100) can be written in terms of the following stochastic integral equation:

$$Q_j(t) = Q_j^{(0)}(t) + \int_0^t h_j(t-\tau)\,f_j(\tau)\,d\tau \tag{4-102}$$

where the initial value response $Q_j^{(0)}(t)$ is given by, cf. Eqs. (4-7), (4-8), (4-9):

$$Q_j^{(0)}(t) = a_{j,0}(t)\,Q_{j,0} + a_{j,1}(t)\,\dot{Q}_{j,0} \tag{4-103}$$

$$a_{j,0}(t) = c_j\,h_j(t) + m_j\,\dot{h}_j(t) \tag{4-104}$$

$$a_{j,0}(t) = m_j\,h_j(t) \tag{4-105}$$

For the n-dimensional stochastic integral equation Eq. (4-71), all results indicated in Section 3.2 on stochastic integration of stochastic vector processes can immediately be applied.

The mean value vector function $\boldsymbol{\mu}_{\mathbf{X}}(t)$ of the displacement vector process is related to the mean value vector function $\boldsymbol{\mu}_{\mathbf{F}}(t)$ of the load vector process as, cf. Eqs. (3-88), (4-71) and (4-72):

$$\boldsymbol{\mu}_{\mathbf{X}}(t) = \boldsymbol{\mu}_{\mathbf{X}^{(0)}}(t) + \int_0^t \mathbf{h}(t-\tau)\,\boldsymbol{\mu}_{\mathbf{F}}(\tau)\,d\tau \tag{4-106}$$

$$\boldsymbol{\mu}_{\mathbf{X}^{(0)}}(t) = \mathbf{a}_0(t)\,\boldsymbol{\mu}_{\mathbf{X}_0} + \mathbf{a}_1(t)\,\boldsymbol{\mu}_{\dot{\mathbf{X}}_0} \tag{4-107}$$

where $\boldsymbol{\mu}_{\mathbf{X}_0}$ and $\boldsymbol{\mu}_{\dot{\mathbf{X}}_0}$ indicate the mean value vectors of the initial value vectors $\mathbf{X}_0$ and $\dot{\mathbf{X}}_0$, respectively.

The component forms of Eqs. (4-106) and (4-107) are:

$$\mu_{X_j}(t) = \mu_{X_j^{(0)}}(t) + \sum_{k=1}^n \int_0^t h_{jk}(t-\tau)\,\mu_{F_k}(\tau)\,d\tau \tag{4-108}$$

$$\mu_{X_j^{(0)}}(t) = \sum_{k=1}^n a_{jk,0}(t)\,\mu_{X_{k,0}} + \sum_{k=1}^n a_{jk,1}(t)\,\mu_{\dot{X}_{k,0}} \tag{4-109}$$

where $\mu_{X_{k,0}}$, $\mu_{\dot{X}_{k,0}}$ signify the components of the mean value vectors of the initial value vectors.

It is assumed that the initial value vectors $\mathbf{X}_0$, $\dot{\mathbf{X}}_0$ in Eq. (4-72) are stochastically independent of the load vector process $\{\mathbf{F}(t),\, t \in [0, \infty[\}$. Then, the cross-covariance matrix function $\kappa_{\mathbf{XX}}(t_1, t_2)$ of the displacement vector process is related to the cross-covariance matrix functions $\kappa_{\mathbf{X}^{(0)}\mathbf{X}^{(0)}}(t_1, t_2)$ and $\kappa_{\mathbf{FF}}(\tau_1, \tau_2)$ of the initial value vector process and load vector process, by the following matrix equation, cf. Eq. (3-121):

$$\kappa_{\mathbf{XX}}(t_1, t_2) = \kappa_{\mathbf{X}^{(0)}\mathbf{X}^{(0)}}(t_1, t_2) + \int_0^{t_1} \int_0^{t_2} \mathbf{h}(t_1 - \tau_1)\,\kappa_{\mathbf{FF}}(\tau_1, \tau_2)\,\mathbf{h}^T(t_2 - \tau_2)\,d\tau_1 d\tau_2 \qquad (4\text{--}110)$$

The component form of Eq. (4-110) is, cf. Eq. (3-117):

$$\kappa_{X_{j_1} X_{j_2}}(t_1, t_2) = \kappa_{X_{j_1}^{(0)} X_{j_2}^{(0)}}(t_1, t_2) + \sum_{k_1=1}^{n} \sum_{k_2=1}^{n} \int_0^{t_1} \int_0^{t_2} h_{j_1 k_1}(t_1 - \tau_1)\, \kappa_{F_{k_1} F_{k_2}}(\tau_1, \tau_2)\, h_{j_2 k_2}(t_2 - \tau_2)\, d\tau_1 d\tau_2$$

$$(4\text{--}111)$$

where from Eq. (4-72) we have:

$$\kappa_{X_{j_1}^{(0)} X_{j_2}^{(0)}}(t_1, t_2) = \sum_{k_1=1}^{n} \sum_{k_2=1}^{n} a_{j_1 k_1, 0}(t_1)\, a_{j_2 k_2, 0}(t_2)\, \kappa_{X_{k_1,0} X_{k_2,0}}$$

$$+ \sum_{k_1=1}^{n} \sum_{k_2=1}^{n} \left(a_{j_1 k_1, 0}(t_1)\, a_{j_2 k_2, 1}(t_2)\, \kappa_{X_{k_1,0} \dot{X}_{k_2,0}} + a_{j_1 k_1, 1}(t_1)\, a_{j_2 k_2, 0}(t_2)\, \kappa_{X_{k_2,0} \dot{X}_{k_1,0}} \right)$$

$$+ \sum_{k_1=1}^{n} \sum_{k_2=1}^{n} a_{j_1 k_1, 1}(t_1)\, a_{j_2 k_2, 1}(t_2)\, \kappa_{\dot{X}_{k_1,0} \dot{X}_{k_2,0}} \qquad (4\text{--}112)$$

$\kappa_{X_{k_1,0} X_{k_2,0}}$, $\kappa_{\dot{X}_{k_1,0} \dot{X}_{k_2,0}}$ are the components of the covariance matrices, and $\kappa_{X_{k_1,0} \dot{X}_{k_2,0}}$ are the components of the cross-covariance matrix of the initial value vectors ($\mathbf{X}_0$ and $\dot{\mathbf{X}}_0$).

If $[\mathbf{X}_0^T, \dot{\mathbf{X}}_0^T]^T$ is a $2n$-dimensional normal vector and the load vector process $\{\mathbf{F}(t),\, t \in [0, \infty[\}$ is Gaussian, $\{\mathbf{X}(t),\, t \in [0, \infty[\}$ becomes Gaussian as well. Then, $\{\mathbf{X}(t),\, t \in [0, \infty[\}$ can be completely determined by Eqs. (4-106) and (4-110).

If the load vector process $\{\mathbf{F}(t),\, t \in R\}$ is weakly or strictly stationary, the displacement vector process $\{\mathbf{X}(t),\, t \in [0, \infty[\}$ and the related derivative vector processes eventually attain stationarity in the same sense. The mean value vector function, the cross-covariance matrix function and the cross-spectral density matrix function in the stationary state follow from Eqs. (3-120), (3-121) and (3-122):

$$\mu_{\mathbf{X}} = \mathbf{H}(0)\,\mu_{\mathbf{F}} = \mathbf{K}^{-1}\,\mu_{\mathbf{F}} \qquad (4\text{--}113)$$

$$\kappa_{\mathbf{XX}}(\tau) = \int_0^{\infty} \int_0^{\infty} \mathbf{h}(u_1)\,\kappa_{\mathbf{FF}}(\tau + u_1 - u_2)\,\mathbf{h}^T(u_2)\,du_1 du_2 \qquad (4\text{--}114)$$

$$\mathbf{S_{XX}}(\omega) \ = \ \mathbf{H}^*(\omega)\,\mathbf{S_{FF}}(\omega)\,\mathbf{H}^T(\omega) \tag{4–115}$$

It follows from Eq. (4-98) that the cross-covariance function $\kappa_{f_{j_1}f_{j_2}}(\tau)$ and the cross-spectral density function $S_{f_{j_1}f_{j_2}}(\omega)$ of the modal load processes are related to the corresponding functions of the load vector process as:

$$\kappa_{f_{j_1}f_{j_2}}(\tau) \ = \ \sum_{k_1=1}^{n}\sum_{k_2=1}^{n} \Phi_{j_1,k_1}\Phi_{j_2,k_2}\,\kappa_{F_{k_1}F_{k_2}}(\tau), \quad \tau = t_2 - t_1 \tag{4–116}$$

$$S_{f_{j_1}f_{j_2}}(\omega) \ = \ \sum_{k_1=1}^{n}\sum_{k_2=1}^{n} \Phi_{j_1,k_1}\Phi_{j_2,k_2}\,S_{F_{k_1}F_{k_2}}(\omega) \tag{4–117}$$

where $\Phi_{j,k}$ signifies the kth component of $\mathbf{\Phi}_j$.

Similar with the results in Eqs. (3-102) and (3-104), the following relations between the cross-covariance function $\kappa_{Q_{j_1}Q_{j_2}}(\tau)$ and the cross-spectral density function $S_{Q_{j_1}Q_{j_2}}(\omega)$ of the modal coordinates and the corresponding quantities of the modal load processes can be obtained from Eq. (4-102):

$$\kappa_{Q_{j_1}Q_{j_2}}(\tau) \ = \ \int_0^\infty\int_0^\infty h_{j_1}(u_1)\,\kappa_{f_{j_1}f_{j_2}}(\tau + u_1 - u_2)\,h_{j_2}(u_2)\,du_1 du_2, \quad \tau = t_2 - t_1 \tag{4–118}$$

$$S_{Q_{j_1}Q_{j_2}}(\omega) \ = \ H_{j_1}^*(\omega)\,S_{f_{j_1}f_{j_2}}(\omega)\,H_{j_2}(\omega) \tag{4–119}$$

From Eq. (4-95), it follows that the cross-covariance function $\kappa_{X_{j_1}X_{j_2}}(\tau)$ and the cross-spectral density function $S_{X_{j_1}X_{j_2}}(\omega)$ of the displacement vector process are related to the corresponding quantities of the modal coordinate processes as:

$$\kappa_{X_{j_1}X_{j_2}}(\tau) \ = \ \sum_{k_1=1}^{n}\sum_{k_2=1}^{n} \Phi_{k_1,j_1}\Phi_{k_2,j_2}\,\kappa_{Q_{k_1}Q_{k_2}}(\tau), \quad \tau = t_2 - t_1 \tag{4–120}$$

$$S_{X_{j_1}X_{j_2}}(\omega) \ = \ \sum_{k_1=1}^{n}\sum_{k_2=1}^{n} \Phi_{k_1,j_1}\Phi_{k_2,j_2}\,S_{Q_{k_1}Q_{k_2}}(\omega) \tag{4–121}$$

In Section 4.1, the concept of equivalent white noise approximation was introduced for evaluating the variance of a narrow-banded SDOF system without performing numerical integration. In this section, a similar approach is to be introduced, which provides an approximate determination of the cross-covariance matrix function and the cross-spectral density matrix function of $\{\mathbf{X}(t),\ t \in [0, \infty[\}$ in the stationary state.

Applying the first Wiener-Khintchine relation in Eq. (3-23) into Eq. (4-119), we have:

$$\kappa_{Q_{j_1}Q_{j_2}}(\tau) \ = \ \int_{-\infty}^{\infty} e^{i\omega\tau}S_{Q_{j_1}Q_{j_2}}(\omega)\,d\tau \ = \ \int_{-\infty}^{\infty} e^{i\omega\tau}H_{j_1}^*(\omega)H_{j_2}(\omega)\,S_{f_{j_1}f_{j_2}}(\omega)\,d\omega \tag{4–122}$$

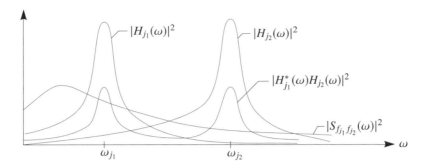

Fig. 4–9 Frequency overlap in a lightly damped system with well separated angular eigenfrequencies.

The basic assumption is that all component processes $\{F_k(t), \ t \in R\}$ of the load vector process may be considered broad-banded. Then, the cross-spectral density function $S_{f_{j_1} f_{j_2}}(\omega)$ between the j_1th and j_2th modal load processes, as expressed by Eq. (4-117), also becomes broad-banded.

Further, the system is assumed to be lightly damped with well separated angular eigenfrequencies. Hence, the following results hold for $\omega_{j_1} < \omega_{j_2}$:

$$\left. \begin{array}{ll} \zeta_j \ll 1 & , \quad j = 1,\dots,n \\ \omega_{j_1}\left(1 + a\,\zeta_{j_1}\right) < \omega_{j_2}\left(1 - a\,\zeta_{j_2}\right), & a \sim 2 - 3 \end{array} \right\} \tag{4–123}$$

Notice that the conditions in Eq. (4-123) are also the sufficient conditions to insure modal decoupling in deterministic linear structural dynamics.[1]

Then, Eq. (4-122) can be approximated as:

$$\kappa_{Q_{j_1} Q_{j_2}}(\tau) \simeq 2\,\mathrm{Re}\Bigg(S_{f_{j_1} f_{j_2}}(\omega_{j_1}) \int_{\omega_{j_1}(1 - a\zeta_{j_1})}^{\omega_{j_1}(1 + a\zeta_{j_1})} e^{i\omega\tau}\, H_{j_1}^*(\omega) H_{j_2}(\omega)\, d\omega$$

$$+ S_{f_{j_1} f_{j_2}}(\omega_{j_2}) \int_{\omega_{j_2}(1 - a\zeta_{j_2})}^{\omega_{j_2}(1 + a\zeta_{j_2})} e^{i\omega\tau}\, H_{j_1}^*(\omega) H_{j_2}(\omega)\, d\omega \Bigg) \tag{4–124}$$

Under the conditions specified in Eq. (4-123), $|H_{j_1}(\omega)|^2$ and $|H_{j_2}(\omega)|^2$ have a marked peak at $\omega = \omega_{j_1}$ and $\omega = \omega_{j_2}$, respectively. They are much larger than the peaks of $|H_{j_1}^*(\omega) H_{j_2}(\omega)|$ at the same angular eigenfrequencies, as shown in Fig. 4-9. As an example, if ω is set to be $\omega = \omega_{j_1}$, the following equation follows from Eq. (4-88):

$$\frac{|H_{j_1}^*(\omega_{j_1}) H_{j_2}(\omega_{j_1})|}{|H_{j_1}(\omega_{j_1})|^2} = \frac{|H_{j_2}(\omega_{j_1})|}{|H_{j_1}(\omega_{j_1})|} = \frac{\dfrac{1}{m_{j_2}} \dfrac{1}{\sqrt{(\omega_{j_2}^2 - \omega_{j_1}^2)^2 + 4\zeta_{j_1}^2 \omega_{j_1}^2 \omega_{j_2}^2}}}{\dfrac{1}{m_{j_1}} \dfrac{1}{2\zeta_{j_1} \omega_{j_1}^2}}$$

$$= \frac{m_{j_1}}{m_{j_2}} \frac{2\omega_{j_1}^2}{|\omega_{j_2}^2 - \omega_{j_1}^2|} \zeta_{j_1} \ll 1 \tag{4–125}$$

[1]S.R.K. Nielsen. *Vibration Theory, Vol. 1: Linear Vibration Theory*. Aalborg University Press, Aalborg, 2004.

Therefore, for lightly damped structures with well separated angular eigenfrequencies, $\kappa_{Q_{j_1}Q_{j_2}}(\tau)$ can be neglected for $j_1 \neq j_2$, so the cross-terms in the double sums in Eqs. (4-120) and (4-121) can be neglected.

Furthermore, for the diagonal terms ($j_1 = j_2 = j$), the equivalent white noise approximation is applied to Eq. (4-122), cf. Eq. (4-55):

$$\kappa_{Q_j Q_j}(\tau) = \int_{-\infty}^{\infty} e^{i\omega\tau} |H_j(\omega)|^2 S_{f_j f_j}(\omega)\, d\omega \simeq \sigma_{Q_j}^2\, \rho_{Q_j Q_j}(\tau) \tag{4–126}$$

$$\sigma_{Q_j}^2 = \frac{\pi S_{f_j f_j}(\omega_j)}{2\zeta_j \omega_j^3 m_j^2} \tag{4–127}$$

$$\rho_{Q_j Q_j}(\tau) = e^{-\zeta_j \omega_j |\tau|} \left(\cos\left(\omega_{d,j}\tau\right) + \frac{\zeta_j}{\sqrt{1 - \zeta_j^2}} \sin\left(\omega_{d,j}|\tau|\right) \right) \tag{4–128}$$

Then, Eqs. (4-120) and (4-121) can approximately be written as:

$$\kappa_{X_{j_1} X_{j_2}}(\tau) \simeq \sum_{k=1}^{n} \Phi_{k,j_1} \Phi_{k,j_2}\, \kappa_{Q_k Q_k}(\tau) \tag{4–129}$$

$$S_{X_{j_1} X_{j_2}}(\omega) \simeq \sum_{k=1}^{n} \Phi_{k,j_1} \Phi_{k,j_2}\, S_{Q_k Q_k}(\omega) = \sum_{k=1}^{n} \Phi_{k,j_1} \Phi_{k,j_2}\, |H_k(\omega)|^2\, S_{f_k f_k}(\omega) \tag{4–130}$$

It follows from Eqs. (4-126) and (4-129) that merely the auto-spectral density $S_{f_k f_k}(\omega_k)$ of the modal load process at the angular eigenfrequency ω_k is needed for the calculation of the cross-covariance functions of the displacement vector process. Moreover, the sums in Eqs. (4-129) and (4-130) may often be truncated after a few terms.

Since $\kappa_{Q_k Q_k}(\tau) = \kappa_{Q_k Q_k}(-\tau)$, it follows from Eq. (4-129) that the indicated approximations imply that the cross-covariance functions of the displacement vector process become even functions of τ as well. Correspondingly, Eq. (4-130) shows that the corresponding cross-spectral density functions become real and symmetric functions of ω.

The condition for ignoring the cross-terms in Eq. (4-120) may alternatively be stated as:

$$\left| \kappa_{Q_{j_1} Q_{j_2}}(\tau) \right| \ll \min \left(\kappa_{Q_{j_1} Q_{j_1}}(\tau), \kappa_{Q_{j_2} Q_{j_2}}(\tau) \right), \quad j_1 \neq j_2 \tag{4–131}$$

implying that $Q_{j_1}(t)$ and $Q_{j_2}(t + \tau)$ must be uncorrelated for all τ. Hence, for lightly damped structures with well separated angular eigenfrequencies, the modal coordinate processes $\{Q_{j_1}(t),\, t \in R\}$ and $\{Q_{j_2}(t),\, t \in R\}$ are uncorrelated.

Fig. 4-10a shows a MDOF system subjected to a scalar load process $\{F(t),\, t \in R\}$ acting indirectly on the degrees of freedom $\mathbf{X}(t) = [X_1(t), X_2(t)]^T$. $F(t)$ is statically equivalent to a

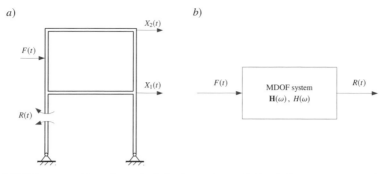

Fig. 4–10 a) A MDOF system subjected to a single load component. b) Symbolic representation of the single-input single-output in a MDOF system.

system of nodal forces $\mathbf{F}(t)$ work-conjugated with $\mathbf{X}(t)$. Due to the linearity, this relation can be written as:

$$\mathbf{F}(t) = \mathbf{b}\,F(t) \tag{4–132}$$

A certain response quantity $R(t)$ is considered, which may represent a stress component, a bending moment, a displacement component, etc. Again, $R(t)$ is linearly dependent on $\mathbf{X}(t)$ due to the linearity of the structure, i.e.:

$$R(t) = \mathbf{a}^T \mathbf{X}(t) \tag{4–133}$$

where $\mathbf{a}$ and $\mathbf{b}$ are both deterministic n-dimensional vectors.

Assuming the external load to be harmonically varying:

$$F(t) = \mathrm{Re}\!\left(F_0\,e^{i\omega t}\right) \tag{4–134}$$

The stationary displacement vector $\mathbf{X}(t)$ and the response $R(t)$ become harmonically varying as well:

$$\mathbf{X}(t) = \mathrm{Re}\!\left(\mathbf{X}_0\,e^{i\omega t}\right) = \mathrm{Re}\!\left(\mathbf{H}(\omega)\,\mathbf{b}\,F_0\,e^{i\omega t}\right) \tag{4–135}$$

$$R(t) = \mathrm{Re}\!\left(R_0\,e^{i\omega t}\right) = \mathrm{Re}\!\left(\mathbf{a}^T \mathbf{X}_0\,e^{i\omega t}\right) = \mathrm{Re}\!\left(\mathbf{a}^T \mathbf{H}(\omega)\,\mathbf{b}\,F_0\,e^{i\omega t}\right) \tag{4–136}$$

where $\mathbf{H}(\omega)$ denotes the frequency response matrix of the MDOF system.

Then, the frequency response function relating the load $F(t)$ to the response $R(t)$ becomes:

$$H(\omega) = \mathbf{a}^T \mathbf{H}(\omega)\,\mathbf{b} = \sum_{j=1}^{n} \frac{c_j}{z^2 + 2\,\zeta_j\,\omega_j\,z + \omega_j^2}, \quad z = i\omega \tag{4–137}$$

$$c_j = \frac{1}{m_j}\,\mathbf{a}^T \mathbf{\Phi}_j \mathbf{\Phi}_j^T \mathbf{b} \tag{4–138}$$

where the modal expansion for the frequency response matrix $\mathbf{H}(\omega)$ as given by Eqs. (4-87) and (4-88) has been used.

If the terms on the right hand side of Eq. (4-137) are brought on a common fraction line, the result may be written as a rational function:

$$H(\omega) = \frac{P(z)}{Q(z)}, \quad z = i\omega \tag{4–139}$$

$$\left. \begin{aligned} P(z) &= p_0 z^r + p_1 z^{r-1} + \cdots + p_{r-1} z + p_r \\ Q(z) &= z^s + q_1 z^{s-1} + \cdots + q_{s-1} z + q_s \end{aligned} \right\} \tag{4–140}$$

where in general:

$$s = 2n, \quad r = 2n - 2 \tag{4–141}$$

Eq. (4-137) is merely a decomposition of Eq. (4-139) into partial fractions.

From Eqs. (4-139) and (4-140), it follows that the response $R(t)$ is related to the load $F(t)$ by the following stochastic differential equations, cf. Eq. (3-157):

$$\left. \begin{aligned} R(t) &= p_0 Y^{(r)} + p_1 Y^{(r-1)} + \ldots + p_{r-1} \dot{Y} + p_r Y \\ Y^{(s)} &+ q_1 Y^{(s-1)} + q_2 Y^{(s-2)} + \ldots + q_{s-1} \dot{Y} + q_s Y = F(t) \end{aligned} \right\} \tag{4–142}$$

Especially, if $\{F(t), t \in R\}$ is a white noise process with the auto-spectral density function S_0, the auto-covariance function of the stationary Gaussian response process $\{R(t), t \in R\}$ becomes, cf. Eq. (3-59):

$$\kappa_{RR}(\tau) = -\pi S_0 \sum_{j=1}^{2n} \frac{e^{z_j |\tau|}}{z_j} \frac{P(z_j) P(-z_j)}{\displaystyle\prod_{\substack{k=1 \\ k \neq j}}^{s} \left(z_k^2 - z_j^2 \right)} \tag{4–143}$$

From Eq. (4-137), it is seen that the poles z_j fulfill:

$$Q(z) = \prod_{j=1}^{n} \left(z^2 + 2\zeta_j \omega_j z + \omega_j^2 \right) = \prod_{j=1}^{n} \left(z - z_{j,1} \right)\left(z - z_{j,2} \right) \tag{4–144}$$

$$\left. \begin{aligned} z_{j,1} \\ z_{j,2} \end{aligned} \right\} = \omega_j \left(-\zeta_j \pm i \sqrt{1 - \zeta_j^2} \right), \quad j = 1, \ldots, n \tag{4–145}$$

The calculation of the coefficients $p_0, p_1, \ldots, p_{r-1}, p_r$ in $P(z)$ is obtained by bringing the partial fractions in Eq. (4-137) on a common fraction line, and collecting terms of equal power

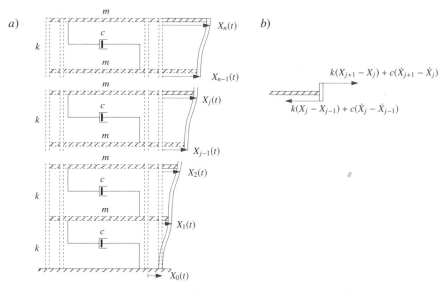

Fig. 4–11 a) An n-storey planar shear frame subjected to a horizontal earthquake excitation. b) Forces on the free storey beam.

of z. However for calculating $\kappa_{RR}(\tau)$ based on Eq. (4-143), we merely need the function values $P(z)$, which can be calculated without explicit knowledge of the coefficients of the polynomial:

$$
P(z) = H(z)\,Q(z) = \left(\sum_{j=1}^{n} \frac{c_j}{z^2 + 2\,\zeta_j\,\omega_j\,z + \omega_j^2} \right) \prod_{k=1}^{n} \left(z^2 + 2\zeta_k\omega_k\,z + \omega_k^2 \right)
$$

$$
= \sum_{j=1}^{n} c_j \prod_{\substack{k=1 \\ k \neq j}}^{n} \left(z^2 + 2\zeta_k\omega_k\,z + \omega_k^2 \right) \tag{4–146}
$$

For $n = 2$, this expression attains the form:

$$
P(z) = c_1 \left(z^2 + 2\zeta_2\omega_2\,z + \omega_2^2 \right) + c_2 \left(z^2 + 2\zeta_1\omega_1\,z + \omega_1^2 \right) \tag{4–147}
$$

Example 4-5: Stochastic response of a multi-storey shear frame exposed to a horizontal earthquake excitation

Fig. 4-11 shows an n-storey *shear frame structure*, where all storeys have the same mass m, and all inter-storeys have the same linear shear stiffness k. The dissipation in the shear columns is modelled by linear viscous damping elements which all have the damping constant c. The ground surface displacement is denoted $X_0(t)$, and the displacement of storey j relative to the ground surface is denoted $X_j(t)$. Thus, the total displacement of storey j becomes $X_0(t) + X_j(t)$.

The storey beams are cut free from the shear columns and damping elements, and the elastic shear forces and the damping forces are applied to the free storey masses. As shown in Fig. 4-11b, the shear force between storey beams $j - 1$ and j becomes $k(X_j - X_{j-1}) + c(\dot{X}_j - \dot{X}_{j-1})$, and the shear force between storey beams j and $j + 1$ becomes $k(X_{j+1} - X_j) + c(\dot{X}_{j+1} - \dot{X}_j)$. The former is acting in opposite direction of the displacement $X_j(t)$, and the latter is co-directional to the displacement. Applying Newton's 2nd law of motion to all n free storey beams, the following equations of motion are obtained:

$$
\left.
\begin{aligned}
m\left(\ddot{X}_0 + \ddot{X}_1\right) &= -\, k\, X_1 + k\left(X_2 - X_1\right) - c\, \dot{X}_1 + c\left(\dot{X}_2 - \dot{X}_1\right) \\
&\vdots \\
m\left(\ddot{X}_0 + \ddot{X}_j\right) &= -\, k\left(X_j - X_{j-1}\right) + k\left(X_{j+1} - X_j\right) \\
&\quad -\, c\left(\dot{X}_j - \dot{X}_{j-1}\right) + c\left(\dot{X}_{j+1} - \dot{X}_j\right) \\
&\vdots \\
m\left(\ddot{X}_0 + \ddot{X}_n\right) &= -\, k\left(X_n - X_{n-1}\right) - c\left(\dot{X}_n - \dot{X}_{n-1}\right)
\end{aligned}
\right\}
\tag{4-148}
$$

After all equations of motion in Eq. (4-148) are divided by the storey mass m, the system of equations can be written in the following matrix form:

$$
\ddot{\mathbf{X}}(t) + 2\zeta_0\omega_0\, \mathbf{k}\, \dot{\mathbf{X}}(t) + \omega_0^2\, \mathbf{k}\, \mathbf{X}(t) = \mathbf{b}\, \ddot{X}_0(t)
\tag{4-149}
$$

$$
\mathbf{X}(t) = \begin{bmatrix} X_1(t) \\ X_2(t) \\ \vdots \\ X_n(t) \end{bmatrix}, \quad \mathbf{b} = -\begin{bmatrix} 1 \\ 1 \\ \vdots \\ 1 \end{bmatrix}, \quad \mathbf{k} = \begin{bmatrix} 2 & -1 & 0 & 0 & 0 & \cdots & 0 & 0 \\ -1 & 2 & -1 & 0 & 0 & \cdots & 0 & 0 \\ 0 & -1 & 2 & -1 & 0 & \cdots & 0 & 0 \\ \vdots & \vdots & \vdots & \vdots & \vdots & \ddots & \vdots & \vdots \\ 0 & 0 & 0 & 0 & 0 & \cdots & -1 & 1 \end{bmatrix}
\tag{4-150}
$$

$$
\omega_0 = \sqrt{\frac{k}{m}}, \quad \zeta_0 = \frac{c}{2\sqrt{km}}
\tag{4-151}
$$

where ω_0 and ζ_0 should merely be considered as parameters in the system, and can not directly be related to the angular eigenfrequencies and the modal damping ratios of the structure.

The angular eigenfrequencies and the mode shapes of the shear frame are:[1]

$$
\omega_j = \omega_0 \sqrt{2 - 2\cos\left(\frac{2j-1}{2n+1}\pi\right)}, \quad j = 1,\ldots,n
\tag{4-152}
$$

$$
\Phi_{j,k} = \sin\left(j\,\frac{2k-1}{2n+1}\pi\right), \quad j = 1,\ldots,n
\tag{4-153}
$$

Corresponding to the normalization used in Eq. (4-81), the modal masses become:

$$
m_j = m\,\mathbf{\Phi}_j^T \mathbf{I} \mathbf{\Phi}_j = m\sum_{k=1}^{n}\Phi_{j,k}^2 = m\sum_{k=1}^{n}\sin^2\left(j\,\frac{2k-1}{2n+1}\pi\right) = \frac{2n+1}{4}\,m
\tag{4-154}
$$

The modal damping ratios become:

$$
2\,\zeta_j\,\omega_j\,m_j = \mathbf{\Phi}_j^T \mathbf{C} \mathbf{\Phi}_j = \frac{c}{k}\,\mathbf{\Phi}_j^T \mathbf{K} \mathbf{\Phi}_j = \frac{c}{k}\,\omega_j^2\,m_j \quad\Rightarrow
$$

$$
\zeta_j = \frac{1}{2}\frac{c}{k}\,\omega_j = \zeta_0 \sqrt{2 - 2\cos\left(\frac{2j-1}{2n+1}\pi\right)}, \quad j = 1,\ldots,n
\tag{4-155}
$$

---

[1]M. Gerardin and D. Rixen. *Mechanical Vibrations: Theory and Application to Structural Dynamics*, 2nd Edition. John Wiley & Sons, Inc., Chichester, 1997.

We may be interested in the displacement response $X_n(t)$ of the top-storey relative to the ground surface:

$$X_n(t) = \mathbf{a}^T \mathbf{X}(t), \quad \mathbf{a} = \begin{bmatrix} 0 \\ 0 \\ \vdots \\ 0 \\ 1 \end{bmatrix} \tag{4–156}$$

Then, the frequency response function $H(\omega)$ relating the ground surface acceleration $\ddot{X}_0(t)$ to the top-storey displacement $X_n(t)$ becomes, cf. Eqs. (4-137) and (4-138):

$$H(\omega) = \sum_{j=1}^{n} \frac{c_j}{z^2 + 2\zeta_j \omega_j z + \omega_j^2} \tag{4–157}$$

$$c_j = \frac{1}{m_j} \mathbf{a}^T \mathbf{\Phi}_j \mathbf{\Phi}_j^T \mathbf{b} = -\frac{4}{(2n+1)m} \sin\left(j \frac{2n-1}{2n+1} \pi\right) \frac{\sin^2\left(j \frac{n}{2n+1} \pi\right)}{\sin\left(j \frac{1}{2n+1} \pi\right)} \tag{4–158}$$

Upon bringing the terms within the summation in Eq. (4-157) on a common fraction line, the parameters of the polynomial $P(z)$ and $Q(z)$ in Eq. (4-139) may be identified. Next, $X_n(t)$ is obtained as the output process of the stochastic differential in Eq. (4-142) driven by the input process $F(t) = \ddot{X}_0(t)$.

Eq. (4-149) may be written as the following equivalent stochastic state vector equation, cf. Eq. (4-69):

$$\left. \begin{aligned} \dot{\mathbf{Y}}(t) &= \mathbf{A}\,\mathbf{Y}(t) + \mathbf{B}\,\ddot{X}_0(t), \quad t > 0 \\ \mathbf{Y}(0) &= \mathbf{Y}_0 \end{aligned} \right\} \tag{4–159}$$

where:

$$\mathbf{Y}(t) = \begin{bmatrix} \mathbf{X}(t) \\ \dot{\mathbf{X}}(t) \end{bmatrix}, \quad \mathbf{Y}_0 = \begin{bmatrix} \mathbf{X}(0) \\ \dot{\mathbf{X}}(0) \end{bmatrix} = \begin{bmatrix} \mathbf{0} \\ \mathbf{0} \end{bmatrix}, \quad \mathbf{A} = \begin{bmatrix} \mathbf{0} & \mathbf{I} \\ -\omega_0^2 \mathbf{k} & -2\zeta_0 \omega_0 \mathbf{k} \end{bmatrix}, \quad \mathbf{B} = \begin{bmatrix} \mathbf{0} \\ \mathbf{b} \end{bmatrix} \tag{4–160}$$

It should be noted that $\mathbf{0}$ in $\mathbf{A}$ signifies an $n \times n$-dimensional zero matrix, and $\mathbf{0}$ in $\mathbf{Y}_0$ and $\mathbf{B}$ are n-dimensional zero vectors.

Next, the non-stationary ground surface acceleration $\{\ddot{X}_0(t), t \in [0, \infty[\}$ is modelled as a *time-modulated white noise*, i.e., $\ddot{X}_0(t)$ is modelled as a zero-mean Gaussian process with the auto-covariance function:

$$\kappa_{\ddot{X}_0 \ddot{X}_0}(t, t + \tau) = 2\pi S_0\, I(t)\, \delta(\tau) \tag{4–161}$$

where $I(t)$ is a deterministic *intensity function*. $\{\ddot{X}_0(t), t \in [0, \infty[\}$ in Eq. (4-161) can be interpreted as a special white noise process of which the auto-spectral density function $S_0 I(t)$ is slowly varying with time. Eq. (4-161) is often applied in earthquake engineering as a way of introducing non-stationarity into the problem.

For the intensity function, the following expression will be used:[1]

---

[1]M. Shinozuka and Y. Sato. Simulation of nonstationary random process. *Journal of Engineering Mechanics, ASCE* 1967; **94** (1): 11-40.

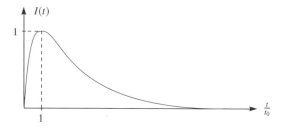

Fig. 4–12 Intensity function $I(t)$. $t_1 = 2.0t_0$, $t_2 = 0.5t_0$.

$$I(t) = \frac{e^{-\frac{t}{t_1}} - e^{-\frac{t}{t_2}}}{e^{-\frac{t_0}{t_1}} - e^{-\frac{t_0}{t_2}}}, \quad t_0 = \frac{t_1 t_2}{t_2 - t_1} \ln \frac{t_2}{t_1} \tag{4–162}$$

It is seen that $I(t)$ increases from 0 to its maximum value 1 at the time t_0, and decreases asymptotically to 0 as $t \to \infty$. The parameters t_1 and t_2 control the rise and fall of $I(t)$, as shown in Fig. 4-12. Since the structure is linear and the input is Gaussian, the state vector process $\{\mathbf{Y}(t), t \in [0, \infty[\}$ becomes Gaussian as well. Hence, the stochastic state vector process is completely determined by the mean value vector function $\boldsymbol{\mu}_\mathbf{Y}(t)$ and the cross-covariance matrix function $\boldsymbol{\kappa}_{\mathbf{YY}}(t_1, t_2)$.

The frame structure is assumed to be at rest when $t = 0$. Then, the mean value vector function is determined from, cf. Eq. (3-187):

$$\frac{d}{dt} \boldsymbol{\mu}_\mathbf{Y}(t) = \mathbf{A} \boldsymbol{\mu}_\mathbf{Y}(t), \quad \boldsymbol{\mu}_\mathbf{Y}(0) = \mathbf{0} \quad \Rightarrow$$

$$\boldsymbol{\mu}_\mathbf{Y}(t) \equiv \mathbf{0} \tag{4–163}$$

The cross-covariance matrix function at zero time-lag, $\boldsymbol{\kappa}_{\mathbf{YY}}(t, t)$, becomes exactly the zero time-lag covariance matrix $\mathbf{C}_{\mathbf{YY}}(t)$, i.e. $\boldsymbol{\kappa}_{\mathbf{YY}}(t, t) = \mathbf{C}_{\mathbf{YY}}(t)$. For the present example, the matrix differential in Eq. (3-211) attain the form:

$$\frac{d}{dt} \mathbf{C}_{\mathbf{YY}}(t) = \mathbf{A}\mathbf{C}_{\mathbf{YY}}(t) + \mathbf{C}_{\mathbf{YY}}(t)\mathbf{A}^T + 2\pi S_0 I(t) \mathbf{B}\mathbf{B}^T, \quad \mathbf{C}_{\mathbf{YY}}(0) = \mathbf{0} \tag{4–164}$$

$$\mathbf{C}_{\mathbf{YY}}(t) = \begin{bmatrix} E\left[\mathbf{X}(t)\mathbf{X}^T(t)\right] & E\left[\mathbf{X}(t)\dot{\mathbf{X}}^T(t)\right] \\ E\left[\dot{\mathbf{X}}(t)\mathbf{X}^T(t)\right] & E\left[\dot{\mathbf{X}}(t)\dot{\mathbf{X}}^T(t)\right] \end{bmatrix} \tag{4–165}$$

As seen from Eq. (4-165), $\mathbf{C}_{\mathbf{YY}}(t)$ provides the covariance between any components of the displacement and velocity vectors, $\mathbf{X}(t)$ and $\dot{\mathbf{X}}(t)$.

The inter-storey displacements are defined as:

$$\left. \begin{aligned} Z_1(t) &= X_1(t) \\ Z_2(t) &= X_2(t) - X_1(t) \\ &\vdots \\ Z_n(t) &= X_n(t) - X_{n-1}(t) \end{aligned} \right\} \tag{4–166}$$

Eq. (4-166) can be rewritten in the matrix form:

$$\mathbf{Z}(t) = \mathbf{a}\,\mathbf{X}(t), \quad \mathbf{a} = \begin{bmatrix} 1 & 0 & 0 & \cdots & 0 & 0 \\ -1 & 1 & 0 & \cdots & 0 & 0 \\ \vdots & \vdots & \vdots & \ddots & \vdots & \vdots \\ 0 & 0 & 0 & \cdots & -1 & 1 \end{bmatrix} \tag{4–167}$$

Correspondingly, the zero time-lag covariance matrices, as well as the zero time-lag cross-covariance matrix of $\mathbf{Z}(t)$ and $\dot{\mathbf{Z}}(t)$ become:

$$\left. \begin{aligned} \mathbf{C}_{\mathbf{ZZ}}(t) &= \mathbf{a}\,\mathbf{C}_{\mathbf{XX}}(t)\,\mathbf{a}^T \\ \mathbf{C}_{\dot{\mathbf{Z}}\dot{\mathbf{Z}}}(t) &= \mathbf{a}\,\mathbf{C}_{\dot{\mathbf{X}}\dot{\mathbf{X}}}(t)\,\mathbf{a}^T \\ \mathbf{C}_{\mathbf{Z}\dot{\mathbf{Z}}}(t) &= \mathbf{a}\,\mathbf{C}_{\mathbf{X}\dot{\mathbf{X}}}(t)\,\mathbf{a}^T \end{aligned} \right\} \tag{4–168}$$

As an example, a five-storey building ($n = 5$) is considered. The fundamental angular eigenfrequency as given by Eq. (4-152) becomes $\omega_1 = 0.2846\,\omega_0$, and the time will be normalized with respect to the corresponding fundamental eigenperiod $T_1 = \frac{2\pi}{\omega_1} = 22.08\,\frac{1}{\omega_0}$. Further, the following system parameters have been selected:

$$\left. \begin{aligned} \omega_0 &= 10\,\text{rad/s}, \quad \zeta_0 = 0.05 \\ t_0 &= 2.0\,T_1 \quad, \quad t_1 = 2.0\,t_0 = 4.0\,T_1, \quad t_2 = 0.5\,t_0 = T_1 \end{aligned} \right\} \tag{4–169}$$

The matrix differential equation in Eq. (4-164) has been solved numerically by means of the 4th order Runge-Kutta scheme using the time step $\Delta t = \frac{1}{160}T_1$.

Fig. 4-13 shows the time variation of the standard deviation $\sigma_{X_5}(t)$ of the top-storey displacement $X_5(t)$, where the time has been normalized with respect to T_1, and the standard deviation has been normalized with respect to the quantity $\sigma_0 = \sqrt{\frac{\pi S_0}{2\zeta_0\omega_0^3 m^2}}$. σ_0 represents the stationary standard deviation of a single-storey frame with the storey mass m, the angular eigenfrequency ω_0 and the damping ratio ζ_0, exposed to a white noise ground surface acceleration process with the auto-spectral density S_0.

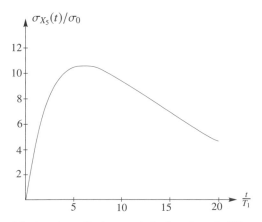

Fig. 4–13 Standard deviation of the top-storey displacement of a five-storey building subjected to a non-stationary horizontal earthquake excitation.

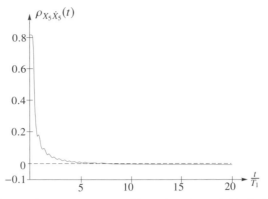

Fig. 4–14 Correlation coefficient of the displacement $X_5(t)$ and the velocity $\dot{X}_5(t)$ of a five-storey building subjected to a non-stationary horizontal earthquake excitation.

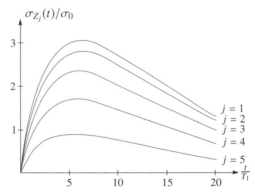

Fig. 4–15 Standard deviations of the inter-storey displacements of a five-storey building subjected to a non-stationary horizontal earthquake excitation.

Fig. 4-14 shows the time variation of the correlation coefficient $\rho_{X_5\dot{X}_5}(t)$ of $X_5(t)$ and the velocity $\dot{X}_5(t)$, calculated as $\rho_{X_5\dot{X}_5}(t) = \kappa_{X_5\dot{X}_5}(t,t)/\big(\sigma_{X_5}(t)\sigma_{\dot{X}_5}(t)\big)$. As seen, $X_5(t)$ and $\dot{X}_5(t)$ become practically uncorrelated for $t > 3T_1$, similar as the case under stationary excitations.

Fig. 4-15 shows the time variation of the standard deviations $\sigma_{Z_j}(t)$ of the inter-storey displacements, normalized with respect to σ_0. The largest and smallest inter-storey displacements take place in the lowest and top-storey, respectively. These results display a general property of a building with uniform stiffness distribution exposed to broad-banded excitations such as earthquakes. For such structures, the damage will be located in the lower storeys. Further, upon comparison with Fig. 4-13, we have:

$$\sigma_{X_5}(t) \simeq \sum_{j=1}^{5} \sigma_{Z_j}(t) \tag{4–170}$$

Eq. (4-170) can be explained by the fact that the structural response is dominated by the lowest eigenmode. Again, this is the case for a lightly damped structure with well separated angular eigenfrequencies, when it is exposed to broad-banded excitations.

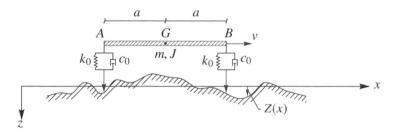

Fig. 4–16 2DOF model of a vehicle moving on an irregular surface.

Example 4-6: 2-degree-of-freedom model of a vehicle moving on an irregular surface

Fig. 4-16 shows a 2DOF model of a vehicle moving with a constant speed v on an irregular surface. The frame of the vehicle is modelled as a rigid plane beam AB of the length $2a$, with a constant mass distribution. The total mass of the beam is m, and the mass moment of inertia around a horizontal line through the center of gravity G is J. At the end points A and B, the vehicle is supported on vertical linear elastic springs with the spring constant k_0 in parallel to linear viscous damping elements with the damping constant c_0. It is assumed that the suspension system is in full contact with the surface at all times.

The *surface irregularity* is modelled by a weakly homogeneous process $\{Z(x), \ x \in R\}$. The irregularities are specified in an (x, z)-coordinate system, where the x-axis is horizontal and placed along the mean value function of the surface irregularity, and the z-axis is orientated downwards. Therefore, it is obvious that:

$$E\big[Z(x)\big] \equiv 0 \tag{4--171}$$

The correlation structure of the surface irregularity is specified by the one-sided auto-spectral density function:

$$S_Z(k) = \begin{cases} 0 & , \quad k \notin \big[k_1, k_2\big] \\ \sigma_Z^2 \dfrac{k_1 k_2}{k_2 - k_1} \dfrac{1}{k^2}, & k \in \big[k_1, k_2\big] \end{cases} \tag{4--172}$$

where σ_Z is the standard deviation of the surface irregularity. k is a wave number, and k_1 and k_2 denote the cut-in and cut-out wave numbers of the spectrum. The indicated auto-spectral density function has a slope of 1:2 in double logarithmic mapping, which is a well-verified property for road surfaces. The spectrum is valid for wavelengths in the interval [0.1m, 10.0 m], which means that $k_1 = \frac{2\pi}{10}$ m^{-1} and $k_2 = \frac{2\pi}{0.1}$ m^{-1}.

The motion of the beam AB is described by the vertical displacement $X_1(t)$ of the mass center of gravity G in the z-direction, and the clock-wise rotation $X_2(t)$ of G, as shown in Fig. 4-17. $X_1(t)$ is measured from a static equilibrium state, where the vehicle is at rest with the support points of the suspension system in the mean level position of the surface irregularities. The vertical displacements of AB are assumed to be small compared with its length $2a$.

The surface irregularities of the points A and B at the time t are designated $Z_A(t)$ and $Z_B(t)$, respectively. Then, the compressions of the springs at A and B become $X_1(t) - aX_2(t) - Z_A(t)$ and $X_1(t) + aX_2(t) - Z_B(t)$, respectively, and the related compression rates become $\dot{X}_1(t) - a\dot{X}_2(t) - \dot{Z}_A(t)$ and $\dot{X}_1(t) + a\dot{X}_2(t) - \dot{Z}_B(t)$. The vehicle beam is cut free from the suspension system, and the forces from the indicated deformations

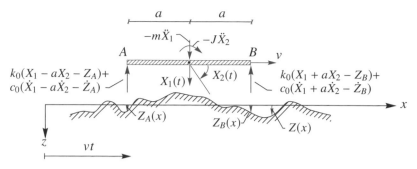

Fig. 4–17 2DOF model of a vehicle moving on an irregular surface. Forces on the free rigid beam structure.

are applied as external forces, as shown in Fig. 4-17. Applying Newton's 2nd law leads to the following equations of motion:

$$m\ddot{X}_1(t) = -k_0\left(2X_1(t) - Z_A(t) - Z_B(t)\right) - c_0\left(2\dot{X}_1(t) - \dot{Z}_A(t) - \dot{Z}_B(t)\right)$$

$$J\ddot{X}_2(t) = -ak_0\left(2aX_2(t) + Z_A(t) - Z_B(t)\right) - ac_0\left(2a\dot{X}_2(t) + \dot{Z}_A(t) - \dot{Z}_B(t)\right) \tag{4-173}$$

It is seen that the selected degrees of freedom are decoupled, and hence $X_1(t)$ and $X_2(t)$ are modal coordinates. This is due to the fact that the displacement degree of freedom has been defined at the mass center of gravity. Eq. (4-173) can be rewritten in the matrix form:

$$\mathbf{M}\ddot{\mathbf{X}}(t) + \mathbf{C}\dot{\mathbf{X}}(t) + \mathbf{K}\mathbf{X}(t) = \mathbf{F}(t)$$

$$\mathbf{F}(t) = \mathbf{P}_0\dot{\mathbf{Z}}(t) + \mathbf{P}_1\mathbf{Z}(t) \tag{4-174}$$

where:

$$\mathbf{X}(t) = \begin{bmatrix} X_1(t) \\ X_2(t) \end{bmatrix}, \quad \mathbf{F}(t) = \begin{bmatrix} F_1(t) \\ F_2(t) \end{bmatrix}, \quad \mathbf{Z}(t) = \begin{bmatrix} Z_A(t) \\ Z_B(t) \end{bmatrix}$$

$$\mathbf{M} = \begin{bmatrix} m & 0 \\ 0 & J \end{bmatrix}, \quad \mathbf{C} = 2c_0\begin{bmatrix} 1 & 0 \\ 0 & a^2 \end{bmatrix}, \quad \mathbf{K} = 2k_0\begin{bmatrix} 1 & 0 \\ 0 & a^2 \end{bmatrix} \tag{4-175}$$

$$\mathbf{P}_0 = c_0\begin{bmatrix} 1 & 1 \\ -a & a \end{bmatrix}, \quad \mathbf{P}_1 = k_0\begin{bmatrix} 1 & 1 \\ -a & a \end{bmatrix}$$

The cross-spectral density matrix function and the cross-covariance matrix function become, cf. Eqs. (4-115) and (3-23):

$$\mathbf{S}_{\mathbf{XX}}(\omega) = \mathbf{H}^*(\omega)\,\mathbf{S}_{\mathbf{FF}}(\omega)\mathbf{H}^T(\omega) \tag{4-176}$$

$$\kappa_{\mathbf{XX}}(\tau) = \int_{-\infty}^{\infty} e^{i\omega\tau}\,\mathbf{H}^*(\omega)\,\mathbf{S}_{\mathbf{FF}}(\omega)\mathbf{H}^T(\omega)\,d\omega \tag{4-177}$$

where the frequency response matrix is given by, cf. Eq. (4-87):

$$\mathbf{H}(\omega) = \left(\mathbf{M}z^2 + \mathbf{C}z + \mathbf{K}\right)^{-1} = \begin{bmatrix} mz^2 + 2c_0z + 2k_0 & 0 \\ 0 & Jz^2 + 2c_0a^2z + 2k_0a^2 \end{bmatrix}^{-1}$$

$$= \frac{1}{mz^2 + 2c_0z + 2k_0}\,\boldsymbol{\Phi}_1\boldsymbol{\Phi}_1^T + \frac{1}{Jz^2 + 2c_0a^2z + 2k_0a^2}\,\boldsymbol{\Phi}_2\boldsymbol{\Phi}_2^T, \quad z = i\omega \tag{4-178}$$

$$\Phi_1 = \begin{bmatrix} 1 \\ 0 \end{bmatrix}, \quad \Phi_2 = \begin{bmatrix} 0 \\ 1 \end{bmatrix} \tag{4-179}$$

Vehicles are strongly damped structures, so the equivalent white noise approximations cannot be used at the evaluation of the Fourier transform in Eq. (4-177), which consequently needs to be performed numerically.

The remaining part of this example deals with the determination of $\mathbf{S_{FF}}(\omega)$.

It is assumed that point A is at the location $x = 0$ when $t = 0$. Hence, the abscissas of A and B at the time t are given as $x_A = vt$ and $x_B = vt + 2a$, respectively. The elevations of the surface irregularity at points A and B at the time t become:

$$\left. \begin{array}{l} Z_A(t) = Z(vt) \\ Z_B(t) = Z(vt + 2a) \end{array} \right\} \tag{4-180}$$

The auto- and cross-covariance functions of $\{Z_A(t),\ t \in R\}$ and $\{Z_B(t),\ t \in R\}$ become:

$$\left. \begin{array}{ll} \kappa_{Z_A Z_A}(\tau) = E\big[Z(vt)Z(v(t+\tau))\big] & = \kappa_{ZZ}(v\tau) \\ \kappa_{Z_B Z_B}(\tau) = E\big[Z(vt+2a)Z(v(t+\tau)+2a)\big] & = \kappa_{ZZ}(v\tau) \\ \kappa_{Z_A Z_B}(\tau) = E\big[Z(vt)Z(v(t+\tau)+2a)\big] & = \kappa_{ZZ}(v\tau + 2a) \end{array} \right\} \tag{4-181}$$

Then, the corresponding auto- and cross-spectral density functions become, cf. Eq. (3-23):

$$\left. \begin{array}{l} S_{Z_A Z_A}(\omega) = S_{Z_B Z_B}(\omega) = \dfrac{1}{2\pi} \displaystyle\int_{-\infty}^{\infty} e^{-i\omega\tau}\, \kappa_{ZZ}(v\tau)\, d\tau \\[3mm] \qquad = \dfrac{1}{v}\dfrac{1}{2\pi}\displaystyle\int_{-\infty}^{\infty} e^{-i\frac{\omega}{v}u}\, \kappa_{ZZ}(u)\, du = \dfrac{1}{v} S_{ZZ}\left(\dfrac{\omega}{v}\right) \\[5mm] S_{Z_A Z_B}(\omega) = \dfrac{1}{2\pi}\displaystyle\int_{-\infty}^{\infty} e^{-i\omega\tau}\, \kappa_{ZZ}(v\tau + 2a)\, d\tau \\[3mm] \qquad = \dfrac{1}{v}\, e^{i2a\frac{\omega}{v}} \dfrac{1}{2\pi}\displaystyle\int_{-\infty}^{\infty} e^{-i\frac{\omega}{v}u}\, \kappa_{ZZ}(u)\, du = \dfrac{1}{v}\, e^{i2a\frac{\omega}{v}} S_{ZZ}\left(\dfrac{\omega}{v}\right) \end{array} \right\} \tag{4-182}$$

It follows from Eq. (4-182) that wave numbers and angular frequencies are related by the following so-called *linear dispersion relation*:

$$\omega = v\,k \tag{4-183}$$

Defining $U(t)$ and $V(t)$ as follows, the corresponding auto/cross-covariance functions and auto/cross-spectral density functions can be obtained:

$$\left. \begin{array}{l} U(t) = Z_A(t) + Z_B(t) \\ V(t) = Z_A(t) - Z_B(t) \end{array} \right\} \quad \Rightarrow$$

$$\left. \begin{array}{l} \kappa_{UU}(\tau) = \kappa_{Z_A Z_A}(\tau) + \kappa_{Z_A Z_B}(\tau) + \kappa_{Z_B Z_A}(\tau) + \kappa_{Z_B Z_B}(\tau) \\ \kappa_{UV}(\tau) = \kappa_{Z_A Z_A}(\tau) - \kappa_{Z_A Z_B}(\tau) + \kappa_{Z_B Z_A}(\tau) - \kappa_{Z_B Z_B}(\tau) \\ \kappa_{VV}(\tau) = \kappa_{Z_A Z_A}(\tau) - \kappa_{Z_A Z_B}(\tau) - \kappa_{Z_B Z_A}(\tau) + \kappa_{Z_B Z_B}(\tau) \end{array} \right\} \quad \Rightarrow$$

$$\left.\begin{aligned}
S_{UU}(\omega) &= S_{Z_A Z_A}(\omega) + S_{Z_A Z_B}(\omega) + S^*_{Z_A Z_B}(\omega) + S_{Z_B Z_B}(\omega) \\
&= \frac{2}{v}\left(1 + \cos\left(2a\frac{\omega}{v}\right)\right) S_{ZZ}\left(\frac{\omega}{v}\right) \\[2mm]
S_{UV}(\omega) &= S_{Z_A Z_A}(\omega) - S_{Z_A Z_B}(\omega) + S^*_{Z_A Z_B}(\omega) - S_{Z_B Z_B}(\omega) \\
&= -\frac{2i}{v}\sin\left(2a\frac{\omega}{v}\right) S_{ZZ}\left(\frac{\omega}{v}\right) \\[2mm]
S_{VV}(\omega) &= S_{Z_A Z_A}(\omega) - S_{Z_A Z_B}(\omega) - S^*_{Z_A Z_B}(\omega) + S_{Z_B Z_B}(\omega) \\
&= \frac{2}{v}\left(1 - \cos\left(2a\frac{\omega}{v}\right)\right) S_{ZZ}\left(\frac{\omega}{v}\right)
\end{aligned}\right\} \qquad (4\text{–}184)$$

where Eqs. (3-24) and (4-182) have been used.

From Eqs. (4-174), (4-175) and (4-184), it follows that:

$$\left.\begin{aligned}
F_1(t) &= k_0\,U(t) + c_0\,\dot{U}(t) \\
F_2(t) &= -a\,k_0\,V(t) - ac_0\,\dot{V}(t)
\end{aligned}\right\} \qquad \Rightarrow$$

$$\left.\begin{aligned}
\kappa_{F_1 F_1}(\tau) &= k_0^2\ \kappa_{UU}(\tau) + k_0 c_0\ \kappa_{U\dot{U}}(\tau) + c_0 k_0\ \kappa_{\dot{U}U}(\tau) + c_0^2\ \kappa_{\dot{U}\dot{U}}(\tau) \\
\kappa_{F_1 F_2}(\tau) &= -ak_0^2\ \kappa_{UV}(\tau) - ak_0 c_0\ \kappa_{U\dot{V}}(\tau) - ac_0 k_0\ \kappa_{\dot{U}V}(\tau) - ac_0^2\ \kappa_{\dot{U}\dot{V}}(\tau) \\
\kappa_{F_2 F_2}(\tau) &= a^2 k_0^2\ \kappa_{VV}(\tau) + a^2 k_0 c_0\ \kappa_{V\dot{V}}(\tau) + a^2 c_0 k_0\ \kappa_{\dot{V}V}(\tau) + a^2 c_0^2\ \kappa_{\dot{V}\dot{V}}(\tau)
\end{aligned}\right\} \qquad \Rightarrow$$

$$\left.\begin{aligned}
S_{F_1 F_1}(\omega) &= k_0^2\,S_{UU}(\omega) + k_0 c_0\,S_{U\dot{U}}(\omega) + c_0 k_0\,S_{\dot{U}U}(\omega) + c_0^2\,S_{\dot{U}\dot{U}}(\omega) \\
&= \left(k_0 + \omega^2 c_0\right) S_{UU}(\omega) \\[2mm]
S_{F_1 F_2}(\omega) &= -k_0^2\,S_{UV}(\omega) - ak_0 c_0\,S_{U\dot{V}}(\omega) - ac_0 k_0\,S_{\dot{U}V}(\omega) - ac_0^2\,S_{\dot{U}\dot{V}}(\omega) \\
&= -a\left(k_0 + \omega^2 c_0\right) S_{UV}(\omega) \\[2mm]
S_{F_2 F_2}(\omega) &= a^2 k_0^2\,S_{VV}(\omega) + a^2 k_0 c_0\,S_{V\dot{V}}(\omega) + a^2 c_0 k_0\,S_{\dot{V}V}(\omega) + a^2 c_0^2\,S_{\dot{V}\dot{V}}(\omega) \\
&= a^2\left(k_0 + \omega^2 c_0\right) S_{VV}(\omega)
\end{aligned}\right\} \qquad (4\text{–}185)$$

where the following results have been used, cf. Eq. (3-142):

$$\left.\begin{aligned}
S_{U\dot{U}}(\omega) &= -S_{\dot{U}U}(\omega) = i\omega\,S_{UU}(\omega), & S_{\dot{U}\dot{U}}(\omega) &= \omega^2\,S_{UU}(\omega) \\
S_{U\dot{V}}(\omega) &= -S_{\dot{V}U}(\omega) = i\omega\,S_{UV}(\omega), & S_{\dot{U}\dot{V}}(\omega) &= \omega^2\,S_{UV}(\omega) \\
S_{V\dot{V}}(\omega) &= -S_{\dot{V}V}(\omega) = i\omega\,S_{VV}(\omega), & S_{\dot{V}\dot{V}}(\omega) &= \omega^2\,S_{VV}(\omega)
\end{aligned}\right\} \qquad (4\text{–}186)$$

Finally, the cross-spectral density matrix function of the load vector process $\{\mathbf{F}(t),\ t \in R\}$ follows from

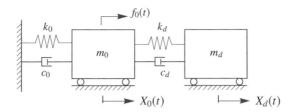

Fig. 4–18 Tuned mass damper with the primary structure exposed to a broad-banded modal load process.

Eqs. (4-184) and (4-185):

$$\mathbf{S_{FF}}(\omega) = \left(k_0 + \omega^2 c_0\right) \begin{bmatrix} S_{UU}(\omega) & -a\, S_{UV}(\omega) \\ -a\, S_{UV}^*(\omega) & a^2\, S_{VV}(\omega) \end{bmatrix}$$

$$= \frac{1}{v}\left(k_0 + \omega^2 c_0\right) \begin{bmatrix} 1 + \cos\left(2a\frac{\omega}{v}\right) & ia\,\sin\left(2a\frac{\omega}{v}\right) \\ -ia\,\sin\left(2a\frac{\omega}{v}\right) & 1 - \cos\left(2a\frac{\omega}{v}\right) \end{bmatrix} S_Z\left(\frac{|\omega|}{v}\right) \tag{4–187}$$

Example 4-7: Tuned mass damper attached to a structure which is exposed to a stationary Gaussian excitation

Fig. 4-18 shows a symbolic representation of a structure with an attached mass damper. The mass damper will be tuned to damp a given mode of the structure, defined by the modal mass m_0, the modal linear viscous damping constant c_0, the modal stiffness k_0 and the modal load $f_0(t)$. The damper mass m_d is connected to the primary structure via a linear elastic spring with the stiffness k_d in parallel to a linear viscous damper with the damping constant c_d. The modal displacement is denoted as $X_0(t)$, and the displacement of the damper from its static referential position is $X_d(t)$. Both displacements are considered positive in the direction of the modal load. The modal load is modelled as a stationary, zero-mean, stochastic Gaussian process $\{f_0(t),\ t \in R\}$ with the auto-spectral density function $S_{f_0 f_0}(\omega)$. The mass damper will be optimally tuned to minimize the variance $\sigma_{X_0}^2$ of the modal displacement.

The stochastic equations of motion are written as:

$$\mathbf{M}\,\ddot{\mathbf{X}}(t) + \mathbf{C}\,\dot{\mathbf{X}}(t) + \mathbf{K}\,\mathbf{X}(t) = \mathbf{b}\,f_0(t) \tag{4–188}$$

$$\left.\begin{aligned} \mathbf{X}(t) &= \begin{bmatrix} X_0(t) \\ X_d(t) \end{bmatrix}, \quad \mathbf{b} = \begin{bmatrix} 1 \\ 0 \end{bmatrix} \\[2mm] \mathbf{M} &= \begin{bmatrix} m_0 & 0 \\ 0 & m_d \end{bmatrix}, \quad \mathbf{C} = \begin{bmatrix} c_0 + c_d & -c_d \\ -c_d & c_d \end{bmatrix}, \quad \mathbf{K} = \begin{bmatrix} k_0 + k_d & -k_d \\ -k_d & k_d \end{bmatrix} \end{aligned}\right\} \tag{4–189}$$

With $\mathbf{a} = \begin{bmatrix} 1, 0 \end{bmatrix}^T$, the frequency response function relating $f_0(t)$ and $X_0(t)$ becomes, cf. (4-137):

$$H(\omega) = \mathbf{a}^T \mathbf{H}(\omega)\,\mathbf{b} = \mathbf{a}^T \left(z^2 \mathbf{M} + z\mathbf{C} + \mathbf{K}\right)^{-1} \mathbf{b}$$

$$= \frac{1}{D(z)}\left(m_d z^2 + c_d z + k_d\right), \quad z = i\omega \tag{4–190}$$

where

$$D(z) = \left(m_0 z^2 + (c_0 + c_d)z + k_0 + k_d\right)\left(m_d z^2 + c_d z + k_d\right) - \left(c_d z + k_d\right)^2 \tag{4–191}$$

Eq. (4-190) may be written as a rational function of the order $(r, s) = (2, 4)$ in the format given by Eqs. (3-54), (3-55) and (3-56):

$$H(\omega) = \frac{P(z)}{Q(z)}, \quad z = i\omega \tag{4–192}$$

$$\left.\begin{aligned}P(z) &= p_0 z^2 + p_1 z + p_2 \\ Q(z) &= z^4 + q_1 z^3 + q_2 z^2 + q_3 z + q_4\end{aligned}\right\} \tag{4–193}$$

$$\left.\begin{aligned}p_0 &= \frac{1}{m_0}, & q_1 &= 2\left(\zeta_0 + \zeta_d\,\alpha\,(1 + \mu)\right)\omega_0 \\ p_1 &= \frac{2\,\zeta_d\,\alpha\,\omega_0}{m_0}, & q_2 &= \left(1 + \alpha^2\,(1 + \mu) + 4\,\zeta_0\,\zeta_d\,\alpha\right)\omega_0^2 \\ p_2 &= \frac{\alpha^2\,\omega_0^2}{m_0}, & q_3 &= 2\left(\zeta_0\,\alpha + \zeta_d\right)\alpha\,\omega_0^3 \\ & & q_4 &= \alpha^2\,\omega_0^4\end{aligned}\right\} \tag{4–194}$$

where:

$$\left.\begin{aligned}\omega_0 &= \sqrt{\frac{k_0}{m_0}}, & \zeta_0 &= \frac{c_0}{2\sqrt{k_0\,m_0}} \\ \omega_d &= \sqrt{\frac{k_d}{m_d}}, & \zeta_d &= \frac{c_d}{2\sqrt{k_d\,m_d}} \\ \mu &= \frac{m_d}{m_0}, & \alpha &= \frac{\omega_d}{\omega_0}\end{aligned}\right\} \tag{4–195}$$

ζ_d, μ and α are denoted the damping ratio, the mass ratio and the *tuning ratio* of the mass damper, respectively. These three non-dimensional parameters are used as design parameters for the mass damper, instead of m_d, c_d and k_d. During the optimization procedure, a specific value of the mass ratio is first chosen, and the optimal values of ζ_d and α are then determined from the following criterion:

$$\min_{\zeta_d, \alpha} \sigma_{X_0}^2 = \int_{-\infty}^{\infty} |H(\omega; \zeta_d, \alpha)|^2 \, S_{f_0 f_0}(\omega)\,d\omega \tag{4–196}$$

It is assumed that the damping of the considered mode is sufficient low even at optimal tuning of the mass damper, so that the modal load process can be replaced by an equivalent white noise with the auto-spectral density $S_0 = S_{f_0 f_0}(\omega_0)$, where ω_0 is the angular eigenfrequency of the considered mode. Fig. 4-19a shows the obtained results for the standard deviation σ_{X_0} as a function of tuning ratio α, for discrete values of the damping ratio, i.e., $\zeta_d = 0.1, 0.2, 0.3, 0.4, 0.5$. The modal damping ratio is $\zeta_0 = 0.01$, and the mass ratio has been chosen to be $\mu = 0.1$. σ_{X_0} has been normalized with respect to the quantity σ_0 defined as:

$$\sigma_0 = \sqrt{\frac{\pi\,S_{f_0 f_0}(\omega_0)}{2\zeta_0\,\omega_0^3\,m_0^2}} \tag{4–197}$$

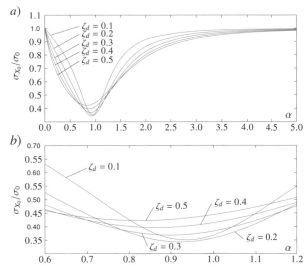

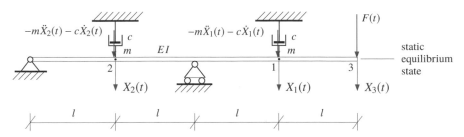

Fig. 4–19 Tuned mass damper attached to a structure, which is exposed to a broad-banded modal load modelled as an equivalent white noise, $\mu = 0.1$, $\zeta_0 = 0.01$. a) Standard deviation of the modal displacement of the primary structure as a function of α and ζ_d. b) Zoomed figure in the vicinity of the minimum value.

Fig. 4–20 2DOF system of a massless Bernoulli-Euler beam with an indirectly acting external dynamic load.

As seen from the zoomed figure in Fig. 4.19b, the minimum variance is achieved when the parameters are $(\alpha_{min}, \zeta_{d,min}) \simeq (0.92, 0.2)$. This result is in good agreement with the corresponding result for an optimal tuned mass damper with the primary structure loaded by a harmonically varying load, for which case $\alpha = \frac{1}{1+\mu} = 0.9091$ and $\zeta_d = \sqrt{\frac{\mu}{2(1+\mu)}} = 0.2032$.[1]

Example 4-8: Planar Bernoulli-Euler beam exposed to an Ornstein-Uhlenbeck load process

The Bernoulli-Euler beam in Example 4-2 is considered again. However, point masses m and linear viscous dampers with the damping constant c are attached to both points 1 and 2 in the present case, as shown in Fig. 4-20.

The load process $\{F(t), t \in R\}$ is obtained by a linear filtration of a white noise $\{W(t), t \in R\}$ (with the auto-spectral density S_0) passing through an Ornstein-Uhlenbeck shaping filter. As

[1]S. Krenk. Frequency analysis of the tuned mass damper. *Journal of Applied Mechanics, ASME* 2005; **72** (6): 936-942.

given by Eq. (4-48), $F(t)$ is obtained from the stochastic differential equation:

$$\dot{F}(t) + \alpha F(t) = \alpha W(t) \tag{4–198}$$

The beam is cut free from the dampers. The damping forces $-c\dot{X}_1(t)$ and $-c\dot{X}_2(t)$, and the inertial forces $-m\ddot{X}_1$ and $-m\ddot{X}_2$ are applied to the beam as equivalent static forces in accordance with d'Alambert's principle with the signs defined in Fig. 4-20. Then, the equations of motion follow from the force method of structural mechanics, cf. Eq. (4-23):

$$\begin{bmatrix} X_1(t) \\ X_2(t) \end{bmatrix} = \mathbf{d}_0\, F(t) + \mathbf{D} \begin{bmatrix} -m\ddot{X}_1(t) - c\dot{X}_1(t) \\ -m\ddot{X}_2(t) - c\dot{X}_2(t) \end{bmatrix} \tag{4–199}$$

$$\mathbf{d}_0 = \begin{bmatrix} \delta_{13} \\ \delta_{23} \end{bmatrix} = \frac{1}{6}\frac{l^3}{EI} \begin{bmatrix} 13 \\ -3 \end{bmatrix}, \quad \mathbf{D} = \begin{bmatrix} \delta_{11} & \delta_{12} \\ \delta_{12} & \delta_{22} \end{bmatrix} = \frac{1}{12}\frac{l^3}{EI} \begin{bmatrix} 12 & -3 \\ -3 & 2 \end{bmatrix} \tag{4–200}$$

where the flexibility coefficients are given by Eq. (4-24), and $\mathbf{D}$ indicates the flexibility matrix for the present system.

Premultiplying Eq. (4-199) with the inverse flexibility matrix yields the following matrix equation of motion:

$$\mathbf{M}\ddot{\mathbf{X}}(t) + \mathbf{C}\dot{\mathbf{X}}(t) + \mathbf{K}\mathbf{X}(t) = \mathbf{F}(t) \tag{4–201}$$

$$\left.\begin{array}{l} \mathbf{X}(t) = \begin{bmatrix} X_1(t) \\ X_2(t) \end{bmatrix}, \quad \mathbf{F}(t) = \mathbf{d}\, F(t), \quad \mathbf{d} = \mathbf{D}^{-1}\mathbf{d}_0 = \frac{1}{15}\begin{bmatrix} 34 \\ 6 \end{bmatrix} \\[4mm] \mathbf{M} = \begin{bmatrix} m & 0 \\ 0 & m \end{bmatrix}, \quad \mathbf{C} = \begin{bmatrix} c & 0 \\ 0 & c \end{bmatrix} \\[4mm] \mathbf{K} = \mathbf{D}^{-1} = \frac{4\,EI}{5\,l^3}\begin{bmatrix} 2 & 3 \\ 3 & 12 \end{bmatrix} \end{array}\right\} \tag{4–202}$$

where $\mathbf{M}$ is the mass matrix, $\mathbf{C}$ is the damping matrix and $\mathbf{K}$ is the stiffness matrix of the system.

Eqs. (4-198) and (4-201) can be combined into the following stochastic state vector differential equation, cf. Eqs. (4-44) and (4-45):

$$\left.\begin{array}{l} \dfrac{d}{dt}\mathbf{Y}(t) = \mathbf{A}\,\mathbf{Y}(t) + \mathbf{b}\,W(t), \quad t > 0 \\[3mm] \mathbf{Y}(0) = \mathbf{y}_0 \end{array}\right\} \tag{4–203}$$

$$\mathbf{Y}(t) = \begin{bmatrix} \mathbf{X}(t) \\ \dot{\mathbf{X}}(t) \\ F(t) \end{bmatrix}, \quad \mathbf{b} = \begin{bmatrix} \mathbf{0} \\ \mathbf{0} \\ \alpha \end{bmatrix}, \quad \mathbf{A} = \begin{bmatrix} \mathbf{0} & \mathbf{I} & 0 \\ -\mathbf{M}^{-1}\mathbf{K} & -\mathbf{M}^{-1}\mathbf{C} & \mathbf{M}^{-1}\mathbf{d} \\ 0 & 0 & -\alpha \end{bmatrix} \tag{4–204}$$

Next, the stationary zero time-lag covariance matrix $\mathbf{C_{YY}}$ of the state vector follows from the Lyapunov equation Eq. (3-197), from which the zero time-lag covariance matrices $\mathbf{C_{XX}}$, $\mathbf{C_{\dot{X}\dot{X}}}$ and the zero time-lag cross-covariance matrix $\mathbf{C_{X\dot{X}}}$ of $\mathbf{X}(t)$ and $\dot{\mathbf{X}}(t)$ can be obtained.

In the following, the system is analyzed based on the equivalent white noise approximation together with the approximate expression given by Eq. (4-129).

Using $\mathbf{M}$ and $\mathbf{K}$ in Eq. (4-202), the undamped angular eigenfrequencies and the related eigenmodes are obtained as solutions to the eigenvalue problem:

$$
\begin{bmatrix} 2 - \lambda_j & 3 \\ 3 & 12 - \lambda_j \end{bmatrix} \begin{bmatrix} \Phi_{j1} \\ \Phi_{j2} \end{bmatrix} = \begin{bmatrix} 0 \\ 0 \end{bmatrix}, \quad \lambda_j = \frac{5}{4} \frac{ml^3}{EI} \omega_j^2, \quad j = 1, 2 \tag{4-205}
$$

$$
\left.
\begin{aligned}
\omega_1 &= \sqrt{\frac{4}{5}} \sqrt{7 - \sqrt{34}} \sqrt{\frac{EI}{ml^3}}, \quad \Phi_1 = \begin{bmatrix} 1 \\ \frac{1}{3}(5 - \sqrt{34}) \end{bmatrix} \\
\omega_2 &= \sqrt{\frac{4}{5}} \sqrt{7 + \sqrt{34}} \sqrt{\frac{EI}{ml^3}}, \quad \Phi_2 = \begin{bmatrix} 1 \\ \frac{1}{3}(5 + \sqrt{34}) \end{bmatrix}
\end{aligned}
\right\} \tag{4-206}
$$

Since $\mathbf{C} = \frac{c}{m}\mathbf{M}$, it follows that $\Phi_1^T \mathbf{C} \Phi_2 = 0$, cf. Eq. (4-83). Hence, the modal equations of motion decouple. Recalling that modal damping ratios are given by $\zeta_j = \frac{c_j}{2\sqrt{m_j k_j}} = \frac{1}{2\omega_j}\frac{c}{m}$, the modal masses and the modal damping ratios in the present case become:

$$
\left.
\begin{aligned}
m_1 &= \frac{1}{9}\left(68 - 10\sqrt{34}\right)m, \quad \zeta_1 = \sqrt{\frac{5}{112 - 16\sqrt{34}}} \sqrt{\frac{c^2}{m}\frac{l^3}{EI}} \\
m_2 &= \frac{1}{9}\left(68 + 10\sqrt{34}\right)m, \quad \zeta_2 = \sqrt{\frac{5}{112 + 16\sqrt{34}}} \sqrt{\frac{c^2}{m}\frac{l^3}{EI}}
\end{aligned}
\right\} \tag{4-207}
$$

The modal loads become:

$$
f_j(t) = \Phi_j^T \mathbf{F}(t) = a_j F(t), \quad a_j = \Phi_j^T \mathbf{d} \tag{4-208}
$$

where from Eqs. (4-202) and (4-206) we have:

$$
\left.
\begin{aligned}
a_1 &= \frac{1}{15}\left(44 - 2\sqrt{34}\right) \\
a_2 &= \frac{1}{15}\left(44 + 2\sqrt{34}\right)
\end{aligned}
\right\} \tag{4-209}
$$

The cross-spectral density functions of the modal load processes are given by, cf. Eqs. (3-203) and (4-208):

$$
S_{f_j f_k}(\omega) = a_j a_k \frac{\alpha^2}{\alpha^2 + \omega^2} S_0 \tag{4-210}
$$

As seen from Eq. (4-206), the angular eigenfrequencies ω_1 and ω_2 are well separated. Further, in order to apply the equivalent white noise approximation at variance calculations, it is necessary to assume light damping of the system, i.e., $\zeta_1 \ll 1 \wedge \zeta_2 \ll 1$. Then, the stationary variances of the displacement processes $\{X_1(t),\, t \in R\}$ and $\{X_2(t),\, t \in R\}$ become, cf. Eq. (4-129):

$$\left.\begin{aligned}
\sigma_{X_1}^2 &\simeq \Phi_{1,1}^2\,\sigma_{Q_1}^2 + \Phi_{2,1}^2\,\sigma_{Q_2}^2 = \sigma_{Q_1}^2 + \sigma_{Q_2}^2 \\
\sigma_{X_2}^2 &\simeq \Phi_{1,2}^2\,\sigma_{Q_1}^2 + \Phi_{2,2}^2\,\sigma_{Q_2}^2 = \frac{1}{9}\left(59 - 10\sqrt{34}\right)\sigma_{Q_1}^2 + \frac{1}{9}\left(59 + 10\sqrt{34}\right)\sigma_{Q_2}^2
\end{aligned}\right\} \qquad (4\text{–}211)$$

where the variances of the modal coordinate processes can be calculated from Eq. (4-127):

$$\left.\begin{aligned}
\sigma_{Q_1}^2 &\simeq \frac{\pi\,S_{f_1 f_1}(\omega_1)}{2\,\zeta_1\,\omega_1^3\,m_1^2} = \frac{\pi\,a_1^2}{2\,\zeta_1\,\omega_1^3\,m_1^2}\,\frac{\alpha^2}{\alpha^2 + \omega_1^2}\,S_0 = 13.4669\,\frac{\alpha^2}{\alpha^2 + \omega_1^2}\,S_0\,\frac{l^3}{c\,EI} \\
\sigma_{Q_2}^2 &\simeq \frac{\pi\,S_{f_2 f_2}(\omega_2)}{2\,\zeta_2\,\omega_2^3\,m_2^2} = \frac{\pi\,a_2^2}{2\,\zeta_2\,\omega_2^3\,m_2^2}\,\frac{\alpha^2}{\alpha^2 + \omega_2^2}\,S_0 = 0.0214\,\frac{\alpha^2}{\alpha^2 + \omega_2^2}\,S_0\,\frac{l^3}{c\,EI}
\end{aligned}\right\} \qquad (4\text{–}212)$$

It is noticed from Eq. (4-212) that the dynamic response of the beam is dominated by the 1st mode.

Finally, Eqs. (4-211) and (4-212) provide the following results of the stationary variances of $\{X_1(t),\, t \in R\}$ and $\{X_2(t),\, t \in R\}$:

$$\left.\begin{aligned}
\sigma_{X_1}^2 &\simeq \left(\frac{13.4669}{\alpha^2 + \omega_1^2} + \frac{0.0214}{\alpha^2 + \omega_2^2}\right)\alpha^2\,S_0\,\frac{l^3}{c\,EI} \\
\sigma_{X_2}^2 &\simeq \left(\frac{1.0332}{\alpha^2 + \omega_1^2} + \frac{0.2789}{\alpha^2 + \omega_2^2}\right)\alpha^2\,S_0\,\frac{l^3}{c\,EI}
\end{aligned}\right\} \qquad (4\text{–}213)$$

4.3 Continuous systems

Fig. 4-21 shows a structure occupying the domain D, which is in the static equilibrium state at $t = 0$. Material particles of the structure are identified by their Cartesian coordinates $\mathbf{x} = [x_1, x_2, x_3]^T$ in the static equilibrium state. The *mass density* is denoted $\rho(\mathbf{x})$, and the damping coefficient per unit volume of a distributed linear viscous damping force is denoted $c(\mathbf{x})$. The surface Γ of the domain D is divided into two disjoint subsurfaces Γ_1 and Γ_2, which prescribe the *mechanical* and *kinematic boundary conditions*, respectively.

The displacement of the material particle from static equilibrium is modelled as a 3-dimensional vector process $\{\mathbf{U}(\mathbf{x}, t),\, (\mathbf{x}, t) \in (D \cup \Gamma_1) \times [0, \infty[\}$. In what follows, it is assumed that the kinematic boundary conditions are homogenous, i.e., the prescribed displacements on Γ_2 are all zero. The stochastic initial displacement and velocity fields are denoted as $\mathbf{U}_0(\mathbf{x}) = \mathbf{U}(\mathbf{x}, 0)$ and $\dot{\mathbf{U}}_0(\mathbf{x}) = \dot{\mathbf{U}}(\mathbf{x}, 0)$, respectively. The displacements are caused by the following vector processes: The initial value processes $\{\mathbf{U}_0(\mathbf{x}),\, \mathbf{x} \in D \cup \Gamma_1\}$ and $\{\dot{\mathbf{U}}_0(\mathbf{x}),\, \mathbf{x} \in D \cup \Gamma_1\}$, the load process

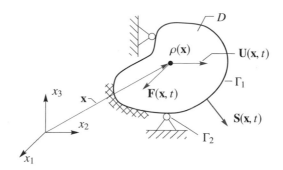

Fig. 4–21 Continuous structure.

per unit volume $\{\mathbf{F}(\mathbf{x}, t), (\mathbf{x}, t) \in D \times [0, \infty[\}$, and the *surface traction* process per unit area $\{\mathbf{S}(\mathbf{x}, t), (\mathbf{x}, t) \in \Gamma_1 \times [0, \infty[\}$.

Then, the displacement field admits the following eigenmode expansion:[1]

$$\mathbf{U}(\mathbf{x}, t) = \sum_{j=1}^{\infty} Q_j(t)\, \mathbf{\Phi}_j(\mathbf{x}) \tag{4-214}$$

where $Q_j(t)$ denote the modal coordinates of the continuous system.

The eigenfunctions $\mathbf{\Phi}_j(\mathbf{x})$ are determined as solutions to a linear eigenvalue problem. The differential operator $\mathbf{L}(\mathbf{x})$ of the eigenvalue problem is self-adjoint, and the following orthogonality properties hold:

$$\int_D \rho(\mathbf{x})\, \mathbf{\Phi}_j^T(\mathbf{x})\, \mathbf{\Phi}_k(\mathbf{x})\, d\mathbf{x} = \begin{cases} 0\,, & j \neq k \\ m_j\,, & j = k \end{cases} \tag{4-215}$$

$$\int_D \left(\mathbf{L}(\mathbf{x})\mathbf{\Phi}_j(\mathbf{x})\right)^T \mathbf{\Phi}_k(\mathbf{x})\, d\mathbf{x} = \int_D \mathbf{\Phi}_j^T(\mathbf{x}) \left(\mathbf{L}(\mathbf{x})\mathbf{\Phi}_k(\mathbf{x})\right) d\mathbf{x} = \begin{cases} 0\,, & j \neq k \\ k_j\,, & j = k \end{cases} \tag{4-216}$$

where m_j and k_j signify the jth modal mass and modal stiffness, respectively. They are related as:

$$k_j = \omega_j^2 m_j \tag{4-217}$$

where ω_j indicates the undamped angular eigenfrequency of the jth mode.

Further, it is assumed that the distributed viscous damping coefficient per unit volume $c(\mathbf{x})$ fulfils the *decoupling condition*:

$$\int_D c(\mathbf{x})\, \mathbf{\Phi}_j^T(\mathbf{x})\, \mathbf{\Phi}_k(\mathbf{x})\, d\mathbf{x} = \begin{cases} 0\,, & j \neq k \\ c_j\,, & j = k \end{cases} \tag{4-218}$$

---

[1]S.R.K. Nielsen. *Vibration Theory, Vol. 1: Linear Vibration Theory.* Aalborg University Press, Aalborg, 2004.

where $c_j = 2\zeta_j\omega_j m_j$ is the modal damping coefficient, ζ_j indicates the modal damping ratio of the jth mode.

From Eqs. (4-215), (4-216) and (4-218), the following decoupled stochastic equations of motion may be derived for the modal coordinate processes $\{Q_j(t), t \in [0, \infty[\}$:[1]

$$\left.\begin{aligned} \ddot{Q}_j(t) + 2\,\zeta_j\,\omega_j\,\dot{Q}_j(t) + \omega_j^2\,Q_j(t) &= \frac{f_j(t)}{m_j}, \quad j = 1, 2, \ldots, \quad t > 0 \\ Q_j(0) &= Q_{j,0}, \quad \dot{Q}_j(0) = \dot{Q}_{j,0} \end{aligned}\right\} \tag{4–219}$$

where $f_j(t)$ is the *modal load process* in the jth mode, expressed as:

$$f_j(t) = \int_{\Gamma_1} \mathbf{\Phi}_j^T(\mathbf{x})\,\mathbf{S}(\mathbf{x}, t)\,dA + \int_D \mathbf{\Phi}_j^T(\mathbf{x})\,\mathbf{F}(\mathbf{x}, t)\,d\mathbf{x} \tag{4–220}$$

Eq. (4-219) is identical to Eq. (4-101). Therefore, its solution is provided by Eq. (4-102) with the modal impulse response function $h_j(t)$ formally given by Eq. (4-91).

The initial values $Q_{j,0}$, $\dot{Q}_{j,0}$ can be written as:

$$\left.\begin{aligned} Q_{j,0} &= \frac{1}{m_j} \int_D \rho(\mathbf{x})\,\mathbf{\Phi}_j^T(\mathbf{x})\,\mathbf{U}_0(\mathbf{x})\,d\mathbf{x} \\ \dot{Q}_{j,0} &= \frac{1}{m_j} \int_D \rho(\mathbf{x})\,\mathbf{\Phi}_j^T(\mathbf{x})\,\dot{\mathbf{U}}_0(\mathbf{x})\,d\mathbf{x} \end{aligned}\right\} \tag{4–221}$$

In order to prove the results in Eq. (4-221), Eq. (4-214) is considered at time $t = 0$:

$$\mathbf{U}_0(\mathbf{x}) = \mathbf{U}(\mathbf{x}, 0) = \sum_{k=1}^{\infty} Q_j(0)\,\mathbf{\Phi}_k(\mathbf{x}) \tag{4–222}$$

Eq. (4-222) is pre-multiplied by $\rho(\mathbf{x})\mathbf{\Phi}_j^T(\mathbf{x})$, and integration is performed over the volume D. Then, the first equation of (4-221) follows from the orthogonality property in Eq. (4-215). The second initial value in Eq. (4-221) can be proved in the same way.

The solution to Eq. (4-219) reads, cf. Eqs. (4-102), (4-103), (4-104) and (4-105):

$$Q_j(t) = \left(c_j\,h_j(t) + m_j\,\dot{h}_j(t)\right) Q_{j,0} + m_j\,h_j(t)\,\dot{Q}_{j,0} + \int_0^t h_j(t - \tau)\,f_j(\tau)\,d\tau \tag{4–223}$$

Inserting Eq. (4-223) into Eq. (4-214) and using Eqs. (4-220) and (4-221), the following stochastic integral representation for the displacement vector stochastic process $\{\mathbf{U}(\mathbf{x}, t), (\mathbf{x}, t) \in (D \cup \Gamma_1) \times [0, \infty[\}$ is obtained:

$$\begin{aligned} \mathbf{U}(\mathbf{x}, t) = &\int_D \mathbf{a}_0(\mathbf{x}, \mathbf{y}, t)\,\mathbf{U}_0(\mathbf{y})\,d\mathbf{y} + \int_D \mathbf{a}_1(\mathbf{x}, \mathbf{y}, t)\,\dot{\mathbf{U}}_0(\mathbf{y})\,d\mathbf{y} \\ &+ \int_0^t \int_D \mathbf{h}(\mathbf{x}, \mathbf{y}, t - \tau)\,\mathbf{F}(\mathbf{y}, \tau)\,d\mathbf{y}\,d\tau + \int_0^t \int_{\Gamma_1} \mathbf{h}(\mathbf{x}, \mathbf{y}, t - \tau)\,\mathbf{S}(\mathbf{y}, \tau)\,dA\,d\tau \end{aligned} \tag{4–224}$$

---

[1] S.R.K. Nielsen. *Vibration Theory, Vol. 1: Linear Vibration Theory.* Aalborg University Press, Aalborg, 2004.

where the kernel functions in the stochastic integrals are given by:

$$\mathbf{h}(\mathbf{x}, \mathbf{y}, t) = \sum_{j=1}^{\infty} h_j(t)\, \mathbf{\Phi}_j(\mathbf{x})\mathbf{\Phi}_j^T(\mathbf{y}) \tag{4–225}$$

$$\mathbf{a}_0(\mathbf{x}, \mathbf{y}, t) = \left(\sum_{j=1}^{\infty} \left(\dot{h}_j(t) + 2\zeta_j\omega_j h_j(t) \right) \mathbf{\Phi}_j(\mathbf{x})\mathbf{\Phi}_j^T(\mathbf{y}) \right) \rho(\mathbf{y}) \tag{4–226}$$

$$\mathbf{a}_1(\mathbf{x}, \mathbf{y}, t) = \left(\sum_{j=1}^{\infty} h_j(t)\, \mathbf{\Phi}_j(\mathbf{x})\mathbf{\Phi}_j^T(\mathbf{y}) \right) \rho(\mathbf{y}) \tag{4–227}$$

Eqs. (4-225), (4-226) and (4-227) represent the continuous versions of the corresponding results in Eqs. (4-90), (4-93) and (4-94) for discrete systems.

Eq. (4-224) determines the displacement field $\mathbf{U}(\mathbf{x}, t)$ as a linear combination of all external excitations such as the prescribed initial values and the previous differential impulses $\mathbf{F}(\mathbf{y}, \tau)d\mathbf{y}d\tau$ and $\mathbf{S}(\mathbf{y}, \tau)dAd\tau$. Eq. (4-224) needs to have the indicated form due the linearity of the system and the consequent validity of the superposition theorem. A further consequence of the linearity is that the displacement process $\{\mathbf{U}(\mathbf{x}, t), (\mathbf{x}, t) \in (D \cup \Gamma_1) \times [0, \infty[\}$ becomes Gaussian, if the initial value processes and the external load processes are Gaussian stochastic processes.

Next, joint statistical moments of the displacement process $\{\mathbf{U}(\mathbf{x}, t), (\mathbf{x}, t) \in (D \cup \Gamma_1) \times [0, \infty[\}$ can be determined by expressions similar to Eqs. (4-106), (4-110), (4-113), (4-114) and (4-115), which become somewhat more involved due to the increased number of source terms on the right-hand side of Eq. (4-224).

Example 4-9: Bernoulli-Euler beams

Fig. 4-22 shows a Bernoulli-Euler beam of the length l. The cross-sections are identified by the coordinate x measured from the left-end side. The beam is non-homogeneous with varying bending stiffness $EI(x)$, *mass per unit length* $\mu(x)$ and linear viscous damping coefficient per unit length $c(x)$.

The beam is loaded with a stochastic dynamic load per unit length $F(x, t)$. Additionally, external stochastic moments $M_0(t)$, $M_1(t)$ and external stochastic shear forces $Q_0(t)$, $Q_1(t)$ with signs defined in Fig. 4-22 may be acting at the end-sections of the beam.

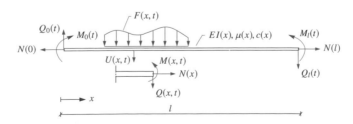

Fig. 4–22 Bernoulli-Euler beam.

The displacement field from the static equilibrium state is modelled by the stochastic process $\{U(x, t), (x, t) \in [0, l] \times [0, \infty[\}$, where the displacement $U(x, t)$ is considered positive in the same direction as $F(x, t)$. $U(x, t)$ is caused by the indicated external loads, as well as the initial value fields $\{U_0(x), x \in [0, l]\}$ and $\{\dot{U}_0(x), x \in [0, l]\}$.

The section forces are made up by the normal force $N(x)$, the shear force $Q(x, t)$ and the bending moment $M(x, t)$ with signs defined in Fig. 4-22. Since no dynamic external load or motion are present in the axial direction, the normal force is constant in time. $M(x, t)$ and $Q(x, t)$ are related to the displacement $U(x, t)$ as:

$$M(x, t) = -EI(x)\frac{\partial^2 U(x, t)}{\partial x^2} \tag{4–228}$$

$$Q(x, t) = -\frac{\partial}{\partial x}\left(EI(x)\frac{\partial^2 U(x, t)}{\partial x^2}\right) + N(x)\frac{\partial U(x, t)}{\partial x} \tag{4–229}$$

Then, $U(x, t)$ is determined from the following initial and boundary value problem:

Stochastic partial differential equation :

$$\frac{\partial^2}{\partial x^2}\left(EI(x)\frac{\partial^2 U}{\partial x^2}\right) - \frac{\partial}{\partial x}\left(N(x)\frac{\partial U}{\partial x}\right) + c(x)\frac{\partial U}{\partial t} + \mu(x)\frac{\partial^2 U}{\partial t^2} = F(x, t)$$

Stochastic initial value fields :

$$U(x, 0) = U_0(x), \quad \frac{\partial U(x, 0)}{\partial t} = \dot{U}_0(x)$$

Geometric boundary conditions :

$$U(0, t) = 0, \quad \frac{\partial U(0, t)}{\partial x} = 0$$

$$U(l, t) = 0, \quad \frac{\partial U(l, t)}{\partial x} = 0 \tag{4–230}$$

Stochastic mechanical boundary conditions :

$$-EI(0)\frac{\partial^2 U(0, t)}{\partial x^2} = M_0(t)$$

$$-\frac{\partial}{\partial x}\left(EI(0)\frac{\partial^2 U(0, t)}{\partial x^2}\right) + N(0)\frac{\partial U(0, t)}{\partial x} = Q_0(t)$$

$$-EI(l)\frac{\partial^2 U(l, t)}{\partial x^2} = M_l(t)$$

$$-\frac{\partial}{\partial x}\left(EI(l)\frac{\partial^2 U(l, t)}{\partial x^2}\right) + N(l)\frac{\partial U(l, t)}{\partial x} = Q_l(t)$$

For ease, the kinematic (geometric) boundary conditions, indicating the displacements or rotations at the end-sections, have been assumed to be homogeneous.

Two boundary conditions need to be specified at $x = 0$ and $x = l$, respectively. They could be both kinematic or both mechanical or a mixture of these.

Undamped eigenvibrations are obtained as non-trivial solutions to Eq. (4-230) for $c(x) = 0$ and $F(x, t) = M_0(t) = Q_0(t) = M_1(t) = Q_1(t) \equiv 0$. Eigenvibrations are searched in the form $U(x, t) = \text{Re}(\Phi(x)\, e^{i\omega t})$.

Insertion this form into Eq. (4-230) provides the following linear eigenvalue problem for the determination of the eigenfunctions $\Phi(x)$ and the related angular eigenfrequency ω:

Differential equation :

$$\frac{d^2}{dx^2}\left(EI(x)\frac{d^2\Phi(x)}{dx^2}\right) - \frac{d}{dx}\left(N(x)\frac{d\Phi(x)}{dx}\right) - \omega^2\mu(x)\,\Phi(x) = 0$$

Geometric boundary conditions :

$$\Phi(0) = 0, \quad \frac{d\Phi(0)}{dx} = 0$$

$$\Phi(l) = 0, \quad \frac{d\Phi(l)}{dx} = 0$$

Mechanical boundary conditions : .   (4–231)

$$-EI(0)\frac{d^2\Phi(0)}{dx^2} = 0$$

$$-\frac{d}{dx}\left(EI(0)\frac{d^2\Phi(0)}{dx^2}\right) + N(0)\frac{d\Phi(0)}{dx} = 0$$

$$-EI(l)\frac{d^2\Phi(l)}{dx^2} = 0$$

$$-\frac{d}{dx}\left(EI(l)\frac{d^2\Phi(l)}{dx^2}\right) + N(l)\frac{d\Phi(l)}{dx} = 0$$

The orthogonality properties of the eigenfunctions $\Phi_j(x)$ can be written as:

$$\int_0^l \mu(x)\,\Phi_j(x)\,\Phi_k(x)\,dx = \begin{cases} 0, & j \neq k \\ m_j, & j = k \end{cases} \tag{4–232}$$

$$\int_0^l \left(L(x)\Phi_j(x)\right)\Phi_k(x)\,dx = \int_0^l \Phi_j(x)\left(L(x)\Phi_k(x)\right)dx$$

$$= \int_0^l \left(EI(x)\frac{d^2\Phi_j(x)}{dx^2}\frac{d^2\Phi_k(x)}{dx^2} + N(x)\frac{d\Phi_j(x)}{dx}\frac{d\Phi_k(x)}{dx}\right)dx = \begin{cases} 0, & j \neq k \\ k_j, & j = k \end{cases} \tag{4–233}$$

where $k_j = \omega_j^2 m_j$. $L(x)$ is the self-adjoint differential operator of the eigenvalue problem and in the present case is given by:

$$L(x)\Phi(x) = \frac{d^2}{dx^2}\left(EI(x)\frac{d^2\Phi(x)}{dx^2}\right) - \frac{d}{dx}\left(N(x)\frac{d\Phi(x)}{dx}\right) \tag{4–234}$$

The modal load is given by, cf. Eq. (4-220):

$$f_j(t) = \int_0^l \Phi_j(x)\,F(x,t)\,dx + \Phi_j(0)\,Q_0(t) - \frac{d\Phi_j(0)}{dx}\,M_0(t) - \Phi_j(l)\,Q_l(t) + \frac{d\Phi_j(l)}{dx}\,M_l(t) \tag{4–235}$$

In the following, the geometric stiffness from the normal force is ignored, corresponding to $N(x) = 0$. Then, for a simply supported homogeneous beam, Eq. (4-231) is simplified as:

Differential equation :

$$\frac{d^4}{d\xi^4}\Phi(\xi) - \lambda^4\,\Phi(\xi) = 0, \quad \xi = \frac{x}{l}, \quad \lambda^4 = \omega^2\frac{\mu\,l^4}{EI}$$

Geometric boundary conditions :

$$\Phi(0) = 0, \quad \Phi(1) = 0$$

Mechanical boundary conditions :

$$\frac{d^2\Phi(0)}{d\xi^2} = 0, \quad \frac{d^2\Phi(1)}{d\xi^2} = 0$$

(4–236)

The eigenvalues λ_j in this case become:

$$\lambda_j = j\pi, \quad j = 1, 2, \ldots \tag{4–237}$$

The related eigenfunctions $\Phi_j(x)$, the modal masses m_j and the angular eigenfrequencies ω_j become:

$$\Phi_j(x) = \sin\left(j\pi\frac{x}{l}\right)$$

$$m_j = \frac{1}{2}\mu\,l$$

$$\omega_j = j^2\pi^2\sqrt{\frac{EI}{\mu\,l^4}}$$

(4–238)

Next, the eigenvalue problem for a homogeneous cantilever beam clamped at $x = 0$ can be written as:

Differential equation :

$$\frac{d^4}{d\xi^4}\Phi(\xi) - \lambda^4\,\Phi(\xi) = 0, \quad \xi = \frac{x}{l}, \quad \lambda^4 = \omega^2\frac{\mu\,l^4}{EI}$$

Geometric boundary conditions :

$$\Phi(0) = 0, \quad \frac{d\Phi(0)}{d\xi} = 0$$

Mechanical boundary conditions :

$$\frac{d^2\Phi(1)}{d\xi^2} = 0, \quad \frac{d^3\Phi(1)}{d\xi^3} = 0$$

(4–239)

The eigenvalues λ_j in this case are solutions to the transcendental equation:

$$\cos(\lambda_j)\cosh(\lambda_j) + 1 = 0 \qquad \Rightarrow$$

$$\lambda_1 = 1.8751$$
$$\lambda_2 = 4.6941$$
$$\lambda_3 = 7.8548$$

(4–240)

$$\vdots$$

The related eigenfunctions $\Phi_j(x)$ and angular eigenfrequencies ω_j become:

$$
\left.
\begin{aligned}
\Phi_j(x) &= \left(\sin\left(\lambda_j \frac{x}{l}\right) - \sinh\left(\lambda_j \frac{x}{l}\right)\right)\left(\cos(\lambda_j) + \cosh(\lambda_j)\right) \\
&\quad - \left(\cos\left(\lambda_j \frac{x}{l}\right) - \cosh\left(\lambda_j \frac{x}{l}\right)\right)\left(\sin(\lambda_j) + \sinh(\lambda_j)\right) \\
\omega_j &= \lambda_j^2 \sqrt{\frac{EI}{\mu l^4}}
\end{aligned}
\right\}
\qquad (4\text{--}241)
$$

Example 4-10: Simply supported homogeneous beam excited by a moving vehicle with stochastic velocity and load magnitude

Fig. 4-23 shows a bridge modelled as a simply supported homogenous Bernoulli-Euler beam with the length l, the bending stiffness EI, the mass per unit length μ, and the linear viscous damping coefficient per unit length c.

At $t = 0$ the bridge is at rest, and a vehicle with the load P is entering the bridge with a time-independent velocity V. P and V are assumed to be independent stochastic variables. Generally, the suspension system of the vehicle will cause dynamic responses of the system, for which reason the reaction force between the wheels of the suspension system and the bridge surface will vary with time during the passage. However, this dynamic effect of the reaction force is ignored in the following, so the reaction force on the bridge is constantly equal to P. The problem is to determine the mean value function $\mu_M(x, t)$ and the cross-covariance function $\kappa_{MM}(x_1, t_1; x_2, t_2)$ of the bending moment stochastic process $\{M(x, t), (x, t) \in [0, l] \times [0, \infty[\}$.

The displacement process $\{U(x, t), (x, t) \in [0, l] \times [0, \infty[\}$ is given by the modal expansion, cf. Eqs. (4-214) and (4-238):

$$
U(x, t) = \sum_{j=1}^{\infty} Q_j(t)\,\Phi_j(x) = \sum_{j=1}^{\infty} Q_j(t)\,\sin\left(j\pi\frac{x}{l}\right)
\qquad (4\text{--}242)
$$

As long as the vehicle is on the bridge, the load per unit length can formally be written as:

$$
F(x, t) = P\,\delta(x - Vt)
\qquad (4\text{--}243)
$$

where $\delta(\cdot)$ indicates the *Dirach's delta function*. Of course, $F(x, t) \equiv 0$ as soon as the vehicle has left the

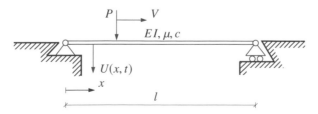

Fig. 4–23 Simply supported homogeneous beam excited by a moving vehicle with stochastic velocity and load magnitude.

bridge at the time $t = \frac{l}{V}$. Then, the modal load $f_j(t)$ becomes, cf. Eq. (4-235):

$$f_j(t) = \int_0^l \sin\left(j\pi \frac{x}{l}\right) F(x,t)\, dx = \begin{cases} P \sin\left(j\pi \dfrac{Vt}{l}\right), & t \in \left[0, \dfrac{l}{V}\right] \\[2ex] 0 & , \quad t \in \left]\dfrac{l}{V}, \infty\right[\end{cases} \tag{4-244}$$

The modal coordinate $Q_j(t)$ can be obtained from Eq. (4-223) with the initial conditions $Q_j(0) = \dot{Q}_j(0) = 0$, together with the modal load given by Eq. (4-244):

$$Q_j(t) = P \int_0^{\min\left(t, \frac{l}{V}\right)} h_j(t - \tau) \sin\left(j\pi \frac{V\tau}{l}\right) d\tau \tag{4-245}$$

where $h_j(t)$ is the modal impulse response function as given by Eq. (4-91).

Next, the bending moment process is also determined by the modal expansion, cf. Eq. (4-228):

$$M(x,t) = -EI \sum_{j=1}^{\infty} Q_j(t) \frac{d^2}{dx^2} \Phi_j(x) = P \sum_{j=1}^{\infty} M_j(t, V) \sin\left(j\pi \frac{x}{l}\right) \tag{4-246}$$

$$M_j(t, V) = EI \left(\frac{j\pi}{l}\right)^2 \int_0^{\min\left(t, \frac{l}{V}\right)} h_j(t - \tau) \sin\left(j\pi \frac{V\tau}{l}\right) d\tau \tag{4-247}$$

The stochastic independence of P and V, and hence of P and $M_j(t, V)$, yields the following expressions for the mean value function and cross-covariance function of $\{M(x,t), (x,t) \in [0,l] \times [0, \infty[\}$:

$$\mu_M(x,t) = \mu_P \sum_{j=1}^{\infty} \mu_{M_j}(t) \sin\left(j\pi \frac{x}{l}\right) \tag{4-248}$$

$$\kappa_{MM}(x_1, t_1; x_2, t_2) = \sigma_P^2 \sum_{j=1}^{\infty} \sum_{k=1}^{\infty} \kappa_{M_j M_k}(t_1, t_2) \sin\left(j\pi \frac{x_1}{l}\right) \sin\left(k\pi \frac{x_2}{l}\right) \tag{4-249}$$

where μ_P and σ_P^2 indicate the mean value and the variance of P, respectively. $\mu_{M_j}(t)$ and $\kappa_{M_j M_k}(t_1, t_2)$ are the mean value function and the cross-covariance function of the stochastic process $\{M_j(t, V), t \in [0, \infty[\}$, respectively.

Eq. (4-247) can not be evaluated analytically due to its non-linear dependence on V, and hence $\mu_{M_j}(t)$ and $\kappa_{M_j M_k}(t_1, t_2)$ need to be obtained by numerical calculation of one-dimensional quadratures over the sample space of V. At least 40 terms are needed in the summations of Eqs. (4-248) and (4-249) in order to achieve convergence.

It should be noticed that due to the non-linear dependence on V, the bending moment process is not Gaussian even if P and V are normally distributed stochastic variables. Hence, $\mu_{M_j}(t)$ and $\kappa_{M_j M_k}(t_1, t_2)$ do not provide the full description of the stochastic structure of the bending moment process.

CHAPTER 5
RELIABILITY THEORY OF DYNAMICALLY EXCITED STRUCTURES

5.1 First-passage failure problem

Fig. 5-1 shows realizations of a scalar stochastic process $\{X(t), t \in [0, \infty[\}$, as well as a deterministic upper *limit state function* $b(t)$ that separates unsafe and safe states of the process. The *safe domain* at the time t is given by:

$$S_t = \big\{x \mid -\infty < x < b(t)\big\} \tag{5-1}$$

Consequently, the limit state function $b(t)$ itself is considered as part of the unsafe domain. Next, consider N realizations (N is a sufficiently large number). Among all these realizations, N_0 realizations originate from the safe domain S_0 at the time $t = 0$, and the remaining $N - N_0$ realizations originate from the unsafe domain, as shown in Fig. 5-1. During the time interval $]t, t + \Delta t]$, a certain number ΔN of realizations are leaving the safe domain. Among these, ΔN_0 realizations originate from the safe domain at the time $t = 0$, whereas the remaining $\Delta N - \Delta N_0$

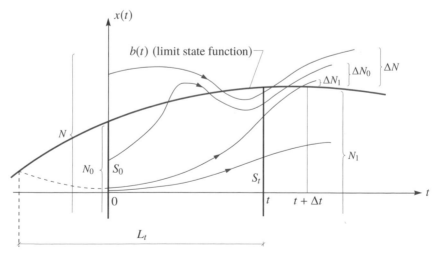

Fig. 5–1 First passages and out-crossings.

realizations originate from the unsafe domain. A further distinction will be made among the ΔN_0 realizations. Let ΔN_1 of them cross out for the first time during the interval $]t, t + \Delta t]$, and the remaining $\Delta N_0 - \Delta N_1$ realizations have at least one previous *out-crossing* in the interval $]0, t]$. ΔN_1 realizations are designated *first passages* in the interval $]t, t + \Delta t]$. Obviously, $\Delta N_1 \leq \Delta N_0 \leq \Delta N$.

The probability of an out-crossing in $]t, t + \Delta t]$ can be estimated as the fraction $\frac{\Delta N}{N}$. It is assumed that this fraction becomes proportional to Δt as $\Delta t \to 0$, i.e.:

$$\frac{\Delta N}{N} = f_1(t)\Delta t + O(\Delta t^2) \tag{5-2}$$

where the *Landau symbol* $O(g(x))$ means that $O(g(x)) \leq A|g(x)|$, with A being a positive constant. $f_1(t)$ is designated the *out-crossing frequency*.

Further, it is assumed that the probability of two or more out-crossings in $]t, t + \Delta t]$ is negligible compared with the probability of exactly one out-crossing if Δt is sufficiently small. The indicated probability is at most of the magnitude of $O(\Delta t^2)$. Then, the *expected number of out-crossings per unit time* at the time t becomes:

$$\lim_{\Delta t \to 0} \frac{1}{\Delta t} \sum_{n=0}^{\infty} n \cdot P(n \text{ out-crossings in}]t, t + \Delta t])$$

$$= \lim_{\Delta t \to 0} \frac{1}{\Delta t} \Big(0 \cdot P(0 \text{ out-crossing}) + 1 \cdot P(1 \text{ out-crossing}) + 2 \cdot P(2 \text{ out-crossings}) + \cdots \Big)$$

$$= \lim_{\Delta t \to 0} \frac{1}{\Delta t} \Big(0 + f_1(t)\Delta t + O(\Delta t^2) + 2 \cdot O(\Delta t^2) + \cdots \Big) = f_1(t) \tag{5-3}$$

Therefore, $f_1(t)$ may alternatively be interpreted as the expected (or mean) number of out-crossings per unit time.

In the same way, the fraction $\frac{\Delta N_1}{N_0}$ is assumed to be proportional to Δt as $\Delta t \to 0$, i.e.:

$$\frac{\Delta N_1}{N_0} = f_T(t)\Delta t + O(\Delta t^2) \tag{5-4}$$

where $f_T(t)$ is termed the *first-passage probability density function*.

Next, the interval $]0, t]$ is divided into n disjoint subintervals $]t_{i-1}, t_i]$, $i = 1, \ldots, n$. $\Delta N_{1,i}$ signifies the number of first passages in the interval $]t_{i-1}, t_{i-1} + \Delta t_i]$, $\Delta t_i = t_i - t_{i-1}$, as shown in Fig. 5-2. Then, the total number of first passages in $]0, t]$ is given by $\sum_{i=1}^{n} \Delta N_{1,i}$. The *probability of failure* in $]0, t]$ on condition of being in the safe domain at the time $t = 0$ is designated $P_f(]0, t] | X(0) \in S_0)$. This conditional probability can be expressed as the proportion of the number of first passages $\sum_{i=1}^{n} \Delta N_{1,i}$ in the interval $]0, t]$ to the total number N_0 of realizations originating from the safe domain S_0 ($t = 0$). Using Eq. (5-4), we obtain the following result:

$$P_f(]0, t] | X(0) \in S_0) = \lim_{n \to \infty} \frac{\sum_{i=1}^{n} \Delta N_{1,i}}{N_0} = \lim_{n \to \infty} \frac{N_0 \sum_{i=1}^{n} f_T(t_{i-1})\Delta t_i}{N_0} = \int_0^t f_T(\tau)d\tau = F_T(t) \tag{5-5}$$

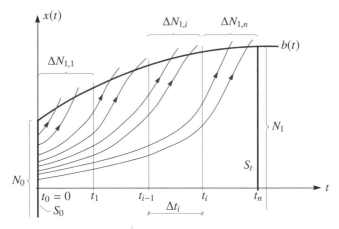

Fig. 5–2 First passages in disjoint intervals.

The limit operation in Eq. (5-5) is performed in such a way that the maximum interval length $\Delta t = \max(\Delta t_1, \ldots, \Delta t_n)$ goes to zero. The function $F_T(t)$ is termed the *first-passage distribution function*. The reason for the naming will be explained below.

Let T denote the random elapsed time when the first out-crossing of the safe domain takes place on condition of a start in the safe domain S_0 at $t = 0$. Obviously, the system has been failed at the time t if and only if $T \leq t$. Hence, the conditional probability of failure $P_f\left(\,]0, t]\mid X(0) \in S_0\right)$ can be written as:

$$P_f\left(\,]0, t]\mid X(0) \in S_0\right) = P(T \leq t) = F_T(t) \tag{5–6}$$

Next, the *unconditional probability of failure* in $[0, t]$, $P_f([0, t])$, can be expressed as:

$$P_f([0, t]) = 1 - P\left(X(0) \in S_0\right) + P_f\left(\,]0, t]\mid X(0) \in S_0\right) P\left(X(0) \in S_0\right)$$

$$= 1 - P\left(X(0) \in S_0\right)\left(1 - F_T(t)\right) \tag{5–7}$$

where

$$P\left(X(0) \in S_0\right) = P\left(X(0) \leq b(0)\right) = F_X\left(b(0), 0\right) \tag{5–8}$$

$1 - P\left(X(0) \in S_0\right)$ indicates the probability of initial failure, i.e., realizations originating from the unsafe domain at the time $t = 0$. $F_X(x, t)$ is the 1st order probability distribution function of the process $\{X(t), \, t \in [0, \infty[\}$.

Among the N_0 realizations, a certain number N_1 of them have not left the safe domain in the interval $[0, t]$, as shown in Fig. 5-1. Based on these realizations, the *hazard rate* $h(t)$ is defined as:

$$h(t)\Delta t + \mathrm{O}\left(\Delta t^2\right) = \frac{\Delta N_1}{N_1} \tag{5–9}$$

The right-hand side of Eq. (5-9) signifies the number of first passages in $]t, t + \Delta t]$ relative to the number of samples which have not been failed in the preceding interval $[0, t]$. From Eqs. (5-4) and (5-9), we have:

$$
h(t)\Delta t + O(\Delta t^2) = \frac{\Delta N_1}{N_0} \frac{N_0}{N_1} = \frac{f_T(t)\Delta t}{1 - F_T(t)} \qquad \Rightarrow
$$

$$
h(t) = \frac{f_T(t)}{1 - F_T(t)} \tag{5–10}
$$

where the result $\frac{N_1}{N_0} = 1 - F_T(t)$ has been applied. Eq. (5-10) provides the following solution for $F_T(t)$:

$$
F_T(t) = 1 - \exp\left(-\int_0^t h(\tau)\, d\tau\right) \tag{5–11}
$$

Eq. (5-11) can be proved by insertion into Eq. (5-10), and using $f_T(t) = \frac{d}{dt} F_T(t)$.

Inserting Eq. (5-11) into Eq. (5-7) provides the following expression for $P_f([0, t])$ in terms of the hazard rate:

$$
P_f([0, t]) = 1 - P(X(0) \in S_0) \exp\left(-\int_0^t h(\tau)\, d\tau\right) \tag{5–12}
$$

No exact solutions to $h(\tau)$ are available for even the simplest system of engineering significance. Hence, the subject of reliability theory for dynamically excited structures is mainly to specify suitable approximations for the hazard rate for the considered system. Before doing this, one further useful expression for the first-passage probability density function will be derived. From Eqs. (5-2) and (5-4), an alternative expression for the first-passage probability density function $f_T(t)$ can be obtained as:

$$
\begin{aligned}
f_T(t) + O(\Delta t) &= \frac{\Delta N_1}{\Delta t\, N_0} = \frac{\Delta N_1}{\Delta N} \frac{N}{N_0} \frac{\Delta N}{\Delta t\, N} = \frac{\Delta N_1}{\Delta N} \frac{N}{N_0} \left(f_1(t) + O(\Delta t)\right) \\
&= \frac{f_1(t) + O(\Delta t)}{P(X(0) \in S_0)} \frac{\Delta N_1}{\Delta N}
\end{aligned} \tag{5–13}
$$

where $P(X(0) \in S_0) = \frac{N_0}{N}$ has been introduced.

In Fig. 5-1, the ΔN_1-realization has been artificially prolonged (with the dashed line) backwards until the time prior to $t = 0$, when it reaches the limit state function. Let L_t signify the time length that is spent in the safe domain before an out-crossing takes place at the time t. Among the total number ΔN of out-crossings in $]t, t + \Delta t]$, ΔN_1 realizations are then distinguished by the property that $L_t > t$. Hence, the fraction $\frac{\Delta N_1}{\Delta N}$ can be interpreted as an estimate of the probability $P(L_t > t)$, i.e., the time length L_t spent in the safe domain prior to an out-crossing at the time t is larger than t. In the limit of $\Delta t \to 0$, Eq. (5-13) can be written as:

$$
f_T(t) = \frac{f_1(t)}{P(X(0) \in S_0)} P(L_t > t) = \frac{f_1(t)}{P(X(0) \in S_0)} \left(1 - F_{L_t}(t)\right) \tag{5–14}
$$

where $F_{L_t}(l)$ is the probability distribution function of L_t, with l being the variable of the function.

If $\{X(t), t \in R\}$ is a stationary process and the limit state function $b(t)$ is constant with time, both the out-crossing frequency $f_1(t)$ and the probability distribution function $F_{L_t}(l)$ become independent of t. Introducing the designation $F_L(l)$ for the time-invariant probability distribution function, Eq. (5-14) can be simplified to:

$$f_T(t) = \frac{f_1}{P(X(0) \in S)} \left(1 - F_L(t)\right) \tag{5–15}$$

Integration of Eq. (5-15) from 0 to ∞ and using integration by parts yields:

$$1 = \int_0^\infty f_T(\tau) \, d\tau = \frac{f_1}{P(X(0) \in S)} \int_0^\infty \left(1 - F_L(\tau)\right) d\tau$$

$$= \frac{f_1}{P(X(0) \in S)} \left(\left[\tau\left(1 - F_L(\tau)\right)\right]_0^\infty + \int_0^\infty \tau f_L(\tau) \, d\tau \right) = \frac{f_1}{P(X(0) \in S)} E[L] \tag{5–16}$$

where $E[L] = \int_0^\infty \tau f_L(\tau) \, d\tau$ is the expected value of L, with $f_L(\tau)$ being the probability density function of L. Then, the first-passage probability density function in Eq. (5-15) can be written as:

$$f_T(t) = \frac{1}{E[L]} \left(1 - F_L(t)\right) \tag{5–17}$$

In section 6.4, it will be demonstrated how $P(X(0) \in S_0)$, $E[L]$ and $F_L(l)$ can be determined by ergodic sampling of a single sufficiently long time-series. Hence, Eq. (5-17) provides a solution for $f_T(t)$. Next, $F_T(t)$ can be determined by numerical integration of $f_T(t)$, and the (unconditional) probability of failure follows from Eq. (5-7). It should be noticed that the probability thus obtained is interpreted as the failure probability of a system with random initial value $X(0)$. If $X(0)$ is known, i.e., $X(0)$ is deterministic, $\{X(t), t \in [0, \infty[\}$ becomes non-stationary during the transient phase and Eq. (5-17) is no longer valid.

At $t = 0$, all out-crossings are first passages. Hence, we have $\Delta N = \Delta N_1$ in the first interval $[0, \Delta t]$. It follows directly from Eq. (5-13) that:

$$f_T(0) = \frac{f_1(0)}{P(X(0) \in S_0)} \tag{5–18}$$

Since the out-crossing frequency $f_1(0)$ and the initial *reliability* $P(X(0) \in S_0)$ under certain conditions can be evaluated analytically, $f_T(0)$ is the only point on the first-passage probability density curve that can be determined without Monte Carlo simulation.

5.2   Calculation of out-crossing frequency

As shown in Fig. 5-3a, the out-crossing frequency $f_1(t)$ can alternatively be expressed as:

$$f_1(t)\Delta t + O(\Delta t^2) = P(X(t) < b(t) \wedge X(t + \Delta t) \geq b(t + \Delta t))$$

$$= P\left(b(t) + (\dot{b}(t) - \dot{X}(t))\Delta t + O(\Delta t^2) \leq X(t) < b(t)\right) \tag{5–19}$$

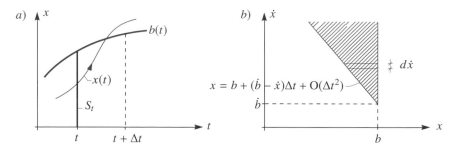

Fig. 5–3 a) Out-crossing in interval $]t, t + \Delta t]$. b) Calculation of out-crossing frequency.

where the 1st order Taylor expansions $X(t + \Delta t) = X(t) + \dot{X}(t)\Delta t + O(\Delta t^2)$ and $b(t + \Delta t) = b(t) + \dot{b}(t)\Delta t + O(\Delta t^2)$ have been applied. Based on the joint probability density function $f_{X\dot{X}}(x, \dot{x}, t)$ of $X(t)$ and $\dot{X}(t)$, the right-hand side of Eq. (5-19) can be evaluated as shown in Fig. 5-3b:

$$f_1(t)\Delta t + O(\Delta t^2) = \int_{b}^{\infty} \int_{b+(\dot{b}-\dot{x})\Delta t+O(\Delta t^2)}^{b} f_{X\dot{X}}(x, \dot{x}, t)\, dx d\dot{x}$$

$$= \int_{b}^{\infty} (\dot{x} - \dot{b})\Delta t\, f_{X\dot{X}}(b(t), \dot{x}, t)\, d\dot{x} + O(\Delta t^2)$$

$$\Rightarrow \quad f_1(t) = \int_{\dot{b}(t)}^{\infty} (\dot{x} - \dot{b}(t)) f_{X\dot{X}}(b(t), \dot{x}, t)\, d\dot{x} \tag{5–20}$$

Within an error $O(\Delta t^2)$, $X(t)$ may be fixed at the value $b(t)$ at the limit state function, from which the 2nd statement of Eq. (5-20) follows. Eq. (5-20) is known as *Rice's formula*.[1]

For the so-called *two-barrier problem* illustrated in Fig. 5-4, the safe domain at the time t is written as:

$$S_t = \{x \mid a(t) < x < b(t)\} \tag{5–21}$$

Since the realizations $x(t)$ are continuous, simultaneous out-crossings at the upper and lower barriers during the interval $]t, t+\Delta t]$ as $\Delta t \to 0$ cannot take place. Hence, these two out-crossing events must be independent, and the total out-crossing frequency is obtained as the sum of out-crossing frequencies at the upper and at the lower barrier. In this case, Eq. (5-20) is generalized to:

$$f_1(t) = \int_{\dot{b}(t)}^{\infty} (\dot{x} - \dot{b}(t)) f_{X\dot{X}}(b(t), \dot{x}, t)\, d\dot{x} + \int_{-\infty}^{\dot{a}(t)} (\dot{a}(t) - \dot{x}) f_{X\dot{X}}(a(t), \dot{x}, t)\, d\dot{x} \tag{5–22}$$

---

[1] S.O. Rice. Mathematical analysis of random noise. Reprinted in *Selected papers on noise and stochastic processes*. Dover Publications, Inc., New York, 133-294, 1954.

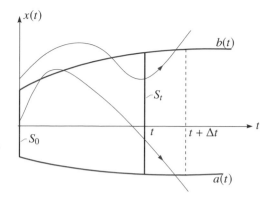

Fig. 5–4 Out-crossings in the two-barrier problem.

The safe domain for a stochastic vector process $\{\mathbf{X}(t), \ t \in [0, \infty[\}$ is defined by a failure surface Γ_t, which may expand or contract deterministically with time as indicated by the subscript t. At time t, a differential area element dA on the surface is considered, which is defined by the position vector $\mathbf{b}(t)$ and the unit normal vector in the outward direction $\mathbf{n}(t)$. $\mathbf{X}(t)$ is assumed in the vicinity of $\mathbf{b}(t)$ at the time t. As shown in Fig. 5-5, $\mathbf{X}(t)$ is in the safe domain S_t if $\left(\mathbf{X}(t) - \mathbf{b}(t)\right)^T \mathbf{n}(t) < 0$. Correspondingly, if an out-crossing through the area element dA takes place during the interval $]t, t + \Delta t]$, we have $\left(\mathbf{X}(t + \Delta t) - \mathbf{b}(t + \Delta t)\right)^T \mathbf{n}(t) > 0$ at the time $t + \Delta t$. Hence, the probability of an out-crossing through dA during the time interval $]t, t + \Delta t]$ can be written as:

$$df_1 \Delta t = P\left(\left(\mathbf{X}(t) - \mathbf{b}(t)\right)^T \mathbf{n}(t) < 0 \ \wedge \ \left(\mathbf{X}(t + \Delta t) - \mathbf{b}(t + \Delta t)\right)^T \mathbf{n}(t) > 0\right)$$

$$= P\left(\mathbf{b}(t)^T \mathbf{n}(t) - \Delta t \dot{X}_n(t) + O(\Delta t^2) \le \mathbf{X}(t)^T \mathbf{n}(t) < \mathbf{b}(t)^T \mathbf{n}(t)\right) \tag{5–23}$$

where

$$\dot{X}_n(t) = \left(\dot{\mathbf{X}}(t) - \dot{\mathbf{b}}(t)\right)^T \mathbf{n}(t) \tag{5–24}$$

At the derivation of Eq. (5-23), the 1st order Taylor expansion $\left(\mathbf{X}(t + \Delta t) - \mathbf{b}(t + \Delta t)\right)^T \mathbf{n}(t) = \left(\mathbf{X}(t) - \mathbf{b}(t)\right)^T \mathbf{n}(t) + \Delta t \left(\dot{\mathbf{X}}(t) - \dot{\mathbf{b}}(t)\right)^T \mathbf{n}(t) + O(\Delta t^2)$ has been applied. $\dot{X}_n(t)$ in Eq. (5-24) signifies

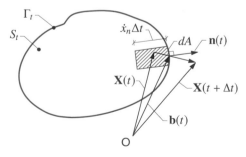

Fig. 5–5 Out-crossing of a stochastic vector process from time-varying safe domain.

the velocity of $\{\mathbf{X}(t), t \in [0, \infty[\}$ relative to the expanding/contracting failure surface in the direction of the normal vector $\mathbf{n}(t)$. From the last statement of Eq. (5-23), it is seen that the considered out-crossing event on condition of $\dot{X}_n(t) = \dot{x}_n > 0$ is tantamount to the case, where $\mathbf{X}(t)$ lies within a cylinder with the depth $\Delta t \dot{x}_n$ and the volume $dV = \Delta t \, \dot{x}_n \, dA$, corresponding to the hatched domain in Fig. 5-5. Within an error $O(\Delta t^2)$, $\mathbf{X}(t)$ may be fixed at the value $\mathbf{b}(t)$ at the failure surface when evaluating the probability integral over the volume dV. Hence, the out-crossing probability on condition of $\dot{X}_n(t) = \dot{x}_n$ is written as $dV f_{\mathbf{X}|\dot{X}_n}(\mathbf{b}(t) \mid \dot{x}_n, t) + O(\Delta t^2)$, where $f_{\mathbf{X}|\dot{X}_n}(\mathbf{x} \mid \dot{x}_n, t)$ is the joint probability density function of $\mathbf{X}(t)$ on condition of $\dot{X}_n(t) = \dot{x}_n$. Then, the unconditional probability of an out-crossing is given by the following integral with $\dot{x}_n$ taken from 0 to ∞:

$$
df_1 \Delta t + O(\Delta t^2) = \int_0^\infty dV \, f_{\mathbf{X}|\dot{X}_n}(\mathbf{b}(t) \mid \dot{x}_n, t) \, f_{\dot{X}_n}(\dot{x}_n, t) \, d\dot{x}_n
$$

$$
= \Delta t \int_0^\infty \dot{x}_n \, f_{\mathbf{X}\dot{X}_n}(\mathbf{b}(t), \dot{x}_n, t) \, d\dot{x}_n dA \tag{5-25}
$$

where $f_{\mathbf{X}\dot{X}_n}(\mathbf{x}, \dot{x}_n, t)$ signifies the joint probability density function of $\mathbf{X}(t)$ and $\dot{X}_n(t)$. The out-crossings during $]t, t + \Delta t]$ from disjoint differential area elements dA_1 and dA_2 are independent events. Hence, the probability of out-crossings from either dA_1 or dA_2 is the sum of probabilities of the out-crossing as given by Eq. (5-25). It follows that the probability of an out-crossing through the failure surface during $]t, t + \Delta t]$ can be obtained by the surface integral of the differential contributions in Eq. (5-25), which leads to the following expression for the out-crossing frequency of the stochastic vector process, known as *Belyaev's formula*:[1]

$$
f_1(t)\Delta t + O(\Delta t^2) = \Delta t \int_{\Gamma_t} \left(\int_0^\infty \dot{x}_n \, f_{\mathbf{X}\dot{X}_n}(\mathbf{b}(t), \dot{x}_n, t) \, d\dot{x}_n \right) dA
$$

$$
\Rightarrow \quad f_1(t) = \int_{\Gamma_t} \left(\int_0^\infty \dot{x}_n \, f_{\mathbf{X}\dot{X}_n}(\mathbf{b}(t), \dot{x}_n, t) \, d\dot{x}_n \right) dA \tag{5-26}
$$

Example 5-1: Out-crossing frequency of a stationary Gaussian process

The barrier (limit state function) b is assumed to be constant, and $\{X(t), \, t \in R\}$ is assumed to be a stationary Gaussian process with $f_{X\dot{X}}(x, \dot{x})$ given by Eq. (3-154). Since $\dot{b} \equiv 0$, Eq. (5-20) becomes:

$$
f_1 = \int_0^\infty \dot{x} \frac{1}{\sigma_X} \varphi \left(\frac{b - \mu_X}{\sigma_X} \right) \frac{1}{\sigma_{\dot{X}}} \varphi \left(\frac{\dot{x}}{\sigma_{\dot{X}}} \right) d\dot{x} = \frac{\sigma_{\dot{X}}}{\sigma_X} \varphi \left(\frac{b - \mu_X}{\sigma_X} \right) \int_0^\infty u \, \varphi(u) \, du
$$

$$
= \frac{1}{2\pi} \frac{\sigma_{\dot{X}}}{\sigma_X} \exp\left(-\frac{1}{2} \left(\frac{b - \mu_X}{\sigma_X} \right)^2 \right) \tag{5-27}
$$

Therefore, f_1 depends on the statistical moments μ_X, σ_X and $\sigma_{\dot{X}}$. Especially for the special case of $b = \mu_X$, the expected number of up-crossings per unit time with respect to the mean value μ_X, designated as ν_0, is obtained. Introducing the spectral moments in Eq. (3-48), we have:

$$
\nu_0 = \frac{1}{2\pi} \frac{\sigma_{\dot{X}}}{\sigma_X} = \frac{1}{2\pi} \sqrt{\frac{\lambda_2}{\lambda_0}} \tag{5-28}
$$

---

[1] Y.K. Belyaev. On the number of exits across the boundary of a region by a vector stochastic process. *Theory of Probability and its Applications* 1968; **13**: 320-324.

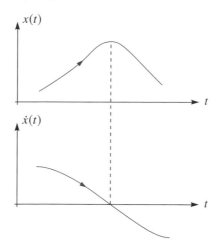

Fig. 5–6 Local maximum of the displacement process and simultaneous down-crossing of the velocity process.

Example 5-2: Expected number of local maxima per unit time

A *local maximum* of the process $\{X(t), t \in [0, \infty[\}$ is characterized by a down-crossing of the constant barrier $a \equiv 0$ of the velocity process $\{\dot{X}(t), t \in [0, \infty[\}$, so $\dot{X}(t) = 0$ at the local maximum point, as shown in Fig. 5-6. The expected number of local maxima per unit time, designated as $\mu_0(t)$, can thus be calculated by Eq. (5-22), replacing $[X(t), \dot{X}(t)]$ with $[\dot{X}(t), \ddot{X}(t)]$, and using $b(t) = \infty$ and $a(t) \equiv 0$:

$$\mu_0(t) = -\int_{-\infty}^{0} \ddot{x} \, f_{\dot{X}\ddot{X}}(0, \ddot{x}, t) \, d\ddot{x} \qquad (5\text{–}29)$$

If $\{X(t), t \in R\}$ is stationary and Gaussian, the joint probability density function becomes:

$$f_{\dot{X}\ddot{X}}(\dot{x}, \ddot{x}) = \frac{1}{\sigma_{\dot{X}}} \varphi\left(\frac{\dot{x}}{\sigma_{\dot{X}}}\right) \frac{1}{\sigma_{\ddot{X}}} \varphi\left(\frac{\ddot{x}}{\sigma_{\ddot{X}}}\right) \qquad (5\text{–}30)$$

In this case, $\mu_0(t)$ becomes constant, which by complete analogy with Eq. (5-28) can be written as:

$$\mu_0 = \frac{1}{2\pi} \frac{\sigma_{\ddot{X}}}{\sigma_{\dot{X}}} = \frac{1}{2\pi} \sqrt{\frac{\lambda_4}{\lambda_2}} \qquad (5\text{–}31)$$

Next, the bandwidth parameter γ and ε as given by Eqs. (3-46) and (3-47) can be calculated analytically using Eqs. (5-28) and (5-31):

$$\gamma = \frac{\mu_0}{\nu_0} = \sqrt{\frac{\lambda_0 \lambda_4}{\lambda_2^2}} \qquad (5\text{–}32)$$

$$\varepsilon = \sqrt{1 - \frac{\lambda_2^2}{\lambda_0 \lambda_4}} \qquad (5\text{–}33)$$

It should be noted that Eqs. (5-32) and (5-33) can only be used for stationary Gaussian processes for which the acceleration process exists. In contrast, Vanmarcke's bandwidth parameter in Eq. (3-49) can be applied to arbitrary stationary processes for which the velocity process exists (i.e. $\lambda_2 < \infty$).

Example 5-3: Out-crossing frequency from a rectangular time-invariant safe domain

As shown in Fig. 5-7, the safe domain is written as:

$$S = \left\{ (x_1, x_2) \mid a_1 < x_1 < b_1 \wedge a_2 < x_2 < b_2 \right\} \tag{5–34}$$

where a_1, b_1, a_2, b_2 are time-invariant barriers. The out-crossing frequency of the stochastic vector process $\{\mathbf{X}(t), \, t \in [0, \infty[\}$, $\mathbf{X}(t) = \left[X_1(t), X_2(t) \right]^T$, with respect to the indicated safe domain, is to be specified.

The surface Γ of S is divided into four sub-surfaces as shown in Fig. 5-7. The out-crossing contributions from each of these four sub-surfaces are determined below.

Sub-surface 1:

$$\left. \begin{aligned} \mathbf{n} &= \begin{bmatrix} 1 \\ 0 \end{bmatrix}, \ \mathbf{b} = \begin{bmatrix} b_1 \\ x_2 \end{bmatrix}, \ \dot{X}_n(t) = \mathbf{n}^T \dot{\mathbf{X}}(t) = \dot{X}_1(t) \\ f_{\mathbf{X}\dot{X}_n}(\mathbf{b}, \dot{x}_n, t) &= f_{X_1 X_2 \dot{X}_1}(b_1, x_2, \dot{x}_n, t) \end{aligned} \right\} \tag{5-35a}$$

Sub-surface 2:

$$\left. \begin{aligned} \mathbf{n} &= \begin{bmatrix} -1 \\ 0 \end{bmatrix}, \ \mathbf{b} = \begin{bmatrix} a_1 \\ x_2 \end{bmatrix}, \ \dot{X}_n(t) = \mathbf{n}^T \dot{\mathbf{X}}(t) = -\dot{X}_1(t) \\ f_{\mathbf{X}\dot{X}_n}(\mathbf{b}, \dot{x}_n, t) &= f_{X_1 X_2 \dot{X}_1}(a_1, x_2, -\dot{x}_n, t) \end{aligned} \right\} \tag{5-35b}$$

Sub-surface 3:

$$\left. \begin{aligned} \mathbf{n} &= \begin{bmatrix} 0 \\ 1 \end{bmatrix}, \ \mathbf{b} = \begin{bmatrix} x_1 \\ b_2 \end{bmatrix}, \ \dot{X}_n(t) = \mathbf{n}^T \dot{\mathbf{X}}(t) = \dot{X}_2(t) \\ f_{\mathbf{X}\dot{X}_n}(\mathbf{b}, \dot{x}_n, t) &= f_{X_1 X_2 \dot{X}_2}(x_1, b_2, \dot{x}_n, t) \end{aligned} \right\} \tag{5-35c}$$

Sub-surface 4:

$$\left. \begin{aligned} \mathbf{n} &= \begin{bmatrix} 0 \\ -1 \end{bmatrix}, \ \mathbf{b} = \begin{bmatrix} x_1 \\ a_2 \end{bmatrix}, \ \dot{X}_n(t) = \mathbf{n}^T \dot{\mathbf{X}}(t) = -\dot{X}_2(t) \\ f_{\mathbf{X}\dot{X}_n}(\mathbf{b}, \dot{x}_n, t) &= f_{X_1 X_2 \dot{X}_2}(x_1, a_2, -\dot{x}_n, t) \end{aligned} \right\} \tag{5-35d}$$

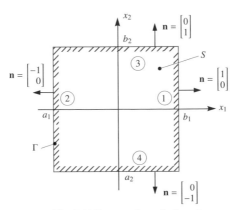

Fig. 5–7 Rectangular safe domain.

In Eqs. (5-35b) and (5-35d), the following identity has been applied to the velocity component:

$$f_{-\dot{X}}(\dot{x}) = f_{\dot{X}}(-\dot{x}) \tag{5–36}$$

By the use of Eq. (5-35), Eq. (5-26) attains the form:

$$
\begin{aligned}
f_1(t) = {} & \int_{a_2}^{b_2} \left(\int_0^\infty \dot{x}_n\, f_{X_1 X_2 \dot{X}_1}(b_1, x_2, \dot{x}_n, t)\, d\dot{x}_n \right) dx_2 \\
& + \int_{a_2}^{b_2} \left(\int_0^\infty \dot{x}_n\, f_{X_1 X_2 \dot{X}_1}(a_1, x_2, -\dot{x}_n, t)\, d\dot{x}_n \right) dx_2 \\
& + \int_{a_1}^{b_1} \left(\int_0^\infty \dot{x}_n\, f_{X_1 X_2 \dot{X}_2}(x_1, b_2, \dot{x}_n, t)\, d\dot{x}_n \right) dx_1 \\
& + \int_{a_1}^{b_1} \left(\int_0^\infty \dot{x}_n\, f_{X_1 X_2 \dot{X}_2}(x_1, a_2, -\dot{x}_n, t)\, d\dot{x}_n \right) dx_1
\end{aligned}
\tag{5–37}
$$

Example 5-4: Out-crossing frequency of a stationary Gaussian vector process from a time-invariant safe domain

Assume that $\{\mathbf{X}(t),\, t \in R\}$ is a stationary Gaussian vector process and the safe domain is time-invariant. Then, $\{\dot{\mathbf{X}}(t),\, t \in R\}$ and $\{\dot{X}_n(t),\, t \in R\}$ become stationary Gaussian as well. As follows from Eq. (5-26), the joint probability density of $\mathbf{X}(t)$ and $\dot{X}_n(t)$ at $\mathbf{x} = \mathbf{b}$ on the failure surface is needed. This can be written as:

$$f_{\mathbf{X}\dot{X}_n}(\mathbf{b}, \dot{x}_n) = f_{\dot{X}_n \mid \mathbf{X}}(\dot{x}_n \mid \mathbf{b})\, f_{\mathbf{X}}(\mathbf{b}) = \frac{1}{\sigma}\, \varphi\!\left(\frac{\dot{x}_n - \mu}{\sigma}\right) f_{\mathbf{X}}(\mathbf{b}) \tag{5–38}$$

$f_{\mathbf{X}}(\mathbf{b})$ is given by, cf. Eq. (2-15):

$$f_{\mathbf{X}}(\mathbf{b}) = \frac{1}{(2\pi)^{\frac{n}{2}} \left(\det(\mathbf{C}_{\mathbf{XX}}) \right)^{\frac{1}{2}}} \exp\!\left(-\frac{1}{2}(\mathbf{b} - \boldsymbol{\mu}_{\mathbf{X}})^T \mathbf{C}_{\mathbf{XX}}^{-1}(\mathbf{b} - \boldsymbol{\mu}_{\mathbf{X}})\right) \tag{5–39}$$

where $\boldsymbol{\mu}_{\mathbf{X}} = E\big[\mathbf{X}(t)\big]$ and $\mathbf{C}_{\mathbf{XX}}$ denote the mean value vector and the zero time-lag covariance matrix of $\mathbf{X}(t)$, respectively.

$\mu = \mu(\mathbf{b})$ and $\sigma = \sigma(\mathbf{b})$ in Eq. (5-38) signify the expected value and the standard deviation of $\dot{X}_n(t)$ on condition of $\mathbf{X}(t) = \mathbf{b}$. Using Eqs. (2-75), (2-76) in combination with $E\big[\dot{X}_n(t)\big] = E\big[\dot{\mathbf{X}}(t) - \mathbf{b}\big]^T \mathbf{n}(\mathbf{b}) = 0$, these two quantities are given by the following expressions:

$$\mu(\mathbf{b}) = \mathbf{n}^T(\mathbf{b}) \mathbf{C}_{\mathbf{X}\dot{\mathbf{X}}}^T \mathbf{C}_{\mathbf{XX}}^{-1}\left(\mathbf{b} - \boldsymbol{\mu}_{\mathbf{X}}\right) \tag{5–40}$$

$$\sigma^2(\mathbf{b}) = \mathbf{n}^T(\mathbf{b})\left(\mathbf{C}_{\dot{\mathbf{X}}\dot{\mathbf{X}}} - \mathbf{C}_{\mathbf{X}\dot{\mathbf{X}}}^T \mathbf{C}_{\mathbf{XX}}^{-1}\mathbf{C}_{\mathbf{X}\dot{\mathbf{X}}}\right)\mathbf{n}(\mathbf{b}) \tag{5–41}$$

where $\mathbf{C}_{\dot{\mathbf{X}}\dot{\mathbf{X}}}$ signifies the zero time-lag covariance matrix of $\dot{\mathbf{X}}(t)$, and $\mathbf{C}_{\mathbf{X}\dot{\mathbf{X}}}$ is the zero time-lag cross-covariance matrix of $\mathbf{X}(t)$ and $\dot{\mathbf{X}}(t)$.

Inserting Eq. (5-38) into Eq. (5-26) provides:

$$f_1 = \int_\Gamma \left(\int_0^\infty \frac{\dot{x}_n}{\sigma}\, \varphi\!\left(\frac{\dot{x}_n - \mu}{\sigma}\right) d\dot{x}_n \right) f_{\mathbf{X}}(\mathbf{b})\, dA \tag{5–42}$$

Using integration by parts, the innermost integral of Eq. (5-42) can be evaluated as:

$$\int_0^\infty \frac{\dot{x}_n}{\sigma} \varphi\left(\frac{\dot{x}_n - \mu}{\sigma}\right) d\dot{x}_n = \sigma\varphi\left(\frac{\mu}{\sigma}\right) + \mu\Phi\left(\frac{\mu}{\sigma}\right) \tag{5-43}$$

Then, Eq. (5-42) becomes:

$$f_1 = \int_\Gamma \left(\sigma(\mathbf{b})\,\varphi\left(\frac{\mu(\mathbf{b})}{\sigma(\mathbf{b})}\right) + \mu(\mathbf{b})\,\Phi\left(\frac{\mu(\mathbf{b})}{\sigma(\mathbf{b})}\right)\right) f_{\mathbf{X}}(\mathbf{b})\, dA \tag{5-44}$$

Because of the complicated dependency of the integrand on the coordinates $\mathbf{b}$ of the surface element dA (on the failure surface), the surface integral Eq. (5-44) has to be evaluated numerically in most cases. Veneziano *et al.*[1] have indicated some analytical results for simple geometries of the failure surface and simplified correlation structures as determined by the matrices $\mathbf{C_{XX}}$, $\mathbf{C_{X\dot{X}}}$, and $\mathbf{C_{\dot{X}\dot{X}}}$.

Example 5-5: Out-crossing frequency of the normal stress process in a Bernoulli-Euler beam

The normal stress $S(t)$ at a certain cross-section of a Bernoulli-Euler beam is determined by *Navier's formula*:

$$S(t) = \frac{N(t)}{A} + \frac{M(t)}{W} \tag{5-45}$$

where A is the area of the cross-section, and W is the *sectional modulus*. The yield stress is assumed to be equal to s_y in both compression and tension. The axial force $N(t)$ and the bending moment $M(t)$ are modelled as stationary Gaussian stochastic processes $\{N(t),\ t \in R\}$ and $\{M(t),\ t \in R\}$, respectively, with the mean values μ_N, μ_M, the auto-covariance functions $\kappa_{NN}(\tau)$, $\kappa_{MM}(\tau)$, and the cross-covariance function $\kappa_{MN}(\tau)$. A typically realization of the normal stress process $\{S(t),\ t \in R\}$ from a lightly damped beam structure has been shown in Fig. 5-8. As seen, the mean value of the stress is negative corresponding to a compressive normal force.

$\{S(t),\ t \in R\}$ becomes a stationary Gaussian process, because $\{N(t),\ t \in R\}$ and $\{M(t),\ t \in R\}$ are stationary Gaussian processes. Then, $S(t) \sim N(\mu_S, \sigma_S^2)$ and $\dot{S}(t) \sim N(0, \sigma_{\dot{S}}^2)$ become independent

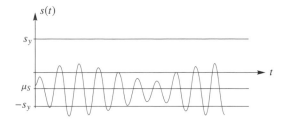

Fig. 5–8 Realization of the narrow-banded normal stress process.

[1]D. Veneziano, M. Grigoriu and C.A. Cornell. Vector-process models for system reliability. *Journal of the Engineering Mechanics Division* 1977; **103** (3): 441-460.

stochastic variables, cf. Eq. (3-154). Further, the involved statistical moments are given by:

$$
\left.\begin{aligned}
\mu_S &= \frac{\mu_N}{A} + \frac{\mu_M}{W} \\
\sigma_S^2 &= \frac{\sigma_N^2}{A^2} + 2\frac{\kappa_{MN}(0)}{AW} + \frac{\sigma_M^2}{W^2} \\
\sigma_{\dot{S}}^2 &= \frac{\sigma_{\dot{N}}^2}{A^2} + 2\frac{\kappa_{\dot{M}\dot{N}}(0)}{AW} + \frac{\sigma_{\dot{M}}^2}{W^2}
\end{aligned}\right\}
\tag{5-46}
$$

where:

$$
\left.\begin{aligned}
\sigma_N^2 &= \kappa_{NN}(0) \\
\sigma_M^2 &= \kappa_{MM}(0) \\
\sigma_{\dot{N}}^2 &= -\frac{d^2}{d\tau^2}\kappa_{NN}(0) \\
\sigma_{\dot{M}}^2 &= -\frac{d^2}{d\tau^2}\kappa_{MM}(0) \\
\kappa_{\dot{M}\dot{N}}(0) &= -\frac{d^2}{d\tau^2}\kappa_{MN}(0)
\end{aligned}\right\}
\tag{5-47}
$$

From Eqs. (5-22) and (5-27), the out-crossing frequency becomes:

$$
f_1 = \frac{1}{2\pi}\frac{\sigma_{\dot{S}}}{\sigma_S}\left(\exp\left(-\frac{1}{2}\left(\frac{s_y - \mu_S}{\sigma_S}\right)^2\right) + \exp\left(-\frac{1}{2}\left(\frac{s_y + \mu_S}{\sigma_S}\right)^2\right)\right)
\tag{5-48}
$$

Next, it is assumed that failure of the beam may also take place due to buckling. The critical buckling load is denoted n_y. The reliability problem is formulated as a out-crossing problem for the vector process $\mathbf{X}(t) = [S(t), N(t)]^T$ with the safe domain shown in Fig. 5-9, which can be interpreted as a rectangular domain with $a_1 = -s_y$, $b_1 = s_y$, $a_2 = -n_y$ and $b_2 = \infty$. From Eq. (5-37), the out-crossing frequency

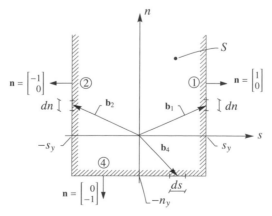

Fig. 5-9 Definition of safe domain. Combined yield limit and buckling failure criterion.

becomes:

$$
\begin{aligned}
f_1(t) = & \int_{-n_y}^{\infty} \left(\int_0^{\infty} \dot{s}\, f_{SN\dot{S}}(s_y, n, \dot{s})\, d\dot{s} \right) dn \\
& + \int_{-n_y}^{\infty} \left(\int_0^{\infty} \dot{s}\, f_{SN\dot{S}}(-s_y, n, -\dot{s})\, d\dot{s} \right) dn \\
& + \int_{-s_y}^{s_y} \left(\int_0^{\infty} \dot{n}\, f_{SN\dot{N}}(s, -n_y, -\dot{n})\, d\dot{n} \right) ds
\end{aligned}
\tag{5-49}
$$

The contribution to the out-crossing frequency from the sub-surface 3 with $b_2 = \infty$ is equal to 0, because $f_{SN\dot{N}}(s, \infty, \dot{n}) = 0$.

In analogy to Eq. (5-38), the joint probability density functions entering Eq. (5-49) are reformulated in the form:

$$
\left.
\begin{aligned}
f_{SN\dot{S}}(s, n, \dot{s}) &= f_{\dot{S}|SN}(\dot{s} \mid s, n)\, f_{SN}(s, n) = \frac{1}{\sigma_1} \varphi\left(\frac{\dot{s} - \mu_1}{\sigma_1} \right) f_{SN}(s, n) \\
f_{SN\dot{N}}(s, n, \dot{n}) &= f_{\dot{N}|SN}(\dot{n} \mid s, n)\, f_{SN}(s, n) = \frac{1}{\sigma_4} \varphi\left(\frac{\dot{n} - \mu_4}{\sigma_4} \right) f_{SN}(s, n)
\end{aligned}
\right\}
\tag{5-50}
$$

Therefore, $[\dot{S}(t), S(t), N(t)]^T \sim N(\boldsymbol{\mu}_1, \mathbf{C}_1)$ and $[\dot{N}(t), S(t), N(t)]^T \sim N(\boldsymbol{\mu}_4, \mathbf{C}_4)$, with the related parameters defined as:

$$
\left.
\begin{aligned}
\boldsymbol{\mu}_1 = \boldsymbol{\mu}_4 &= \begin{bmatrix} 0 \\ \mu_S \\ \mu_N \end{bmatrix} \\[2em]
\mathbf{C}_1 &= \begin{bmatrix} \sigma_{\dot{S}}^2 & 0 & \kappa_{\dot{S}N}(0) \\ 0 & \sigma_S^2 & \kappa_{SN}(0) \\ \kappa_{\dot{S}N}(0) & \kappa_{SN}(0) & \sigma_N^2 \end{bmatrix} \\[2em]
\mathbf{C}_4 &= \begin{bmatrix} \sigma_{\dot{N}}^2 & \kappa_{S\dot{N}}(0) & 0 \\ \kappa_{S\dot{N}}(0) & \sigma_S^2 & \kappa_{SN}(0) \\ 0 & \kappa_{SN}(0) & \sigma_N^2 \end{bmatrix}
\end{aligned}
\right\}
\tag{5-51}
$$

where $\mu_S, \sigma_S^2, \sigma_{\dot{S}}^2$ are given by Eq. (5-46). Further, it has been applied that $E\left[S(t)\dot{S}(t)\right] = E\left[N(t)\dot{N}(t)\right] = 0$ due to the stationarity of the involved stochastic processes. The remaining covariances in Eq. (5-51) can be written as:

$$
\left.
\begin{aligned}
\kappa_{SN}(0) &= E\left[\left(\frac{N(t)}{A} + \frac{M(t)}{W} \right) N(t) \right] = \frac{\sigma_N^2}{A} + \frac{\kappa_{MN}(0)}{W} \\
\kappa_{\dot{S}N}(0) &= E\left[\left(\frac{\dot{N}(t)}{A} + \frac{\dot{M}(t)}{W} \right) N(t) \right] = \frac{\kappa_{\dot{M}N}(0)}{W} = -\frac{1}{W} \frac{d}{d\tau} \kappa_{MN}(0) \\
\kappa_{S\dot{N}}(0) &= E\left[\left(\frac{N(t)}{A} + \frac{M(t)}{W} \right) \dot{N}(t) \right] = \frac{\kappa_{M\dot{N}}(0)}{W} = \frac{1}{W} \frac{d}{d\tau} \kappa_{MN}(0)
\end{aligned}
\right\}
\tag{5-52}
$$

$f_{SN}(s, n)$ is the joint probability density function of the two-dimensional normally distributed stochastic vector $[S(t), N(t)]^T \sim N(\boldsymbol{\mu}, \mathbf{C})$, where:

$$\left. \begin{array}{l} \boldsymbol{\mu} = \begin{bmatrix} \mu_S \\ \mu_N \end{bmatrix} \\[12pt] \mathbf{C} = \begin{bmatrix} \sigma_S^2 & \kappa_{SN}(0) \\ \kappa_{SN}(0) & \sigma_N^2 \end{bmatrix} \end{array} \right\} \tag{5–53}$$

Let $\mu_1 = \mu_1(s, n)$ and σ_1 signify the expected value and the standard deviation of $\dot{S}(t)$ on condition of $[S(t), N(t)] = [s, n]$, and $\mu_4 = \mu_4(s, n)$ and σ_4 are the expected value and the standard deviation of $\dot{N}(t)$ on condition of $[S(t), N(t)] = [s, n]$. From Eqs. (2-75), (2-76) and (5-51), these four parameters can be expressed as:

$$\left. \begin{array}{l} \mu_1(s, n) = \begin{bmatrix} 0 \\ \kappa_{\dot{S}N}(0) \end{bmatrix}^T \begin{bmatrix} \sigma_S^2 & \kappa_{SN}(0) \\ \kappa_{SN}(0) & \sigma_N^2 \end{bmatrix}^{-1} \begin{bmatrix} s - \mu_S \\ n - \mu_N \end{bmatrix} \\[18pt] \sigma_1^2 = \sigma_{\dot{S}}^2 - \begin{bmatrix} 0 \\ \kappa_{\dot{S}N}(0) \end{bmatrix}^T \begin{bmatrix} \sigma_S^2 & \kappa_{SN}(0) \\ \kappa_{SN}(0) & \sigma_N^2 \end{bmatrix}^{-1} \begin{bmatrix} 0 \\ \kappa_{\dot{S}N}(0) \end{bmatrix} \\[18pt] \mu_4(s, n) = \begin{bmatrix} \kappa_{S\dot{N}}(0) \\ 0 \end{bmatrix}^T \begin{bmatrix} \sigma_S^2 & \kappa_{SN}(0) \\ \kappa_{SN}(0) & \sigma_N^2 \end{bmatrix}^{-1} \begin{bmatrix} s - \mu_S \\ n - \mu_N \end{bmatrix} \\[18pt] \sigma_4^2 = \sigma_{\dot{N}}^2 - \begin{bmatrix} \kappa_{S\dot{N}}(0) \\ 0 \end{bmatrix}^T \begin{bmatrix} \sigma_S^2 & \kappa_{SN}(0) \\ \kappa_{SN}(0) & \sigma_N^2 \end{bmatrix}^{-1} \begin{bmatrix} \kappa_{S\dot{N}}(0) \\ 0 \end{bmatrix} \end{array} \right\} \tag{5–54}$$

Then, using the result in Eq. (5-43), Eq. (5-49) can be written in the following form:

$$\begin{aligned} f_1 &= \int_{-n_y}^{\infty} \left(\sigma_1 \varphi\left(\frac{\mu_1(s_y, n)}{\sigma_1}\right) + \mu_1(s_y, n) \Phi\left(\frac{\mu_1(s_y, n)}{\sigma_1}\right) \right) f_{SN}(s_y, n)\, dn \\ &\quad + \int_{-n_y}^{\infty} \left(\sigma_1 \varphi\left(\frac{\mu_1(-s_y, n)}{\sigma_1}\right) - \mu_1(-s_y, n) \Phi\left(-\frac{\mu_1(-s_y, n)}{\sigma_1}\right) \right) f_{SN}(-s_y, n)\, dn \\ &\quad + \int_{-s_y}^{s_y} \left(\sigma_4 \varphi\left(\frac{\mu_4(s, -n_y)}{\sigma_4}\right) - \mu_4(s, -n_y) \Phi\left(-\frac{\mu_4(s, -n_y)}{\sigma_4}\right) \right) f_{SN}(s, -n_y)\, ds \end{aligned} \tag{5–55}$$

The one-dimensional quadratures in Eq. (5-55) must be evaluated numerically.

Finally, the axial force and the bending moment are assumed to be given by the following single term modal expansions:

$$\left. \begin{array}{l} N(t) \simeq \mu_N + N_1 Q_1(t) \\ M(t) \simeq \mu_M + M_1 Q_1(t) \end{array} \right\} \tag{5–56}$$

where the mean values μ_N and μ_M indicate the axial force and the bending moment from the mean value (the static part) of the external load process, respectively. $\{Q_1(t), t \in R\}$ is the first modal coordinate process caused by the zero-mean dynamic part of the load process, which means $\mu_{Q_1} = 0$. N_1 and

M_1 signify the axial force and the bending moment, when the structure is deformed in the first mode shape.

In case of a single mode approximation, the normal stress process becomes:

$$S(t) = \mu_S + S_1 Q_1(t) \tag{5-57}$$

$$S_1 = \frac{N_1}{A} + \frac{M_1}{W} \tag{5-58}$$

where S_1 is interpreted as the normal stress when the structure is deformed in the first mode shape.

The safe domain S with respect to the modal coordinate $Q_1(t)$ is given by:

$$S = \{q \mid -s_y < \mu_S + S_1 q < s_y \ \wedge \ \mu_N + N_1 q > -n_y\} = [a, b] \tag{5-59}$$

where the lower and upper barriers are:

$$a = \begin{cases} \max\left(-\dfrac{s_y + \mu_S}{S_1}, \ -\dfrac{n_y + \mu_N}{N_1}\right), & S_1 > 0 \ \wedge \ N_1 > 0 \\[2mm] -\dfrac{s_y + \mu_S}{S_1}, & S_1 > 0 \ \wedge \ N_1 < 0 \\[2mm] \max\left(\dfrac{s_y - \mu_S}{S_1}, \ -\dfrac{n_y + \mu_N}{N_1}\right), & S_1 < 0 \ \wedge \ N_1 > 0 \\[2mm] \dfrac{s_y - \mu_S}{S_1}, & S_1 < 0 \ \wedge \ N_1 < 0 \end{cases} \tag{5-60}$$

$$b = \begin{cases} \dfrac{s_y - \mu_S}{S_1}, & S_1 > 0 \ \wedge \ N_1 > 0 \\[2mm] \min\left(\dfrac{s_y - \mu_S}{S_1}, \ -\dfrac{n_y + \mu_N}{N_1}\right), & S_1 < 0 \ \wedge \ N_1 > 0 \\[2mm] -\dfrac{s_y + \mu_S}{S_1}, & S_1 > 0 \ \wedge \ N_1 > 0 \\[2mm] \min\left(-\dfrac{s_y + \mu_S}{S_1}, \ -\dfrac{n_y + \mu_N}{N_1}\right), & S_1 < 0 \ \wedge \ N_1 < 0 \end{cases} \tag{5-61}$$

The case where only the stress criterion is active, can be obtained with $n_y = \infty$ in Eqs. (5-60) and (5-61). In any case, the out-crossing frequency can be written as:

$$f_1 = \frac{1}{2\pi} \frac{\sigma_{\dot{Q}_1}}{\sigma_{Q_1}} \left(\exp\left(-\frac{1}{2} \frac{b^2}{\sigma_{Q_1}^2}\right) + \exp\left(-\frac{1}{2} \frac{a^2}{\sigma_{Q_1}^2}\right) \right) \tag{5-62}$$

Generally, when the response of the structure can be described by a single modal coordinate, the out-crossing problem reduces to a one-dimensional crossing problem as illustrated above. Further, if an n-dimensional system can be reduced to an n_1-dimensional system, $n_1 < n_1$, by means of a truncated modal expansion, the out-crossing problem can be reduced to a crossing problem of an n_1-dimensional vector process made up of the retained modal coordinates. The safe domain of the modal vector process is the set in the modal subspace, which is mapped on to the safe domain in the physical coordinate space using the truncated modal matrix as the transformation matrix.

Example 5-6: Out-crossing frequency of an isotropic stationary Gaussian vector process through the surface of an n-dimensional sphere

A stationary Gaussian vector process $\{\mathbf{X}(t), t \in R\}$ is said to be isotropic if the mean values $\mu_{X_j} = 0$ and the cross-covariance functions are written as $\kappa_{X_j X_k}(\tau) = \rho(\tau)\delta_{jk}$, where $\rho(\tau)$ signifies the auto-correlation coefficient function of all components, and δ_{jk} is Kronecker's delta. For ease, all component processes are normalized to have unit variance as shown below, but a similar analysis can be performed when all component processes have the same variance σ_0^2 rather than one. The out-crossing frequency of the process through the surface Γ of an n-dimensional sphere with the radius r is to be determined.

The surface area A_n of an n-dimensional sphere is given by:

$$A_n = 2 \frac{\pi^{\frac{n}{2}}}{\Gamma(\frac{n}{2})} r^{n-1} \tag{5–63}$$

where $\Gamma(x)$ is the *Gamma function*, and r is the radius of an n-dimensional sphere with its centre at $\mathbf{x} = \mathbf{0}$. For any point that lies on the surface, the following expression holds:

$$|\mathbf{x}|^2 = x_1^2 + \cdots + x_n^2 = r^2 \tag{5–64}$$

According to Eqs. (3-136) and (3-138), the following zero time-lag covariances may be calculated as:

$$\left.\begin{aligned}
E\left[X_j(t)\dot{X}_k(t)\right] &= \frac{\partial}{\partial t_2} E\left[X_j(t_1)X_k(t_2)\right]\bigg|_{t_1=t_2=t} = \delta_{jk}\frac{d\rho(\tau)}{d\tau}\bigg|_{\tau=0} = 0 \\
E\left[\dot{X}_j(t)\dot{X}_k(t)\right] &= \frac{\partial^2}{\partial t_1 \partial t_2} E\left[X_j(t_1)X_k(t_2)\right]\bigg|_{t_1=t_2=t} = -\delta_{jk}\frac{d^2\rho(\tau)}{d\tau^2}\bigg|_{\tau=0} = \delta_{jk}\omega_0^2
\end{aligned}\right\} \tag{5–65}$$

where from Eq. (3-50):

$$\omega_0^2 = -\frac{d^2\rho(\tau)}{d\tau^2}\bigg|_{\tau=0} \tag{5–66}$$

Then, the zero time-lag covariance (or cross-covariance) matrices $\mathbf{C_{XX}}$, $\mathbf{C_{X\dot{X}}}$ and $\mathbf{C_{\dot{X}\dot{X}}}$ become:

$$\mathbf{C_{XX}} = \mathbf{I}, \quad \mathbf{C_{X\dot{X}}} = \mathbf{0}, \quad \mathbf{C_{\dot{X}\dot{X}}} = \omega_0^2 \mathbf{I} \tag{5–67}$$

where $\mathbf{I}$ and $\mathbf{0}$ indicate the unit matrix and zero matrix of dimension $n \times n$, respectively.

With $\mathbf{b} = \mathbf{x}$, the components of the outward directed unit vector from the surface Γ become:

$$n_j(\mathbf{x}) = \frac{x_j}{r} \tag{5–68}$$

Eqs. (5-40) and (5-41) can thus be written as:

$$\left.\begin{aligned}
\mu(\mathbf{x}) &= \mathbf{n}^T(\mathbf{x})\, \mathbf{C}_{\mathbf{X\dot{X}}}^T \mathbf{C}_{\mathbf{XX}}^{-1}\left(\mathbf{x} - \mathbf{0}\right) = 0 \\
\sigma^2(\mathbf{x}) &= \mathbf{n}^T(\mathbf{x})\left(\mathbf{C}_{\mathbf{\dot{X}\dot{X}}} - \mathbf{C}_{\mathbf{X\dot{X}}}^T \mathbf{C}_{\mathbf{XX}}^{-1}\mathbf{C}_{\mathbf{X\dot{X}}}\right)\mathbf{n}(\mathbf{x}) = \mathbf{n}^T(\mathbf{x})\,\omega_0^2\,\mathbf{I}\,\mathbf{n}(\mathbf{x}) = \omega_0^2
\end{aligned}\right\} \tag{5–69}$$

Hence, $\sigma^2(\mathbf{x})$ is independent of $\mathbf{x}$, and the distribution of $\dot{X}_n(t)$ on condition of $\mathbf{X}(t) = \mathbf{x}$ becomes $\dot{X}_n(t) \sim N(0, \omega_0^2)$. From Eq. (5-42), we have:

$$f_1 = \int_\Gamma \left(\int_0^\infty \frac{\dot{x}_n}{\omega_0}\,\varphi\left(\frac{\dot{x}_n}{\omega_0}\right) d\dot{x}_n\right) f_{\mathbf{X}}(\mathbf{x})\, dA = \frac{\omega_0}{\sqrt{2\pi}} \int_\Gamma f_{\mathbf{X}}(\mathbf{x})\, dA \tag{5–70}$$

Table 5-1 Normalized surface area and out-crossing frequency as a function of the dimension n.

n	$\Gamma\left(\frac{n}{2}\right)$	$\frac{A_n}{r^{n-1}}$	$\frac{f_1}{\omega_0\exp(-\frac{1}{2}r^2)}$
1	$\sqrt{\pi}$	2	$\frac{\sqrt{2}}{\pi}$
2	1	2π	$\frac{1}{\sqrt{\pi}}\,r$
3	$\frac{1}{2}\sqrt{\pi}$	4π	$\frac{\sqrt{2}}{\pi}\,r^2$
4	1	$2\pi^2$	$\frac{1}{2\sqrt{\pi}}\,r^3$
5	$\frac{3}{4}\sqrt{\pi}$	$\frac{8}{3}\pi^2$	$\frac{\sqrt{2}}{3\pi}\,r^4$
6	2	π^3	$\frac{1}{8\sqrt{\pi}}\,r^5$

Since $X_1(t), \ldots, X_n(t)$ are mutually independent normal variables with the same mean and standard deviation, i.e, $X_i(t) \sim N(0, 1)$, the joint probability density function is given by:

$$f_{\mathbf{X}}(\mathbf{x}) = f_{X_1}(x_1) \cdots f_{X_n}(x_n) = \varphi(x_1) \cdots \varphi(x_n)$$

$$= \frac{1}{\left(\sqrt{2\pi}\right)^n} \exp\left(-\frac{1}{2}\left(x_1^2 + \ldots + x_n^2\right)\right) = \frac{1}{\left(\sqrt{2\pi}\right)^n} \exp\left(-\frac{1}{2}r^2\right) \tag{5–71}$$

Hence, $f_{\mathbf{X}}(\mathbf{x})$ is constant on the surface Γ, and Eq. (5-70) can be written as:

$$f_1 = \frac{\omega_0}{\sqrt{2\pi}} \frac{1}{\left(\sqrt{2\pi}\right)^n} \exp\left(-\frac{1}{2}r^2\right) \int_{\Gamma} dA$$

$$= \frac{\omega_0}{\left(\sqrt{2\pi}\right)^{n+1}} \exp\left(-\frac{1}{2}r^2\right) 2 \frac{\pi^{\frac{n}{2}}}{\Gamma(\frac{n}{2})} r^{n-1} = \frac{r^{n-1}}{2^{\frac{n}{2}-1}\sqrt{\pi}\,\Gamma(\frac{n}{2})} \omega_0 \exp\left(-\frac{1}{2}r^2\right) \tag{5–72}$$

Table 5-1 indicates the surface area A_n normalized with respect to r^{n-1} and the out-crossing frequency f_1 through an n-dimensional sphere, normalized with respect to $\omega_0\exp(-\frac{1}{2}r^2)$, for the low dimensional cases. Fig. 5-10 shows a plot of the normalized out-crossing frequency as a function of n for $r = 2.0$. A remarkable observation, which holds for arbitrary small values of r, is that $f_1 \to 0$ as $n \to \infty$. Hence, the process never crosses out of an infinitely dimensional sphere, no matter how small the radius of the sphere r may be in proportion to the standard deviation $\sigma_{X_j} = \sigma_0 = 1$ of the component processes. This fact is known as the *dimensionality paradox in reliability theory*.

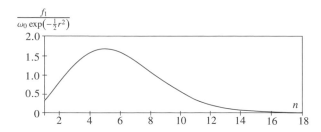

Fig. 5–10 Normalized out-crossing frequency as a function of the dimension n, $r = 2.0$.

5.3 Approximations to probability of failure

If the out-crossings in $]t, t + \Delta t]$ are stochastically independent of the previous out-crossings in $]0, t]$, the hazard rate $h(t)$ defined by Eq. (5-9) may be approximated as:

$$h(t)\Delta t + O(\Delta t^2) = P\Big(1 \text{ out-crossing in }]t, t + \Delta t] \mid \text{no out-crossings in }]0, t]\Big)$$

$$= P\big(1 \text{ out-crossing in }]t, t + \Delta t]\big) = f_1(t)\Delta t + O(\Delta t^2)$$

$$\Rightarrow \quad h(t) = f_1(t) \tag{5–73}$$

Inserting Eq. (5-73) into Eq. (5-12) yields:

$$P_f([0, t]) = 1 - P(X(0) \in S_0) \exp\left(-\int_0^t f_1(\tau)\, d\tau\right) \tag{5–74}$$

For broad-banded processes, the correlation length is of the magnitude of the average zero up-crossings time. The maxima between succeeding zero up-crossings are virtually uncorrelated. Hence, the out-crossings from the safe domain related to these maxima will also be independent events, and Eq. (5-74) is valid in this case.

For narrow-banded processes with low to medium barrier levels, such as b_2 shown in Fig. 5-11, the out-crossings tend to occur in clumps. In this case, the out-crossing events are significantly correlated, and Eq. (5-74) is no longer valid. However, at higher barrier levels such as b_1 shown in Fig. 5-11, only the highest peak in a clump is likely to imply an out-crossing. This suggests that the out-crossings tend to become independent as $b \to \infty$. Actually, this hypothesis can be formally proved for Gaussian processes.[1]

In Fig. 5-11, $e(t)$ indicates the realization of a so-called *envelope process* $\{E(t),\ t \in [0, \infty[\}$. This is defined as a maximum process, which is tangent to the underlying process $\{X(t),\ t \in [0, \infty[\}$ at the local maxima, and has a smooth variation between these maxima. Being a maximum process, it is clear that out-crossings of the envelope process precede out-crossings of $\{X(t),\ t \in [0, \infty[\}$.

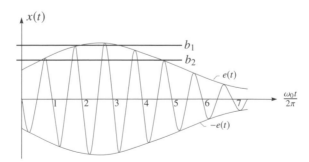

Fig. 5–11 Out-crossings of a narrow-banded process.

---

[1]H. Cramér and M.R. Leadbetter. *Stationary and related stochastic processes: Sample function properties and their applications.* John Wiley & Sons, Inc., New York, 1967.

Hence, the probability of failure $P_{f,E}([0,t])$ and the hazard rate $h_E(t)$ of the envelope process are upper bounds to the corresponding quantities $P_f([0,t])$ and $h(t)$ of $\{X(t),\ t \in [0,\infty[\}$, i.e.:

$$P_f([0,t]) \leq 1 - P\big(X(0) \in S_0\big)\ \exp\left(-\int_0^t h_E(\tau)\,d\tau\right) \tag{5–75}$$

The largest maxima in neighbouring clumps are separated at an time interval comparable to the correlation length of the narrow-banded process, and are consequently uncorrelated. Since the crossings of the envelope process are determined by the magnitude of these largest maxima, the out-crossing events of the envelope process are virtually stochastically independent. Hence, Eq. (5-73) also approximately holds for the envelope process:

$$h_E(t) \simeq f_{1,E}(t) \tag{5–76}$$

where $f_{1,E}(t)$ indicates the out-crossing frequency of the envelope process. If the approximation in Eq. (5-76) is inserted into the right-hand side of Eq. (5-75), a better approximation is obtained for $P_f([0,t])$, which is valid for narrow-banded processes $\{X(t),\ t \in [0,\infty[\}$ with respect to low or medium barrier levels.

Several definitions of the envelope process have been proposed, among which the following definition is most frequently used:

$$E(t) = \sqrt{U^2(t) + V^2(t)} \tag{5–77}$$

$$U(t) = \frac{X(t) - \mu_X}{\sigma_X}, \quad V(t) = \frac{\dot{X}(t)}{\sigma_{\dot{X}}} = \frac{\dot{U}(t)}{\omega_0}, \quad \omega_0 = \frac{\sigma_{\dot{X}}}{\sigma_X} \tag{5–78}$$

$\{U(t),\ t \in [0,\infty[\}$ and $\{V(t),\ t \in [0,\infty[\}$ are standardized versions of $\{X(t),\ t \in [0,\infty[\}$ and $\{\dot{X}(t),\ t \in [0,\infty[\}$, with zero mean value and unit variance. Clearly, $E(t) = U(t)$ at the local maxima of $X(t)$, where $\dot{X}(t) = \dot{U}(t) = V(t) = 0$. Further, if $\{X(t),\ t \in [0,\infty[\}$ is a stationary Gaussian process, $U(t)$ and $V(t)$ become stochastically independent, and ω_0 becomes the angular up-crossing frequency with respect to the mean value level, cf. Eq. (5-28). Because $E^2(t)$ has some resemblance to the mechanical energy of a SDOF oscillator with the displacement $U(t)$ and the velocity $\dot{U}(t)$, Eq. (5-77) is designated the *energy envelope process*.

Out-crossings of $\{X(t),\ t \in [0,\infty[\}$ with respect to the barrier b is equivalent to out-crossings of $\{U(t),\ t \in [0,\infty[\}$ with respect to the normalized barrier β given by:

$$\beta = \frac{b - \mu_X}{\sigma_X} \tag{5–79}$$

For a stationary Gaussian process, it will be shown in Example 5-7 below that the out-crossing frequency of the energy envelope process at the upper barrier β can be written as:

$$f_{1,E}(b) = \alpha_1(b)\, f_1(b) \tag{5–80}$$

$$\alpha_1(b) = \sqrt{\frac{8}{\pi}}\ \sqrt{\frac{\lambda_0 \lambda_4}{\lambda_2^2} - 1}\ \frac{b - \mu_X}{\sigma_X} \tag{5–81}$$

where $f_1(b)$ is given by Eq. (5-27), and the spectral moments λ_j are given by Eq. (3-48). $\alpha_1(b)$ is a reduction factor, which relates the single out-crossing of the envelope process to the multiple out-crossings of $\{X(t), t \in [0, \infty[\}$. Since $\alpha_1(b)$ increases proportional with b, the approximation given by Eqs. (5-75), (5-76) and (5-80) for the failure probability is only useful when the reduction coefficient $\alpha_1(b) < 1$.

Eq. (5-81) requires the existence of the acceleration process (i.e., $\lambda_4 < \infty$), which means that the out-crossing frequency of the energy envelope cannot be calculated for response processes with infinite acceleration. This excludes the important case of a linear SDOF system subjected to a Gaussian white noise process. Even for a velocity process that is differentiable but contains a zero-mean high-frequency component, the acceleration process may have such high variance that the energy envelope process does not have a smooth variation between the local maxima of $\{X(t), t \in [0, \infty]\}$.

In order to apply the concept to acceleration processes with infinite or large variance, Cramér and Leadbetter suggested the following alternative definition of the envelope process:[1]

$$E(t) = \sqrt{U^2(t) + \hat{U}^2(t)} \tag{5-82}$$

where $\hat{U}(t)$ indicates the Hilbert transform of $U(t)$ defined by Eq. (3-107). Actually, the Hilbert transform filters out the rapid variations of the velocity time-series. As explained next to Eq. (3-111), for a harmonic time-series with the angular frequency ω_0, we have $\hat{U}(t) = \frac{\dot{U}(t)}{\omega_0} = V(t)$. In this case, the envelope definitions in Eq. (5-77) and Eq. (5-82) are identical. For narrow-banded processes with finite band-width, the *Cramér and Leadbetter envelope process* no longer attains the value of $U(t)$ at the local maxima of $X(t)$. Instead, a close upper bound is achieved.

As will be shown in Example 5-8 below, the out-crossing frequency of the Cramér-Leadbetter envelope process may still be represented by Eq. (5-80), with a modified reduction factor $\alpha_2(b)$ determined as:

$$\alpha_2(b) = \sqrt{2\pi} \sqrt{1 - \frac{\lambda_1^2}{\lambda_0 \lambda_2} \frac{b - \mu_X}{\sigma_X}} \tag{5-83}$$

It should be noticed that both Eqs. (5-81) and (5-83) have been derived without any resort to narrow-banded assumptions, and can be applied to broad-banded Gaussian processes as well. However, the use of these results in dynamic reliability problems is restricted to narrow-banded response processes. As seen from Eq. (5-81), the out-crossing frequency of the energy envelope process is actually controlled by the bandwidth parameter ε as given by Eq. (5-33). Correspondingly, the out-crossing frequency of the Cramér-Leadbetter envelope process is controlled by Vanmarcke's bandwidth parameter δ as given by Eq. (3-49). In any case, the envelope approximations are only useful if the corresponding band-width parameters indicate narrow-bandedness.

---

[1]H. Cramér and M.R. Leadbetter. *Stationary and related stochastic processes: Sample function properties and their applications.* John Wiley & Sons, Inc., New York, 1967.

For a stationary Gaussian process with a constant upper barrier b, $P(X(0) \in S_0) = \Phi\left(\frac{b-\mu_X}{\sigma_X}\right)$ with $\Phi(\cdot)$ defined in Eq. (2-24). Then, from Eqs. (5-27), (5-75), (5-76) and (5-80), the failure probability in this case can be determined as:

$$P_f([0,t]) \simeq 1 - \Phi\left(\frac{b-\mu_X}{\sigma_X}\right) \exp\left(-\frac{1}{2\pi}\,\alpha(b)\,\frac{\sigma_{\dot{X}}}{\sigma_X}\exp\left(-\frac{1}{2}\left(\frac{b-\mu_X}{\sigma_X}\right)^2\right)t\right) \qquad (5\text{–}84)$$

$$\alpha(b) = \min\left(1, \alpha_j(b)\right), \quad j = 1, 2 \qquad (5\text{–}85)$$

where $\alpha_j(b)$ is given by Eq. (5-81) or (5-83), depending on which envelope definition is applied. $\alpha(b) = 1$ for broad-banded processes or for narrow-banded processes with respect to high barrier levels, whereas $\alpha(b) = \alpha_j(b)$ for narrow-banded processes at low or medium barrier levels.

Fig. 5-12 shows the first-passage probability density curves obtained from Eq. (5-84) for a linear SDOF system with damping ratio $\zeta = 0.01$, which is subjected to a Gaussian white noise. The barrier level is $b = 2.0\sigma_X$, where σ_X indicates the standard deviation of the stationary stochastic response. The abscissa t is normalized with respect to the undamped eigenperiod T_0. Curve a and curve b correspond to $\alpha(b) = 1$ and $\alpha(b) = \alpha_2(b)$ as in Eq. (5-83), respectively. Further, the first-passage probability density curve with random start at $t = 0$, obtained by Monte Carlo simulation using Eq. (5-17), has also been illustrated for comparison. Except for the first period, curves a and b form upper bounds to $f_T(t)$, and hence to the failure probability. As expected, curve b forms a closer (better) upper bound.

For a constant lower barrier a corresponding to the safe domain $S = \{x \mid a < x < \infty\}$, the analogous result to Eq. (5-84) can be written as:

$$P_f([0,t]) \simeq 1 - \Phi\left(\frac{\mu_X-a}{\sigma_X}\right) \exp\left(-\frac{1}{2\pi}\,\alpha(a)\,\frac{\sigma_{\dot{X}}}{\sigma_X}\exp\left(-\frac{1}{2}\left(\frac{\mu_X-a}{\sigma_X}\right)^2\right)t\right) \qquad (5\text{–}86)$$

$$\alpha(a) = \min\left(1, \alpha_j(a)\right), \quad j = 1, 2 \qquad (5\text{–}87)$$

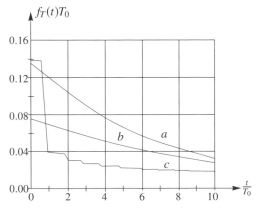

Fig. 5–12 First-passage probability density curves. Linear SDOF system subjected to a Gaussian white noise, $b = 2\sigma_X$, $\zeta = 0.01$. a) $\alpha(b) = 1$. b) $\alpha(b) = \alpha_2(b)$. c) Monte Carlo simulation.

$$\alpha_j(a) = \begin{cases} \sqrt{\dfrac{8}{\pi}} \ \sqrt{\dfrac{\lambda_0 \lambda_4}{\lambda_2^2} - 1} \ \dfrac{\mu_X - a}{\sigma_X}, & j = 1 \\[2em] \sqrt{2\pi} \ \sqrt{1 - \dfrac{\lambda_1^2}{\lambda_0 \lambda_2}} \ \dfrac{\mu_X - a}{\sigma_X}, & j = 2 \end{cases} \tag{5-88}$$

Example 5-7: Out-crossing frequency of the energy envelope process

$\{X(t), t \in R\}$ is a stationary Gaussian process with the mean value μ_X, the standard deviation σ_X and the auto-correlation coefficient function $\rho_{XX}(\tau)$. $X(t)$, $\dot{X}(t)$ and $\ddot{X}(t)$ are standardized to the following stochastic variables with zero mean value and unit variance:

$$U(t) = \frac{X(t) - \mu_X}{\sigma_X}, \quad V(t) = \frac{\dot{X}(t)}{\sigma_{\dot{X}}}, \quad W(t) = \frac{\ddot{X}(t)}{\sigma_{\ddot{X}}} \tag{5-89}$$

where the variances of $\dot{X}(t)$ and $\ddot{X}(t)$ are given by, cf. Eq. (3-152):

$$\sigma_{\dot{X}}^2 = -\rho_{XX}''(0)\,\sigma_X^2, \quad \sigma_{\ddot{X}}^2 = \rho_{XX}^{(4)}(0)\,\sigma_X^2 \tag{5-90}$$

Recalling that $E\left[X(t)\dot{X}(t)\right] = E\left[\dot{X}(t)\ddot{X}(t)\right] = 0 \Rightarrow E\left[U(t)V(t)\right] = E\left[V(t)W(t)\right] = 0$, the (zero time-lag) covariance matrix of the stochastic vector $\mathbf{U} = [U(t), V(t), W(t)]^T$ and its inverse matrix become:

$$\mathbf{C}_{\mathbf{UU}} = \begin{bmatrix} 1 & 0 & r \\ 0 & 1 & 0 \\ r & 0 & 1 \end{bmatrix}, \quad \mathbf{C}_{\mathbf{UU}}^{-1} = \frac{1}{1 - r^2} \begin{bmatrix} 1 & 0 & -r \\ 0 & 1 - r^2 & 0 \\ -r & 0 & 1 \end{bmatrix} \tag{5-91}$$

r indicates the correlation coefficient of $X(t)$ and $\ddot{X}(t)$, calculated as:

$$r = E\left[U(t)W(t)\right] = \frac{E\left[X(t)\ddot{X}(t)\right]}{\sigma_X \sigma_{\ddot{X}}} = \frac{\sigma_X^2 \rho_{XX}''(0)}{\sigma_X \sigma_{\ddot{X}}} = -\frac{\sigma_{\dot{X}}^2}{\sigma_X \sigma_{\ddot{X}}} = -\frac{\lambda_2}{\sqrt{\lambda_0 \lambda_4}} \tag{5-92}$$

where Eqs. (3-152) and (5-90) have been used. $r = -1$ for a harmonic process, and r is slightly larger than -1 for a narrow-banded process.

Then, the joint probability density function of the normal stochastic vector $\mathbf{U}$ becomes, cf. Eq. (2-70):

$$f_{UVW}(u, v, w) = \frac{1}{(2\pi)^{\frac{3}{2}} \sqrt{1 - r^2}} \exp\left(-\frac{1}{2} \frac{u^2 + (1 - r^2)v^2 + w^2 - 2ruw}{1 - r^2}\right)$$

$$= f_{W|UV}(w \,|\, u, v)\, f_{UV}(u, v) \tag{5-93}$$

where:

$$\left. \begin{aligned} f_{UV}(u, v) &= \frac{1}{2\pi} \exp\left(-\frac{u^2 + v^2}{2}\right) \\ f_{W|UV}(w \,|\, u, v) &= \frac{1}{\sqrt{2\pi} \sqrt{1 - r^2}} \exp\left(-\frac{1}{2}\left(\frac{w - ru}{\sqrt{1 - r^2}}\right)^2\right) \end{aligned} \right\} \tag{5-94}$$

$f_{UV}(u, v)$ signifies the marginal joint probability density function of $[U(t), V(t)]^T$, and $f_{W|UV}(w \mid u, v)$ denotes the probability density function of $W(t)$ on condition of $(U(t), V(t)) = (u, v)$. Eq. (5-94) follows from Eqs. (2-75), (2-76) and (2-77) in relation to the covariance matrix $\mathbf{C_{UU}}$.

Next, the following bijective mapping of $(E, \Theta) \in [0, \infty[\times[0, 2\pi[$ onto $(U, V) \in] - \infty, \infty[\times] - \infty, \infty[$ is considered:

$$U(t) = E(t) \cos \Theta(t) \left.\right\}$$
$$V(t) = E(t) \sin \Theta(t) \left.\right.$$
$$\tag{5-95}$$

Eq. (5-95) implies that $E(t) = \sqrt{U^2(t) + V^2(t)}$ holds for all realizations of the phase process $\Theta(t)$. From Eqs. (5-89) and (5-95), we have:

$$V(t) = \frac{\sigma_X}{\sigma_{\dot{X}}} \frac{d}{dt} U(t) = \frac{\sigma_X}{\sigma_{\dot{X}}} (\dot{E} \cos \Theta - E \sin \Theta \cdot \dot{\Theta}) \left.\right\}$$
$$W(t) = \frac{\sigma_{\dot{X}}}{\sigma_{\ddot{X}}} \frac{d}{dt} V(t) = \frac{\sigma_{\dot{X}}}{\sigma_{\ddot{X}}} (\dot{E} \sin \Theta + E \cos \Theta \cdot \dot{\Theta}) \left.\right.$$
$$\tag{5-96}$$

At the same time $V(t) = E(t) \sin \Theta(t)$. Then, comparing with the 1st equation in Eq. (5-96), the following expression for $\dot{\Theta}(t)$ holds:

$$\dot{\Theta}(t) = \frac{\dot{E}(t) \cos \Theta(t)}{E(t) \sin \Theta(t)} - \frac{\sigma_{\dot{X}}}{\sigma_X} \tag{5-97}$$

Elimination of $\dot{\Theta}(t)$ in the 2nd equation of Eq. (5-96) by means of Eq. (5-97) yields:

$$W(t) = \frac{\sigma_{\dot{X}}}{\sigma_{\ddot{X}}} \frac{\dot{E}(t)}{\sin \Theta(t)} - \frac{\sigma_{\dot{X}}^2}{\sigma_X \sigma_{\ddot{X}}} E(t) \cos \Theta(t) = r \left(E(t) \cos \Theta(t) - \frac{1}{\omega_0} \frac{\dot{E}(t)}{\sin \Theta(t)} \right) \tag{5-98}$$

where $\omega_0 = \frac{\sigma_{\dot{X}}}{\sigma_X}$, and r is given by Eq. (5-92).

Eqs. (5-95) and (5-98) provide a bijective mapping of $(E, \dot{E}, \Theta)$ onto (U, V, W). The Jacobian of this transformation becomes:

$$|\det(\mathbf{J})| = \left| \det \left(\begin{bmatrix} \cos \theta & 0 & -e \sin \theta \\ \sin \theta & 0 & e \cos \theta \\ r \cos \theta & -\frac{1}{\omega_0} \frac{r}{\sin \theta} & r \left(-e \sin \theta + \frac{\dot{e}}{\omega_0} \frac{\cos \theta}{\sin^2 \theta} \right) \end{bmatrix} \right) \right| = \frac{e}{\omega_0} \left| \frac{r}{\sin \theta} \right| \tag{5-99}$$

The joint probability density function of $[(E(t), \dot{E}(t), \Theta(t))]^T$ follows from Eqs. (2-30) and (5-93):

$$f_{E\dot{E}\Theta}(e, \dot{e}, \theta) = f_{UVW}(u, v, w) \frac{e}{\omega_0} \left| \frac{r}{\sin \theta} \right| = f_{UV}(u, v) f_{W|UV}(w \mid u, v) \frac{e}{\omega_0} \left| \frac{r}{\sin \theta} \right| \tag{5-100}$$

where the following expressions have been obtained from Eqs. (5-94), (5-95) and (5-98):

$$f_{UV}(u, v) = \frac{1}{2\pi} \exp \left(-\frac{1}{2} (e^2 \cos^2 \theta + e^2 \sin^2 \theta) \right) = \frac{1}{2\pi} \exp \left(-\frac{1}{2} e^2 \right) \tag{5-101}$$

$$f_{W|UV}(w \mid u, v) = \frac{1}{\sqrt{2\pi} \sqrt{1 - r^2}} \exp \left(-\frac{1}{2} \frac{\left(r \left(e \cos \theta - \frac{\dot{e}}{\omega_0 \sin \theta} \right) - r e \cos \theta \right)^2}{1 - r^2} \right)$$

$$= \omega_0 \left| \frac{\sin \theta}{r} \right| \frac{1}{\sqrt{2\pi} \sigma(\theta)} \exp \left(-\frac{1}{2} \frac{\dot{e}^2}{\sigma^2(\theta)} \right) \tag{5-102}$$

with the parameter $\sigma(\theta)$ given by:

$$\sigma(\theta) = \omega_0 \, |\sin\theta| \, \sqrt{\frac{1-r^2}{r^2}} \tag{5–103}$$

From Eqs. (5-100), (5-101) and (5-102), the joint probability density function of $(E, \dot{E}, \Theta)$ can be written as:

$$f_{E\dot{E}\Theta}(e, \dot{e}, \theta) = e \exp\left(-\frac{1}{2}e^2\right) \cdot \frac{1}{2\pi} \cdot \frac{1}{\sqrt{2\pi}\,\sigma(\theta)} \exp\left(-\frac{1}{2}\frac{\dot{e}^2}{\sigma^2(\theta)}\right) \tag{5–104}$$

The marginal probability density function of $\Theta(t)$ becomes:

$$f_\Theta(\theta) = \int_0^\infty \int_{-\infty}^\infty f_{E\dot{E}\Theta}(e, \dot{e}, \theta) \, de \, d\dot{e} = \frac{1}{2\pi} \tag{5–105}$$

Therefore, $\Theta(t) \sim U(0, 2\pi)$. The joint probability density function of $[E, \dot{E}]$ is then given by:

$$f_{E\dot{E}}(e, \dot{e}) = e \exp\left(-\frac{1}{2}e^2\right) \int_0^{2\pi} \frac{1}{\sqrt{2\pi}\,\sigma(\theta)} \exp\left(-\frac{1}{2}\frac{\dot{e}^2}{\sigma^2(\theta)}\right) \frac{1}{2\pi} \, d\theta \tag{5–106}$$

Eq. (5-106) shows that $E(t)$ and $\dot{E}(t)$ are stochastically independent, and $E(t)$ fulfills Rayleigh distribution, i.e., $E(t) \sim R(1)$. The marginal distribution of $\dot{E}(t)$ for the energy envelope process is generally not normally distributed. On the other hand, as will be shown in Example 5-8, $\dot{E}(t)$ for the Cramér-Leadbetter envelope process is indeed normally distributed. This is also the case for the energy envelope process with extreme narrow-bandedness, i.e., when $r = -1$.

Finally, the out-crossing frequency at the upper barrier $e = \beta$ can be calculated as, cf. Eq. (5-20):

$$\begin{aligned}
f_{1,E} &= \int_0^\infty \dot{e} f_{E\dot{E}}(\beta, \dot{e}) \, d\dot{e} \\
&= \beta \exp\left(-\frac{1}{2}\beta^2\right) \int_0^{2\pi} \frac{1}{2\pi} \left(\int_0^\infty \frac{\dot{e}}{\sqrt{2\pi}\sigma(\theta)} \exp\left(-\frac{1}{2}\frac{\dot{e}^2}{\sigma^2(\theta)}\right) d\dot{e}\right) d\theta \\
&= \beta \exp\left(-\frac{1}{2}\beta^2\right) \int_0^{2\pi} \frac{\sigma(\theta)}{2\pi\sqrt{2\pi}} \, d\theta = \beta \exp\left(-\frac{1}{2}\beta^2\right) \frac{\omega_0}{2\pi\sqrt{2\pi}} \sqrt{\frac{1-r^2}{r^2}} \int_0^{2\pi} |\sin(\theta)| \, d\theta \\
&= \sqrt{\frac{8}{\pi}} \sqrt{\frac{1-r^2}{r^2}} \beta \frac{\omega_0}{2\pi} \exp\left(-\frac{1}{2}\beta^2\right) = \alpha_1(b) f_1(b)
\end{aligned} \tag{5–107}$$

where $\alpha_1(b)$ and $f_1(b)$ correspond to Eqs. (5-81) and (5-27), respectively.

Example 5-8: Out-crossing frequency of the Cramér-Leadbetter envelope process

$\{X(t), \, t \in R\}$ is a stationary Gaussian process with the mean value μ_X, the standard deviation σ_X and the auto-spectral density function $S_{XX}(\omega)$. The following stochastic variables normalized to zero mean value and unit variance are introduced:

$$U(t) = \frac{X(t) - \mu_X}{\sigma_X}, \quad \hat{U}(t) = \frac{\hat{X}(t)}{\sigma_X} \tag{5–108}$$

where $\hat{X}(t)$ and $\hat{U}(t)$ indicate the Hilbert transforms of $X(t)$ and $U(t)$ as defined by Eq. (3-107). It follows from Eqs. (3-112) and (3-114) that $\mu_{\hat{X}} = 0$ and $\kappa_{\hat{X}\hat{X}}(\tau) = \kappa_{XX}(\tau) \Rightarrow \sigma_{\hat{X}} = \sigma_X$. Further, according to Eq. (3-115), $U(t)$ and $\hat{U}(t)$ are uncorrelated and hence stochastically independent due to the Gaussianity in the present case.

A harmonically varying input is first considered:

$$U(t) = \mathrm{Re}\left(U_0\, e^{i\omega t}\right) \tag{5–109}$$

Correspondingly, the stationary response of $\dot{U}(t)$, $\hat{U}(t)$ and $\dot{\hat{U}}(t)$ become:

$$\left.\begin{aligned}
\dot{U}(t) &= \mathrm{Re}\left(i\omega\, U_0\, e^{i\omega t}\right) \\
\hat{U}(t) &= \mathrm{Re}\left(H(\omega)\, U_0\, e^{i\omega t}\right) \\
\dot{\hat{U}}(t) &= \mathrm{Re}\left(i\omega\, H(\omega)\, U_0\, e^{i\omega t}\right)
\end{aligned}\right\} \tag{5–110}$$

where $H(\omega)$ indicates the frequency response function of the Hilbert transform as given by Eq. (3-109). Especially, $iH(\omega) = \mathrm{sign}(\omega) = -\mathrm{sign}(-\omega)$.

From Eq. (5-110), it is seen that the frequency response functions of $\dot{U}(t)$, $\hat{U}(t)$ and $\dot{\hat{U}}(t)$ are $i\omega$, $H(\omega)$ and $i\omega H(\omega)$, respectively. Then, the cross-spectral density functions $S_{U\hat{U}}(\omega)$, $S_{U\dot{U}}(\omega)$, $S_{\dot{U}\hat{U}}(\omega)$, $S_{\dot{U}\dot{U}}(\omega)$, $S_{\hat{U}\dot{U}}(\omega)$ and $S_{\dot{U}\dot{U}}(\omega)$ become, cf. Eqs. (3-104) and (3-118):

$$\left.\begin{aligned}
S_{U\hat{U}}(\omega) &= H(\omega)\, S_{UU}(\omega) & &= H(\omega)\, S_{UU}(\omega) \\
S_{U\dot{U}}(\omega) &= i\omega H(\omega)\, S_{UU}(\omega) & &= i\omega H(\omega)\, S_{UU}(\omega) \\
S_{\dot{U}\hat{U}}(\omega) &= (i\omega)^*\, H(\omega)\, S_{UU}(\omega) & &= -i\omega H(\omega)\, S_{UU}(\omega) \\
S_{\dot{U}\dot{U}}(\omega) &= (i\omega)^*\, i\omega H(\omega)\, S_{UU}(\omega) & &= \omega^2 H(\omega)\, S_{UU}(\omega) \\
S_{\hat{U}\dot{U}}(\omega) &= H^*(\omega)\, i\omega H(\omega)\, S_{UU}(\omega) & &= i\omega\, S_{UU}(\omega) \\
S_{\dot{U}\dot{U}}(\omega) &= \left(i\omega H(\omega)\right)^*\, i\omega H(\omega)\, S_{UU}(\omega) &= \omega^2\, S_{UU}(\omega)
\end{aligned}\right\} \tag{5–111}$$

Inserting Eq. (3-109) into Eq. (5-111) yields the following cross-covariances:

$$\kappa_{U\hat{U}}(0) = \int_{-\infty}^{\infty} H(\omega)\, S_{UU}(\omega)\, d\omega = 0$$

$$\kappa_{U\dot{U}}(0) = \int_{-\infty}^{\infty} i\omega H(\omega)\, S_U(\omega)\, d\omega = 2\int_{0}^{\infty} \omega\, S_{UU}(\omega)\, d\omega = c$$

$$\kappa_{\dot{U}\hat{U}}(0) = -\int_{-\infty}^{\infty} i\omega H(\omega)\, S_{UU}(\omega)\, d\omega = -c$$

$$\kappa_{\dot{U}\dot{U}}(0) = \int_{-\infty}^{\infty} \omega^2 H(\omega)\, S_{UU}(\omega)\, d\omega = 0$$

$$\kappa_{\hat{U}\dot{U}}(0) = \int_{-\infty}^{\infty} i\omega\, S_{UU}(\omega)\, d\omega = 0$$

$$\kappa_{\dot{U}\dot{U}}(0) = \int_{-\infty}^{\infty} \omega^2\, S_{UU}(\omega)\, d\omega = d \tag{5–112}$$

where:

$$c = \frac{\lambda_1}{\lambda_0}, \quad d = \frac{\lambda_2}{\lambda_0} \tag{5–113}$$

λ_0, λ_1 and λ_0 indicate the spectral moments of the underlying process $\{X(t), t \in R\}$ as given by Eq. (3-48).

At the evaluation of Eq. (5-112), it has been applied that $i\omega H(\omega)S_{XX}(\omega)$ is an even function of ω, and $H(\omega)S_{XX}(\omega)$, $\omega^2 H(\omega)S_{XX}(\omega)$, $iH(\omega)S_{XX}(\omega)$ are odd functions of ω.

Consider the following stochastic vector:

$$\mathbf{U}(t) = \begin{bmatrix} \mathbf{U}_1(t) \\ \dot{\mathbf{U}}_1(t) \end{bmatrix}, \quad \mathbf{U}_1(t) = \begin{bmatrix} U(t) \\ \hat{U}(t) \end{bmatrix}, \quad \dot{\mathbf{U}}_1(t) = \begin{bmatrix} \dot{U}(t) \\ \dot{\hat{U}}(t) \end{bmatrix} \tag{5–114}$$

From Eq. (5-112), the (zero time-lag) covariance matrix $\mathbf{C_{UU}}$ of the stochastic vector $\mathbf{U}(t)$ may be written as:

$$\mathbf{C_{UU}} = \begin{bmatrix} \mathbf{C_{U_1 U_1}} & \mathbf{C_{U_1 \dot{U}_1}} \\ \mathbf{C_{U_1 \dot{U}_1}^T} & \mathbf{C_{\dot{U}_1 \dot{U}_1}} \end{bmatrix} \tag{5–115}$$

$$\mathbf{C_{U_1 U_1}} = \begin{bmatrix} 1 & 0 \\ 0 & 1 \end{bmatrix}, \quad \mathbf{C_{U_1 \dot{U}_1}} = \begin{bmatrix} 0 & c \\ -c & 0 \end{bmatrix}, \quad \mathbf{C_{\dot{U}_1 \dot{U}_1}} = \begin{bmatrix} d & 0 \\ 0 & d \end{bmatrix} \tag{5–116}$$

Next, the joint probability density function of $\mathbf{U}(t)$ may be written as:

$$f_{\mathbf{U}}(u, \hat{u}, \dot{u}, \dot{\hat{u}}) = f_{\mathbf{U}_1}(u, \hat{u}) \, f_{\dot{\mathbf{U}}_1 | \mathbf{U}_1}(\dot{u}, \dot{\hat{u}} \,|\, u, \hat{u}) \tag{5–117}$$

where $f_{\mathbf{U}_1}(u, \hat{u})$ signifies the marginal joint probability density function of $U(t)$ and $\hat{U}(t)$, given by:

$$f_{\mathbf{U}_1}(u, \hat{u}) = \frac{1}{2\pi} \exp\left(-\frac{1}{2}(u^2 + \hat{u}^2)\right) \tag{5–118}$$

$f_{\dot{\mathbf{U}}_1 | \mathbf{U}_1}(\dot{u}, \dot{\hat{u}} \,|\, u, \hat{u})$ signifies the joint probability density function of $\dot{U}(t)$ and $\dot{\hat{U}}(t)$ on condition that $\mathbf{U}_1 = \mathbf{u}_1 = [u, \hat{u}]^T$. This is a normal probability density function with the conditional mean value vector $\boldsymbol{\mu}_{\dot{\mathbf{U}}_1 | \mathbf{U}_1}$ and the conditional covariance matrix $\mathbf{C}_{\dot{\mathbf{U}}_1 \dot{\mathbf{U}}_1 | \mathbf{U}_1}$ obtained from Eqs. (2-75) and (2-76):

$$\begin{aligned} \boldsymbol{\mu}_{\dot{\mathbf{U}}_1 | \mathbf{U}_1} &= \boldsymbol{\mu}_{\dot{\mathbf{U}}_1} + \mathbf{C}_{\mathbf{U}_1 \dot{\mathbf{U}}_1}^T \mathbf{C}_{\mathbf{U}_1 \mathbf{U}_1}^{-1} (\mathbf{u}_1 - \boldsymbol{\mu}_{\mathbf{U}_1}) \\ &= \begin{bmatrix} 0 \\ 0 \end{bmatrix} + \begin{bmatrix} 0 & -c \\ c & 0 \end{bmatrix} \begin{bmatrix} 1 & 0 \\ 0 & 1 \end{bmatrix}^{-1} \left(\begin{bmatrix} u \\ \hat{u} \end{bmatrix} - \begin{bmatrix} 0 \\ 0 \end{bmatrix} \right) = c \begin{bmatrix} -\hat{u} \\ u \end{bmatrix} \end{aligned} \tag{5–119}$$

$$\begin{aligned} \mathbf{C}_{\dot{\mathbf{U}}_1 \dot{\mathbf{U}}_1 | \mathbf{U}_1} &= \mathbf{C}_{\dot{\mathbf{U}}_1 \dot{\mathbf{U}}_1} - \mathbf{C}_{\mathbf{U}_1 \dot{\mathbf{U}}_1}^T \mathbf{C}_{\mathbf{U}_1 \mathbf{U}_1}^{-1} \mathbf{C}_{\mathbf{U}_1 \dot{\mathbf{U}}_1} \\ &= \begin{bmatrix} d & 0 \\ 0 & d \end{bmatrix} - \begin{bmatrix} 0 & -c \\ c & 0 \end{bmatrix} \begin{bmatrix} 1 & 0 \\ 0 & 1 \end{bmatrix}^{-1} \begin{bmatrix} 0 & c \\ -c & 0 \end{bmatrix} = (d - c^2) \begin{bmatrix} 1 & 0 \\ 0 & 1 \end{bmatrix} \end{aligned} \tag{5–120}$$

where from Eq. (5-113):

$$d - c^2 = \frac{\lambda_2}{\lambda_0} - \frac{\lambda_1^2}{\lambda_0^2} = \omega_0^2 \left(1 - r^2\right)$$

$$\omega_0 = \sqrt{\frac{\lambda_2}{\lambda_0}} = \frac{\sigma_{\dot{X}}}{\sigma_X}, \quad r^2 = \frac{\lambda_1^2}{\lambda_0 \lambda_2} \tag{5–121}$$

Insertion of Eqs. (5-118), (5-119) and (5-120) into Eq. (5-117) yields:

$$f_{U\hat{U}\dot{U}\dot{\hat{U}}}\left(u, \hat{u}, \dot{u}, \dot{\hat{u}}\right) = \frac{1}{2\pi} \exp\left(-\frac{1}{2}\left(u^2 + \hat{u}^2\right)\right) \frac{1}{2\pi \sqrt{d - c^2}} \exp\left(-\frac{1}{2} \frac{\left(\dot{u} + c\hat{u}\right)^2 + \left(\dot{\hat{u}} - cu\right)^2}{d - c^2}\right) \tag{5–122}$$

Next, consider the following bijective transformation of the stochastic vector $\mathbf{V}(t) = \left[E(t), \dot{E}(t), \Theta(t), \dot{\Theta}(t)\right]^T$ onto $\mathbf{U}(t)$, defined as:

$$\left. \begin{aligned} u &= e \cos\theta \\ \hat{u} &= e \sin\theta \\ \dot{u} &= \dot{e} \cos\theta - e \sin\theta \, \dot{\theta} \\ \dot{\hat{u}} &= \dot{e} \sin\theta + e \cos\theta \, \dot{\theta} \end{aligned} \right\} \quad \begin{aligned} e &\in [0, \infty[\, , \quad \dot{e} \in \,]-\infty, \infty[\\ \theta &\in [0, 2\pi[\, , \quad \dot{\theta} \in \,]-\infty, \infty[\end{aligned} \tag{5–123}$$

The Jacobian of this transformation is:

$$|\det(\mathbf{J})| = \left| \det\left(\begin{bmatrix} \cos\theta & 0 & -e \sin\theta & 0 \\ \sin\theta & 0 & e \cos\theta & 0 \\ -\sin\theta \, \dot{\theta} & \cos\theta & -\dot{e} \sin\theta - e \cos\theta \, \dot{\theta} & -e \sin\theta \\ \cos\theta \, \dot{\theta} & \sin\theta & \dot{e} \cos\theta - e \sin\theta \, \dot{\theta} & e \cos\theta \end{bmatrix} \right) \right| = e^2 \tag{5–124}$$

Then, Eqs. (5-122), (5-123) and (5-124) provide the following solution for the joint probability density function of $\mathbf{V}(t)$, cf. Eq. (2-30):

$$\begin{aligned} f_{E\dot{E}\Theta\dot{\Theta}}(e, \dot{e}, \theta, \dot{\theta}) &= f_{U\hat{U}\dot{U}\dot{\hat{U}}}(u, \hat{u}, \dot{u}, \dot{\hat{u}}) \, e^2 \\ &= \frac{e^2}{2\pi} \exp\left(-\frac{1}{2} e^2\right) \frac{1}{2\pi(d - c^2)} \exp\left(-\frac{1}{2} \frac{\dot{e}^2 + e^2 \dot{\theta}^2 + c^2 e^2 - 2c e^2 \dot{\theta}}{d - c^2}\right) \end{aligned} \tag{5–125}$$

where it has been used that $u^2 + \hat{u}^2 = e^2$, $2c\dot{u}\hat{u} - 2c\dot{\hat{u}}u = -2ce^2\dot{\theta}$, $\dot{u}^2 + \dot{\hat{u}}^2 = \dot{e}^2 + e^2\dot{\theta}^2$. The joint probability density function $f_{E\dot{E}}(e, \dot{e})$ is determined by marginalization of Eq. (5-125):

$$\begin{aligned} f_{E\dot{E}}(e, \dot{e}) &= \int_{-\infty}^{\infty} \int_0^{2\pi} f_{E\dot{E}\Theta\dot{\Theta}}(e, \dot{e}, \theta, \dot{\theta}) \, d\theta d\dot{\theta} \\ &= \frac{e}{2\pi} \exp\left(-\frac{1}{2}e^2\right) \frac{1}{\sqrt{2\pi}\sqrt{d - c^2}} \exp\left(-\frac{1}{2}\frac{\dot{e}^2}{d - c^2}\right) \int_{-\infty}^{\infty} \int_0^{2\pi} \frac{e}{\sqrt{2\pi}\sqrt{d - c^2}} \exp\left(-\frac{1}{2}\frac{e^2(\dot{\theta} - c)^2}{d - c^2}\right) d\theta d\dot{\theta} \\ &= e \exp\left(-\frac{1}{2}e^2\right) \frac{1}{\sqrt{2\pi}\sqrt{d - c^2}} \exp\left(-\frac{1}{2}\frac{\dot{e}^2}{d - c^2}\right) \end{aligned} \tag{5–126}$$

Hence, for the Cramér-Leadbetter envelope process, we have $E(t) \sim R(1)$ and $\dot{E}(t) \sim N(0, d - c^2)$. Moreover, $E(t)$ and $\dot{E}(t)$ are stochastically independent.

The out-crossing frequency of $\{E(t), t \in R\}$ with respect to the normalized barrier $\beta = \frac{b - \mu_X}{\sigma_X}$ becomes, cf. Eqs. (5-20) and (5-121):

$$f_{1,E} = \int_0^\infty \dot{e} f_{E\dot{E}}(\beta, \dot{e}) \, d\dot{e} = \beta \exp\left(-\frac{1}{2}\beta^2\right) \frac{1}{\sqrt{2\pi}} \int_0^\infty \frac{\dot{e}}{\sqrt{d - c^2}} \exp\left(-\frac{1}{2}\frac{\dot{e}^2}{d - c^2}\right) d\dot{e}$$

$$= \beta \exp\left(-\frac{1}{2}\beta^2\right) \frac{\sqrt{d - c^2}}{\sqrt{2\pi}} = \sqrt{2\pi} \sqrt{1 - r^2} \beta \frac{\omega_0}{2\pi} \exp\left(-\frac{1}{2}\beta^2\right) = \alpha_2(b) f_1(b) \qquad (5\text{--}127)$$

where $\alpha_2(b)$ and $f_1(b)$ correspond to Eqs. (5-83) and (5-27), respectively.

The bandwidth parameter ε as given by Eq. (5-33) also controls the probability distribution of the local maxima A in a stationary Gaussian process with finite 4th order spectral moment (i.e. $\lambda_4 < \infty$). The following expression for the probability density function of A has been proposed by Huston and Skopinski:[1]

$$f_A(a) = \frac{1}{\sigma_X}\left(\varepsilon\, \varphi\left(\frac{\alpha}{\varepsilon}\right) + \sqrt{2\pi} \sqrt{1 - \varepsilon^2}\, \alpha\, \Phi\left(\frac{\sqrt{1 - \varepsilon^2}}{\varepsilon}\, \alpha\right) \varphi(\alpha)\right) \qquad (5\text{--}128)$$

$$\alpha = \frac{a - \mu_X}{\sigma_X} \qquad (5\text{--}129)$$

Eq. (5-128) is valid for all Gaussian processes with $\lambda_4 < \infty$, although an additional assumption on certain ergodic properties has been applied during the derivation of the result, as will be shown in Example 5-9.

For any $\alpha > 0$, it follows that $\Phi\left(\frac{\alpha\sqrt{1 - \varepsilon^2}}{\varepsilon}\right) \to \Phi(\infty) = 1$ and $\varphi\left(\frac{\alpha}{\varepsilon}\right) \to 0$ as $\varepsilon \to 0$. Then, in the limits of extreme narrow-bandedness ($\varepsilon = 0$) and extreme broad-bandedness ($\varepsilon = 1$), Eq. (5-128) approaches the following limits:

$$\lim_{\varepsilon \to 0} f_A(a) = \frac{\sqrt{2\pi}\alpha}{\sigma_X} \varphi(\alpha) = \frac{\alpha}{\sigma_X} \exp\left(-\frac{1}{2}\alpha^2\right) = \frac{a - \mu_X}{\sigma_X^2} \exp\left(-\frac{1}{2}\left(\frac{a - \mu_X}{\sigma_X}\right)^2\right) \qquad (5\text{--}130)$$

$$\lim_{\varepsilon \to 1} f_A(a) = \frac{1}{\sigma_X} \varphi(\alpha) = \frac{1}{\sigma_X} \varphi\left(\frac{a - \mu_X}{\sigma_X}\right) \qquad (5\text{--}131)$$

Therefore, in the case of extreme narrow-bandedness, the local maxima A becomes Rayleigh-distributed with the parameter σ_X^2, i.e., $A \sim R(\sigma_X^2)$. In the case of extreme broad-bandedness, A is normally distributed with the mean value μ_X and the variance σ_X^2, i.e., $A \sim N(\mu_X, \sigma_X^2)$. It should be noted that in the narrow-banded case, all samples of A will be larger than μ_X, implying that $f_A(a) = 0$ for $a < \mu_X$ in Eq. (5-130).

Example 5-9: Probability density function of local maxima in a stationary Gaussian process

$\{X(t), t \in R\}$ is a stationary and ergodic Gaussian process with the mean value μ_X, the standard deviation σ_X and the auto-correlation coefficient function $\rho_{XX}(\tau)$. Further, we shall assume that the acceleration

---

[1] W.B. Huston and T.H. Skorpinski. Probability and frequency characteristics of some flight buffet loads. *NACA TN 3733*, August, 1956.

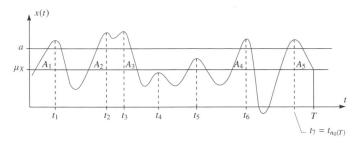

Fig. 5–13 Local maxima above a barrier a.

process $\{\ddot{X}(t),\ t \in R\}$ exists with a finite variance, i.e., $\sigma_{\ddot{X}}^2 < \infty$.

Let $A_1,\ A_2\ \ldots$ denote a sequence of local maxima of $\{X(t),\ t \in R\}$. Due to the stationarity, they become identically distributed and can be denoted as a single stochastic variable A. In what follows, the probability distribution function $F_A(a) = 1 - P(A > a)$ of A is determined in terms of the low order spectral moments of the process $\{X(t),\ t \in R\}$.

Fig. 5-13 shows a realization of the process of the length T. If T is sufficiently large, the probability $P(A > a)$ can be estimated by counting the number $n_a(T)$ of local maxima which fulfill the criteria of being above a in the interval $[0,T]$, in proportion to the total number $n_0(T)$ of local maxima in $[0,T]$. Hence, we have:

$$P(A > a) = \lim_{T \to \infty} \frac{n_a(T)}{n_0(T)} = \lim_{T \to \infty} \frac{\frac{1}{T}\,n_a(T)}{\frac{1}{T}\,n_0(T)} = \frac{\mu_a}{\mu_0} \tag{5–132}$$

where μ_a indicates the expected number of local maxima above a per unit of time, and μ_0 is the expected total number of maxima per unit of time. μ_0 has previously been determined by Eq. (5-31), and a similar explicit solution for μ_a is need for the use of Eq. (5-132).

First, the process is assumed to be narrow-banded. In this case, each up-crossing of the level μ_X is followed by one and only one local maximum. Similarly, each up-crossing of the level $A = a$ is followed by a single local maximum. Hence, the expected number of local maxima per unit of time above the level $A = a$ in proportion to the total expected number of local maxima per unit of time, is equal to the expected number of up-crossings of the level $A = a$ per unit of time in proportion to the expected number of up-crossings of the level μ_X per unit of time. The probability distribution function of A can thus be obtained by using Eq. (5-27):

$$P(A > a) \simeq \frac{f_1(a)}{f_1(\mu_X)} = \frac{\frac{1}{2\pi}\frac{\sigma_{\dot{X}}}{\sigma_X}\exp\left(-\frac{1}{2}\left(\frac{a-\mu_X}{\sigma_X}\right)^2\right)}{\frac{1}{2\pi}\frac{\sigma_{\dot{X}}}{\sigma_X}} = \exp\left(-\frac{1}{2}\left(\frac{a-\mu_X}{\sigma_X}\right)^2\right) \qquad \Rightarrow$$

$$F_A(a) = 1 - \exp\left(-\frac{1}{2}\left(\frac{a-\mu_X}{\sigma_X}\right)^2\right), \quad a \in [\mu_X, \infty[\tag{5–133}$$

It is seen from Eq. (5-133) that the local maxima in a narrow-banded Gaussian process are Rayleigh-distributed, i.e., $A \sim R(\sigma_X^2)$.

Next, an arbitrary stationary Gaussian process $\{X(t),\ t \in R\}$ is considered. Let $d\mu_x$ denote the expected number of local maxima per unit of time in the interval $]x, x + dx]$. This is determined by the following extension of the Rice's formula given by Eq. (5-29):

$$d\mu_x = \int_{-\infty}^{0} -\ddot{x}\, f_{X\dot{X}\ddot{X}}(x, 0, \ddot{x})\, d\ddot{x}dx \tag{5-134}$$

The events of local maxima in disjoint intervals $]x_1, x_1 + dx_1]$ and $]x_2, x_2 + dx_2]$ at the same time are independent, so the probabilities of them can be added linearly. Then, the expected number of local maxima per unit of time in the interval $]a, \infty[$ becomes:

$$\mu_a = \int_{a}^{\infty} \int_{-\infty}^{0} -\ddot{x}\, f_{X\dot{X}\ddot{X}}(x, 0, \ddot{x})\, d\ddot{x}dx \tag{5-135}$$

In order to evaluate the integral on the right-hand side of Eq. (5-135), the joint probability density function of the stochastic vector $\mathbf{X}(t) = [X(t), \dot{X}(t), \ddot{X}(t)]^T$ is needed. The joint probability density function for the related normalized stochastic variables $U(t) = (X(t) - \mu_X)/\sigma_X$, $V(t) = \dot{X}(t)/\sigma_{\dot{X}}$, $W(t) = \ddot{X}(t)/\sigma_{\ddot{X}}$ has been indicated in Eq. (5-93). The corresponding joint probability density function of the underlying stochastic variables reads:

$$f_{X\dot{X}\ddot{X}}(x, \dot{x}, \ddot{x}) = \frac{1}{(2\pi)^{\frac{3}{2}} \sqrt{1 - r^2}\, \sigma_X\, \sigma_{\dot{X}}\sigma_{\ddot{X}}} \exp\left(-\frac{1}{2} \frac{u^2 + (1 - r^2)v^2 + w^2 - 2ruw}{1 - r^2}\right) \tag{5-136}$$

where:

$$u = \frac{x - \mu_X}{\sigma_X}, \quad v = \frac{\dot{x}}{\sigma_{\dot{X}}}, \quad w = \frac{\ddot{x}}{\sigma_{\ddot{X}}} \tag{5-137}$$

r signifies the correlation coefficient of $X(t)$ and $\ddot{X}(t)$ as given by Eq. (5-92).

For $\dot{x} = 0 \Leftrightarrow v = 0$, Eq. (5-136) reduces to:

$$f_{X\dot{X}\ddot{X}}(x, 0, \ddot{x}) = \frac{1}{\sigma_X} \varphi(u) \frac{1}{\sqrt{2\pi}\sigma_{\dot{X}}} \frac{1}{\sigma_{\ddot{X}}\sqrt{1 - r^2}} \varphi\left(\frac{w - ru}{\sqrt{1 - r^2}}\right) \tag{5-138}$$

Inserting Eq. (5-138) into Eq. (5-135) yields:

$$\mu_a = \frac{1}{\sqrt{2\pi}} \frac{\sigma_{\ddot{X}}}{\sigma_{\dot{X}}} \int_{\alpha}^{\infty} \varphi(u) \left(\int_{-\infty}^{0} -\frac{w}{\sqrt{1 - r^2}} \varphi\left(\frac{w - ru}{\sqrt{1 - r^2}}\right) dw\right) du \tag{5-139}$$

where the change of integration variable from $\ddot{x}$ to $w = \ddot{x}/\sigma_{\ddot{X}}$ has bee introduced, and α is given by Eq. (5-129).

The innermost integral in Eq. (5-139) can be written as:

$$\int_{-\infty}^{0} -\frac{w}{\sqrt{1 - r^2}} \varphi\left(\frac{w - ru}{\sqrt{1 - r^2}}\right) dw = \sqrt{1 - r^2}\, \varphi\left(\frac{ru}{\sqrt{1 - r^2}}\right) - ru\, \Phi\left(-\frac{ru}{\sqrt{1 - r^2}}\right) \tag{5-140}$$

From Eqs. (5-31), (5-132), (5-139) and (5-140), it follows that the probability distribution function of A becomes:

$$F_A(a) = 1 - \sqrt{2\pi} \sqrt{1 - r^2} \int_{\alpha}^{\infty} \left(\varphi\left(\frac{ru}{\sqrt{1 - r^2}}\right) - \frac{ru}{\sqrt{1 - r^2}} \Phi\left(-\frac{ru}{\sqrt{1 - r^2}}\right)\right) \varphi(u)\, du \tag{5-141}$$

Table 5-2 Normalized spectral moment and bandwidth parameter ε as functions of the non-dimensional bandwidth parameter ζ.

ζ	$\dfrac{\lambda_2}{\omega_0^2 \lambda_0}$	$\dfrac{\lambda_4}{\omega_0^4 \lambda_0}$	$\varepsilon = \sqrt{1 - \dfrac{\lambda_2^2}{\lambda_0 \lambda_4}}$
0.01	1.0000333	1.000200	0.011546
0.10	1.0033333	1.020200	0.114369
1.00	1.3333333	3.200000	0.666667

The probability density function is then obtained by differentiation with respect to a:

$$f_A(a) = \frac{1}{\sigma_X} \left(\sqrt{1 - r^2} \, \varphi \left(\frac{\alpha}{\sqrt{1 - r^2}} \right) - \sqrt{2\pi} \, r\alpha \, \Phi \left(-\frac{r\alpha}{\sqrt{1 - r^2}} \right) \varphi(\alpha) \right) \tag{5–142}$$

Finally, Eq. (5-128) can be obtained from (5-142) upon elimination of r by the bandwidth parameter $\varepsilon = \sqrt{1 - r^2} \Rightarrow r = -\sqrt{1 - \varepsilon^2}$.

Example 5-10: Probability density function of local maxima in a Gaussian band-limited white noise process

$\{X(t), \, t \in R\}$ is a stationary Gaussian process with the mean value μ_X, the variance σ_X^2, and the band-limited one-sided auto-spectral density function $S_X(\omega)$ as shown in Fig. 3-11. The bandwidth is written as $B = 2\zeta\omega_0$, where ω_0 signifies the angular centre frequency and $\zeta \in]0, 1[$ is a non-dimensional bandwidth parameter. The probability density function of the local maxima is to be calculated as a function of $\alpha = \frac{a - \mu_X}{\sigma_X}$, for $\zeta = 0.01, 0.1, 0.3, 1.0$.

The spectral moments λ_j are given by Eq. (3-52), and they can be represented in the form:

$$\frac{\lambda_j}{\lambda_0 \, \omega_0^j} = \frac{1}{2\zeta (j + 1)} \left((1 + \zeta)^{j+1} - (1 - \zeta)^{j+1} \right) \tag{5–143}$$

Table 5-2 shows the fraction $\lambda_j / \lambda_0 \omega_0^j$ for $j = 2$ and $j = 4$ as a function of ζ, as well as the bandwidth parameter ε as given by Eq. (3-47).

Next, the probability density function of the local maxima A is calculated using Eq. (5-128). Fig. 5-14 shows the result of $\sigma_X f_A(a)$ as a function of $\alpha = \frac{a - \mu_X}{\sigma_X}$ for the considered values of ζ. As seen, the

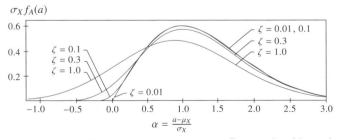

Fig. 5–14 Probability density function of local maxima in a stationary Gaussian band-limited white noise process.

graphs of the probability density functions for $\zeta = 0.01$ and $\zeta = 0.1$ are almost identical, and equal to the one for Rayleigh distribution, $A \sim R(\sigma_X^2)$. In this example, narrow-bandedness can be assumed for $\zeta \leq 0.1$, corresponding to $\varepsilon \leq 0.114$. For $\zeta = 1.0$ ($\varepsilon = \frac{2}{3}$), the probability density function has some resemblance to a normal distribution. However, the mean value of the distribution is still far from that of the limiting normal distribution $A \sim N(\mu_X, \sigma_X^2)$ obtained as $\varepsilon \to 1$, which corresponds to $\alpha = 0$ in Fig. 5-14.

Let's consider a scalar stochastic process $\{X(\tau),\ \tau \in [0, t]\}$. Of specific importance in structural design are the *maximum value* and the *minimum value* of $\{X(\tau),\ \tau \in [0, t]\}$ in the interval $[0, t]$, defined as the following two new stochastic processes:

$$X_{\max}([0, t]) = \max_{\tau \in [0,t]} X(\tau) \tag{5–144}$$

$$X_{\min}([0, t]) = \min_{\tau \in [0,t]} X(\tau) \tag{5–145}$$

$X_{\max}([0, t])$ is smaller than or equal to a certain value x, when all stochastic values in the process $\{X(\tau),\ \tau \in [0, t]\}$ fulfil this criterion. Hence, the probability distribution function of $X_{\max}([0, t])$ can be written as:

$$F_{X_{\max}}(x; [0, t]) = P\Big(X_{\max}([0, t]) \leq x\Big) = P\big(\forall \tau \in [0, t] : X(\tau) \leq x\big) \tag{5–146}$$

The last statement of Eq. (5-146) signifies the *reliability* of the stochastic process $\{X(\tau),\ \tau \in [0, t]\}$ with respect to the safe domain $S = \{\xi \mid -\infty < \xi < x\}$. Clearly, the reliability equals one minus the probability of failure, so does the probability distribution function:

$$F_{X_{\max}}(x; [0, t]) = 1 - P_f([0, t];]-\infty, x[) \tag{5–147}$$

where $P_f([0, t];]-\infty, x[)$ is the probability of failure for the considered safe domain.

Similarly, $X_{\min}([0, t])$ is larger than a certain lower barrier x, when all stochastic variables in $\{X(\tau),\ \tau \in [0, t]\}$ fulfil this criterion. The following equation holds:

$$P\Big(X_{\min}([0, t]) > x\Big) = P\big(\forall \tau \in [0, t] : X(\tau) > x\big) \tag{5–148}$$

The right-hand side represents the reliability of the process $\{X(\tau),\ \tau \in [0, t]\}$ with respect to the safe domain $S = \{\xi \mid x < \xi < \infty\}$. In this case, the probability distribution function equals the probability of failure:

$$F_{X_{\min}}(x; [0, t]) = P\Big(X_{\min}([0, t]) \leq x\Big) = P_f([0, t];]x, \infty[) \tag{5–149}$$

where $P_f([0, t];]x, \infty[)$ is the probability of failure of $\{X(\tau), \tau \in [0, t]\}$ relative to the considered safe domain. Fig. 5-15 shows realizations of the maximum and the minimum stochastic processes, as well as the safe domains.

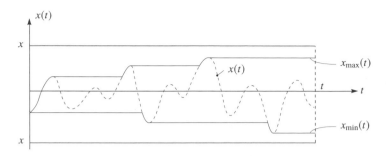

Fig. 5–15 Realizations of the maximum and the minimum stochastic processes.

In Eqs. (5-147) and (5-149), the probability distribution functions of the extremes of a stochastic process $\{X(\tau),\ \tau \in [0,t]\}$ have been expressed in terms of the probability of failure with respect to properly defined time-constant single barrier safe domains. If $\{X(\tau),\ \tau \in [0,t]\}$ is a stationary Gaussian process, $P_f([0,t];\ ]-\infty, x[)$ can be approximated by Eq. (5-84) with $b = x$. Hence, the probability distribution function of the maximum value process becomes:

$$F_{X_{\max}}(x; [0,t]) \simeq \Phi\left(\frac{x - \mu_X}{\sigma_X}\right) \exp\left(-\alpha(x)\, \frac{\sigma_{\dot{X}}}{2\pi\sigma_X}\, \exp\left(-\frac{1}{2}\left(\frac{x - \mu_X}{\sigma_X}\right)^2\right) t\right) \qquad (5\text{–}150)$$

where $\alpha(x)$ is given by Eq. (5-85).

Similarly, $P_f([0,t];\ ]x,\infty[)$ can be approximated by Eqs. (5-86) with $a = x$. Hence, we have:

$$F_{X_{\min}}(x; [0,t]) \simeq 1 - \Phi\left(\frac{\mu_X - x}{\sigma_X}\right) \exp\left(-\alpha(x)\, \frac{\sigma_{\dot{X}}}{2\pi\sigma_X}\, \exp\left((-\frac{1}{2}\left(\frac{\mu_X - x}{\sigma_X}\right)^2)\right) t\right) \qquad (5\text{–}151)$$

where $\alpha(x)$ is given by Eq. (5-87).

If $\{X(\tau),\ \tau \in [0,t]\}$ is a stationary process, the expected value $\mu_{X_{\max}([0,t])}$ and the standard deviation $\sigma_{X_{\max}([0,t])}$ of the maximum value process can be written as:

$$\mu_{X_{\max}([0,t])} = \mu_X + p(t)\,\sigma_X \qquad (5\text{–}152)$$

$$\sigma_{X_{\max}([0,t])} = s(t)\,\sigma_X \qquad (5\text{–}153)$$

where $p(t)$ is termed the *peak factor*.

Furthermore, if $\{X(\tau),\ \tau \in [0,t]\}$ is Gaussian, the probability distribution function of its maximum value is approximated by Eq. (5-150). Let's denote $T_0 = 2\pi\frac{\sigma_X}{\sigma_{\dot{X}}}$ as the average time interval between local maxima, which is the reciprocal of μ_0 given by Eq. (5-31). Then, it can be proved that for a process $\{X(\tau),\ \tau \in [0,t]\}$ with sufficiently large time length, i.e., for $t \gg T_0$, $p(t)$ and $s(t)$ in Eqs. (5-152) and (5-153) have the asymptotic expressions:

$$p(t) \simeq \sqrt{2 \ln n} + \frac{0.5772}{\sqrt{2 \ln n}} \qquad (5\text{–}154)$$

$$s(t) \simeq \frac{\pi}{\sqrt{6}} \frac{1}{\sqrt{2 \ln n}} \tag{5–155}$$

where n denotes the average number of local maxima during the time interval $[0, t]$:

$$n = \frac{t}{T_0} = \frac{1}{2\pi} \frac{\sigma_{\ddot{X}}}{\sigma_{\dot{X}}} t \tag{5–156}$$

Civil engineering structures are usually lightly damped, for which reason the response processes become narrow-banded. Then, T_0 can be approximated as the fundamental eigenperiod of the structure, and n approximately indicates the expected number of up-crossings with respect to the mean value level during the time interval t.

From Eqs. (5-152) to (5-155), it can be deduced that the variational coefficient $V_{X_{\max}([0,t])} \to 0$ as $t \to \infty$. Hence, the sample values of $X_{\max}([0, t])$ will group ever closer at the expected value $\mu_{X_{\max}([0,t])}$ as $t \to \infty$. At sufficiently large excitation intervals, the structural design may be based entirely on $\mu_{X_{\max}([0,t])}$. In the Danish codes of practice for wind engineering, design is based on Eqs. (5-152) and (5-154).

CHAPTER 6
MONTE CARLO SIMULATION TECHNIQUE

For multi-dimensional and nonlinear systems, analytical solutions of the stochastic dynamic response and reliability problems as outlined in Chapters 3, 4, 5 may become cumbersome or even impossible. In such cases, *Monte Carlo simulation* may be the last resort. The method is simple to apply, but with the disadvantage that large computational effort is needed, especially at reliability analyses at high barrier levels.

With reference to the stochastic vector differential equation Eq. (4-67), it is assumed that a stochastic model has been formulated for the load process $\{\mathbf{F}(t),\ t \in [t_0, \infty[\}$ and for the stochastic initial value vector $[\mathbf{X}_0, \dot{\mathbf{X}}_0]$, from which realizations $\mathbf{f}_n(t)$ and $[\mathbf{x}_{0,n}, \dot{\mathbf{x}}_{0,n}]$ may be generated. The corresponding realizations $\mathbf{x}_n(t)$ of the displacement process $\{\mathbf{X}(t),\ t \in [t_0, \infty[\}$ are generated by solving the ordinary differential equations of motion:

$$\left. \begin{array}{l} \mathbf{M}\ddot{\mathbf{x}}_n(t) + \mathbf{C}\dot{\mathbf{x}}_n(t) + \mathbf{K}\mathbf{x}_n(t) = \mathbf{f}_n(t), \quad t \in]t_0, \infty[\\[2mm] \mathbf{x}_n(t_0) = \mathbf{x}_{n,0}, \quad \dot{\mathbf{x}}_n(t_0) = \dot{\mathbf{x}}_{n,0} \end{array} \right\} , \quad n = 1, \ldots, N \qquad (6\text{--}1)$$

Eq. (6-1) needs to be solved numerically by means of an unconditional stable time integration scheme (Newmark, generalized α-method). This has to be performed for each of the N realizations of the load process and the initial values.

Next, unbiased estimates of the mean value functions $\mu_{X_j}(t)$ and the cross-covariance functions $\kappa_{X_j X_k}(t_1, t_2)$ for the component processes are obtained from, cf. Eqs. (3-220) and (3-221):

$$\mu_{X_j}(t) \simeq \frac{1}{N} \sum_{n=1}^{N} x_{nj}(t) \qquad (6\text{--}2)$$

$$\kappa_{X_j X_k}(t_1, t_2) \simeq \frac{1}{N} \sum_{n=1}^{N} x_{nj}(t_1) x_{nk}(t_2) - \mu_{X_j}(t_1)\mu_{X_k}(t_2) \qquad (6\text{--}3)$$

where $x_{nj}(t)$ signifies the jth component of $\mathbf{x}_n(t)$.

Higher order moments can be estimated in a similar way, although generally an increased number of realizations are needed to achieve the same accuracy of the estimates.

Obviously, it is not necessary to store all N realizations of the response in Eqs. (6-2) and (6-3) before the stochastic responses are calculated. Instead, the sums $\sum_{n=1}^{m} x_{nj}(t)$ and $\sum_{n=1}^{m} x_{nj}(t_1)x_{nk}(t_2)$ are updated after each new realization $\mathbf{y}_m(t)$ has been generated. The final estimates are calculated after the Nth realization has been included into the sums.

The indicated Monte Carlo simulation technique using multiple independent realizations is the only approach valid for non-stationary processes. For stationary (ergodic) stochastic processes, ergodic sampling can be carried out from a single realization $\mathbf{x}(t)$ of sufficient length T. In this case, the sampling length T is divided into N equidistant intervals of the length $\Delta t = \frac{T}{N}$. Replacing the time-integrals in Eqs. (3-224) and (3-246) by Riemann sums, the mean values and the cross-covariance functions are estimated as:

$$\mu_{X_j} = \frac{1}{T} \int_0^T x_j(\tau)\,d\tau \;\simeq\; \frac{1}{N} \sum_{n=1}^{N} x_j(t_n) \tag{6–4}$$

$$\kappa_{X_j X_k}(\tau_m) = \frac{1}{T - \tau_m} \int_0^{T-\tau_m} x_j(\tau)x_k(\tau + \tau_m)\,d\tau \;-\; \mu_{X_j}\mu_{X_k}$$

$$\simeq \frac{1}{N-m} \sum_{n=1}^{N-m} x_j(t_n)x_k(t_n + \tau_m) \;-\; \mu_{X_j}\mu_{X_k}, \quad m = 1, \ldots, N-1 \tag{6–5}$$

where:

$$t_n = t_0 + n\,\Delta t, \quad \tau_m = m\,\Delta t \tag{6–6}$$

The realization $\mathbf{x}(t)$ is obtained by numerical integration of Eq. (6-1), using a sufficiently long realization $\mathbf{f}(t)$ of the load process.

6.1   Equivalent white noise processes

In stochastic structural dynamics, the load process is typically obtained by the output of a Gaussian white noise process $\{W(t),\, t \in R\}$ (with the double-sided auto-spectral density function S_0) passing through a rational shaping filter as explained in Eq. (4-43). The white noise process is an abstraction with discontinuous realizations and infinite variance. Hence, for applying Monte Carlo simulations, the white noise process needs to be replaced by a so-called *equivalent white noise process* with finite variance, which effectively produces the same stochastic response of the structure. The underlying idea is based on the equivalent white noise approximation described in Eq. (4-55), where merely the value of the auto-spectral density function at the angular frequency ω_0 is shown to have significant influence on the variance of the response. The equivalent white noise process should be chosen as a broad-banded stationary Gaussian process $\{\tilde{W}(t),\, t \in R\}$, of which the auto-spectral density function is approximately constant at the value S_0 up to a maximum angular frequency ω_m. ω_m should be well above all angular eigenfrequencies of the structural system of importance for the structural response, and above all the poles of the shaping

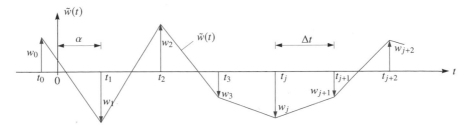

Fig. 6–1 Realization of the broken line process.

filter.

Clough and Penzien[1] suggested to use the so-called *broken line process*, also known as the stochastic finite element model with linear interpolation functions, as an equivalent white noise process. A typical realization of the process is shown in Fig. 6-1. W_0, W_1, W_2, ... is a sequence of mutually independent, identically distributed normal stochastic variables, $W \sim N(0, \sigma_W^2)$, where the variance σ_W^2 will be indicated later. W_j is related to the instant of time $T_j = (\alpha + j - 1)\Delta t$, where α is a stochastic variable uniformly distributed in the interval $[0, 1]$, i.e., $\alpha \sim U(0, 1)$. Additionally, α is stochastically independent of all W_j, $j = 0, 1, 2 \dots$. The interval Δt between the times T_j and T_{j+1}, with the attached stochastic variables W_j and W_{j+1}, will be determined later.

During the interval $[T_j, T_{j+1}[$, the equivalent white noise process is obtained by linear interpolation between the values W_j and W_{j+1}:

$$\tilde{W}(t) = W_j + (W_{j+1} - W_j)\frac{t - T_j}{\Delta t}, \quad t \in [T_j, T_j + \Delta t[\tag{6–7}$$

Since α, and hence T_j, is independent of W_j, the mean value function of $\{\tilde{W}(t), t \in R\}$ becomes:

$$\mu_{\tilde{W}}(t) = E[W_j] + E[W_{j+1} - W_j] E\left[\frac{t - T_j}{\Delta t}\right] = 0 \tag{6–8}$$

As will be shown in Example 6-1, the auto-covariance function of $\{\tilde{W}(t), t \in R\}$ is written as:

$$\kappa_{\tilde{W}\tilde{W}}(\tau) = \begin{cases} \sigma_W^2 \left(\dfrac{2}{3} - \dfrac{\tau^2}{\Delta t^2} + \dfrac{1}{2}\dfrac{|\tau|^3}{\Delta t^3}\right) & , \quad \dfrac{|\tau|}{\Delta t} \in [0, 1] \\[3mm] \sigma_W^2 \left(\dfrac{4}{3} - 2\dfrac{|\tau|}{\Delta t} + \dfrac{\tau^2}{\Delta t^2} - \dfrac{1}{6}\dfrac{|\tau|^3}{\Delta t^3}\right), & \dfrac{|\tau|}{\Delta t} \in]1, 2] \\[3mm] 0 & , \quad \dfrac{|\tau|}{\Delta t} \in]2, \infty[\end{cases} \tag{6–9}$$

It is seen from Eq. (6-9) that the covariance between $\tilde{W}(t)$ and $\tilde{W}(t + \tau)$ is non-zero when $|\tau| < 2\Delta t$. As $\Delta t \to 0$, the equivalent white noise process tends to become δ-correlated.

---

[1] R.W. Clough and J. Penzien. *Dynamics of Structures*. McGraw-Hill, New York, 1982.

The auto-spectral density function $S_{\tilde{W}\tilde{W}}(\omega)$ follows from the Wiener-Khintchine relation in Eq. (3-28):

$$
\begin{aligned}
S_{\tilde{W}\tilde{W}}(\omega) &= \frac{1}{\pi} \int_0^{\Delta t} \cos(\omega\tau)\, \sigma_W^2 \left(\frac{2}{3} - \frac{\tau^2}{\Delta t^2} + \frac{1}{2} \frac{|\tau|^3}{\Delta t^3} \right) d\tau \\
&+ \frac{1}{\pi} \int_{\Delta t}^{2\Delta t} \cos(\omega\tau)\, \sigma_W^2 \left(\frac{4}{3} - 2\frac{|\tau|}{\Delta t} + \frac{\tau^2}{\Delta t^2} - \frac{1}{6}\frac{|\tau|^3}{\Delta t^3} \right) d\tau \\
&= \frac{\sigma_W^2 \Delta t}{2\pi} \left(\frac{\sin\left(\frac{1}{2}\omega\Delta t\right)}{\frac{1}{2}\omega\Delta t} \right)^4
\end{aligned}
\tag{6–10}
$$

For $\omega\Delta t \to 0$, the following limit of Eq. (6-10) is obtained:

$$
S_{\tilde{W}\tilde{W}}(0) = \frac{\sigma_W^2 \Delta t}{2\pi} = S_0
\tag{6–11}
$$

$S_{\tilde{W}\tilde{W}}(\omega)/S_0$ as given by Eqs. (6-10) and (6-11) is shown in Fig. 6-2 as a function of $\omega\Delta t$.

Calibration of Δt and σ_W^2 is to be carried out. It is decided that the auto-spectral density function remains almost flat in the interval $[0, \omega_m]$, and all values of $S_{\tilde{W}\tilde{W}}(\omega)$ in this interval is no less than $0.99S_0$. Δt can thus be obtained as:

$$
\frac{S_{\tilde{W}\tilde{W}}(\omega_m)}{S_0} \geq 0.99 \qquad \Rightarrow \qquad \Delta t \leq \frac{0.2455}{\omega_m}
\tag{6–12}
$$

Next, σ_W is determined from Eq. (6-11):

$$
\sigma_W = \sqrt{\frac{2\pi}{\Delta t} S_0}
\tag{6–13}
$$

Because of the random factor $\frac{t-T_i}{\Delta t}$, $\{\tilde{W}(t),\, t \in [t_0, \infty[\}$ will not be a Gaussian stochastic process. Hence, the displacement process $\{\mathbf{X}(t),\, t \in [t_0, \infty[\}$ in Eq. (6-1) will not be Gaussian either, even

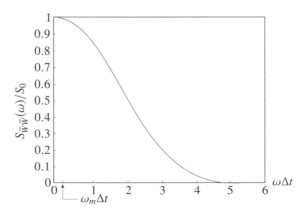

Fig. 6–2 Normalized auto-spectral density function of the broken line process

if the initial values $\mathbf{X}_0$ and $\dot{\mathbf{X}}_0$ are normally distributed and independent of $\{\tilde{W}(t),\, t \in [t_0, \infty[\}$. This may only be of importance for the simulation method using Eqs. (6-2) and (6-3). For the methods based on ergodic sampling, the initial values $\mathbf{X}_0$ and $\dot{\mathbf{X}}_0$, and the parameter α will have fading influence on the structural response. Hence, the response will become Gaussian after a transient phase, which should be passed, before sampling is initiated.

Example 6-1: Derivation of the auto-covariance function of a broken line process

Following a derivation by Ditlevsen[1], the auto-covariance function $\kappa_{\dot{\tilde{W}}\dot{\tilde{W}}}(\tau)$ of the derivative process $\{\dot{\tilde{W}}(t),\, t \in R\}$ is first determined. Next, $\kappa_{\tilde{W}\tilde{W}}(\tau)$ is obtained by a double integration, cf. Eq. (3-136).

From Eq. (6-7), $\{\tilde{W}(t),\, t \in R\}$ can be written as:

$$\dot{\tilde{W}}(t) = \frac{1}{\Delta t}\left(W_{j+1} - W_j\right), \quad t \in [T_j, T_{j+1}[\tag{6–14}$$

On the condition that $\tau \geq 0$, the auto-covariance function of $\{\dot{\tilde{W}}(t),\, t \in R\}$ becomes:

$$\kappa_{\dot{\tilde{W}}\dot{\tilde{W}}}(\tau) = E\left[\dot{\tilde{W}}(0)\dot{\tilde{W}}(\tau)\right]$$

$$= \frac{1}{\Delta t^2} E\left[(W_1 - W_0)^2 \,\Big|\, \frac{\tau}{\Delta t} \leq \alpha\right] P\left(\frac{\tau}{\Delta t} \leq \alpha\right)$$

$$+ \frac{1}{\Delta t^2} E\left[(W_1 - W_0)(W_2 - W_1) \,\Big|\, \alpha < \frac{\tau}{\Delta t} \leq \alpha + 1\right] P\left(\alpha < \frac{\tau}{\Delta t} \leq \alpha + 1\right)$$

$$+ \frac{1}{\Delta t^2} E\left[(W_1 - W_0)(W_3 - W_2) \,\Big|\, \alpha + 1 < \frac{\tau}{\Delta t} \leq \alpha + 2\right] P\left(\alpha + 1 < \frac{\tau}{\Delta t} \leq \alpha + 2\right) + \cdots$$

$$= \frac{1}{\Delta t^2} E\left[(W_1 - W_0)^2\right] P\left(\frac{\tau}{\Delta t} \leq \alpha\right)$$

$$+ \frac{1}{\Delta t^2} E\left[(W_1 - W_0)(W_2 - W_1)\right] P\left(\alpha < \frac{\tau}{\Delta t} \leq \alpha + 1\right)$$

$$+ \frac{1}{\Delta t^2} E\left[(W_1 - W_0)(W_3 - W_2)\right] P\left(\alpha + 1 < \frac{\tau}{\Delta t} \leq \alpha + 2\right) + \cdots$$

$$= \frac{1}{\Delta t^2} E\left[W_1^2 - 2W_0 W_1 + W_0^2\right] P\left(\frac{\tau}{\Delta t} \leq \alpha\right)$$

$$+ \frac{1}{\Delta t^2} E\left[[-W_1^2 + W_1 W_2 - W_0 W_2 + W_0 W_1]\right] P\left(\alpha < \frac{\tau}{\Delta t} \leq \alpha + 1\right)$$

$$= \sigma_W^2 \frac{2}{\Delta t^2} P\left(\frac{\tau}{\Delta t} \leq \alpha\right) - \sigma_W^2 \frac{1}{\Delta t^2} P\left(\alpha < \frac{\tau}{\Delta t} \leq \alpha + 1\right) \tag{6–15}$$

At the derivation of Eq. (6-15), it has been used that all W_i are independent of α, so the conditional expectations equal the corresponding unconditional ones. Further, it has been used that W_i are mutually

---

[1]O. Ditlevsen. *Extremes and First Passage Times with Applications in Civil Engineering: Some Approximate Results in the Theory of Stochastic Processes.* Doctoral Thesis, Technical University of Denmark, Lyngby, Denmark, 1971.

independent zero-mean stochastic variables, so $E[W_0 W_1] = E[W_0]E[W_1] = 0$, etc. Since $\alpha \sim U(0, 1)$, we have $P(a < \alpha < b) = \min(1, b) - \max(0, a)$. Then, Eq. (6-15) becomes:

$$
\kappa_{\dot{\tilde{W}}\dot{\tilde{W}}}(\tau) = \begin{cases}
\sigma_W^2 \dfrac{2}{\Delta t^2}\left(1 - \dfrac{\tau}{\Delta t}\right) - \sigma_W^2 \dfrac{1}{\Delta t^2}\dfrac{\tau}{\Delta t}, & \dfrac{\tau}{\Delta t} \in [0, 1] \\[3mm]
-\sigma_W^2 \dfrac{1}{\Delta t^2}\left(2 - \dfrac{\tau}{\Delta t}\right) & , \quad \dfrac{\tau}{\Delta t} \in]1, 2] \\[3mm]
0 & , \quad \dfrac{\tau}{\Delta t} \in]2, \infty[
\end{cases}
\tag{6–16}
$$

From Eq. (3-136), $\frac{d^2}{d\tau^2}\kappa_{\tilde{W}\tilde{W}}(\tau) = -\kappa_{\dot{\tilde{W}}\dot{\tilde{W}}}(\tau)$. Then, integration of Eq. (6-16) provides;

$$
\frac{d}{d\tau}\kappa_{\tilde{W}\tilde{W}}(\tau) = \begin{cases}
-\sigma_W^2 \dfrac{1}{\Delta t}\left(a_1 + 2\dfrac{\tau}{\Delta t} - \dfrac{3}{2}\dfrac{\tau^2}{\Delta t^2}\right), & \dfrac{\tau}{\Delta t} \in [0, 1] \\[3mm]
\sigma_W^2 \dfrac{1}{\Delta t}\left(b_1 + 2\dfrac{\tau}{\Delta t} - \dfrac{1}{2}\dfrac{\tau^2}{\Delta t^2}\right), & \dfrac{\tau}{\Delta t} \in]1, 2] \\[3mm]
c_1 & , \quad \dfrac{\tau}{\Delta t} \in]2, \infty[
\end{cases}
\tag{6–17}
$$

where a_1, b_1 and c_1 are integration constants. Recalling that $\frac{d}{d\tau}\kappa_{\tilde{W}\tilde{W}}(0) = 0$ as shown in Fig. 3-18, we have $a_1 = 0$. Further, $\frac{d}{d\tau}\kappa_{\tilde{W}\tilde{W}}(\tau) = 0$ for $\tau \geq 2\Delta t$, which implies that $c_1 = 0$. b_1 is determined, so $\frac{d}{d\tau}\kappa_{\tilde{W}\tilde{W}}(\tau)$ is continuous for $\tau = \Delta t$ and $\tau = 2\Delta t$. In both cases, the solution $b_1 = -2$ is obtained. Therefore, Eq.(6-17) attains the form:

$$
\frac{d}{d\tau}\kappa_{\tilde{W}\tilde{W}}(\tau) = \begin{cases}
\sigma_W^2 \dfrac{1}{\Delta t}\left(-2\dfrac{\tau}{\Delta t} + \dfrac{3}{2}\dfrac{\tau^2}{\Delta t^2}\right) & , \quad \dfrac{\tau}{\Delta t} \in [0, 1] \\[3mm]
\sigma_W^2 \dfrac{1}{\Delta t}\left(-2 + 2\dfrac{\tau}{\Delta t} - \dfrac{1}{2}\dfrac{\tau^2}{\Delta t^2}\right), & \dfrac{\tau}{\Delta t} \in]1, 2] \\[3mm]
0 & , \quad \dfrac{\tau}{\Delta t} \in]2, \infty[
\end{cases}
\tag{6–18}
$$

One further integration yields:

$$
\kappa_{\tilde{W}\tilde{W}}(\tau) = \begin{cases}
\sigma_W^2 \left(a_2 - \dfrac{\tau^2}{\Delta t^2} + \dfrac{1}{2}\dfrac{\tau^3}{\Delta t^3}\right) & , \quad \dfrac{\tau}{\Delta t} \in [0, 1] \\[3mm]
\sigma_W^2 \left(b_2 - 2\dfrac{\tau}{\Delta t} + \dfrac{\tau^2}{\Delta t^2} - \dfrac{1}{6}\dfrac{\tau^3}{\Delta t^3}\right), & \dfrac{\tau}{\Delta t} \in]1, 2] \\[3mm]
c_2 & , \quad \dfrac{\tau}{\Delta t} \in]2, \infty[
\end{cases}
\tag{6–19}
$$

where a_2, b_2, c_2 are new integration constants. Again, the requirement $\kappa_{\tilde{W}\tilde{W}}(\tau) = 0$ for $\tau \geq 2\Delta t$ implies that $c_2 = 0$. The continuity of $\kappa_{\tilde{W}\tilde{W}}(\tau)$ for $\tau = 2\Delta t$ implies that $b_2 = \frac{4}{3}$. Further, continuity of $\kappa_{\tilde{W}\tilde{W}}(\tau)$ for $\tau = \Delta t$ implies that $a_2 = \frac{2}{3}$. Insertion of a_2, b_2, c_2 into Eq. (6-19) provides Eq. (6-9) for $\tau > 0$. The final result providing the additional solution for $\tau < 0$, is obtained from the symmetry property, i.e., $\kappa_{\tilde{W}\tilde{W}}(\tau) = \kappa_{\tilde{W}\tilde{W}}(-\tau)$.

Example 6-2: Ergodic sampling of statistical moments of a SDOF oscillator exposed to a white noise excitation

A SDOF oscillator is exposed to a Gaussian white noise excitation. The mean value μ_X and the standard deviation σ_X of the displacement process $\{X(t), t \in R\}$ are to be determined by ergodic sampling.

The stochastic differential equation is given by, cf. Eqs. (4-1), (4-4) and (4-5):

$$\left. \begin{aligned} \ddot{X}(t) + 2\zeta\omega_0 \dot{X}(t) + \omega_0^2 X(t) &= \frac{1}{m} W(t), \quad t > 0 \\ X(0) = 0, \quad \dot{X}(0) &= 0 \end{aligned} \right\} \tag{6-20}$$

where the following values are used for the system parameters:

$$m = 1, \quad \zeta = 0.01, \quad \omega_0 = 1 \tag{6-21}$$

$\{W(t), t \in R\}$ is a unit white noise process with $S_0 = \frac{1}{2\pi}$, cf. Eqs. (3-40). It has the following properties:

$$\left. \begin{aligned} E[W(t)] &= 0 \\ E[W(t)W(t+\tau)] &= \delta(\tau) \end{aligned} \right\} \tag{6-22}$$

From Eq. (4-40), the standard deviation of $X(t)$ when it becomes stationary can be theoretically calculated:

$$\sigma_{X,0} = \sqrt{\frac{\pi S_0}{2\zeta\omega_0^3 m^2}} = 5 \tag{6-23}$$

In the numerical analysis, $\{W(t), t \in R\}$ is replaced by an equivalent white noise process (the broken line process) $\{\tilde{W}(t), t \in R\}$. The cut-out angular frequency is chosen as $\omega_m = 10\omega_0 = 10$. The time step Δt for linear interpolation and the standard deviation $\sigma_{\tilde{W}}^2$ are then obtained from Eqs. (6-12) and (6-13), respectively.

The 4th order Runge-Kutta scheme is used for numerical integration of Eq. (6-20), based on a state vector formulation of the equation of motion, cf. Eqs. (3-168) and (3-205). The time step in the numerical integration has been chosen as $\Delta t_0 = \frac{T_0}{40}$, where $T_0 = \frac{2\pi}{\omega_0}$ is the undamped eigenperiod of the oscillator.

$x_j = x(t_j)$, $t_j = j\Delta t_0$, $j = 0, 1, 2, \ldots$ is determined at the end of each time step Δt_0. Then, the mean value $\mu_{X,k}$ and the standard deviation $\sigma_{X,k}$ after k steps are estimated as:

$$\left. \begin{aligned} \mu_{X,k} &= \frac{1}{k} \sum_{j=1}^{k} x_j \\ \sigma_{X,k} &= \sqrt{\frac{1}{k} \sum_{j=1}^{k} x_j^2 - \mu_{X,k}^2} \end{aligned} \right\} \tag{6-24}$$

Figs. 6-3 and 6-4 show the development of $\mu_{X,k}$ and $\sigma_{X,k}$ as a function of the sampling length t normalized with respect to the eigenperiod T_0. It is seen that convergence demands a rather long sampling length. Especially, the non-zero mean value estimate at the end of 4000 periods brings to the need for an even longer sampling length.

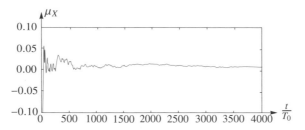

Fig. 6–3 Convergence study for the mean value μ_X.

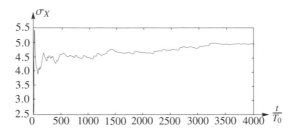

Fig. 6–4 Convergence study for the standard deviation σ_X.

Example 6-3: Simulation of realizations of the sea-surface elevation process by solving linear stochastic differential equations

The auto-spectral density function of the stationary sea-surface elevation process $\{\eta(t),\, t \in R\}$ can be described by the *JONSWAP wave spectrum*. The double-sided spectrum has the following form:[1]

$$S_{\eta\eta}(\omega) = \beta\,\frac{5}{32}\,\frac{H_s^2}{\omega_p}\,\frac{\gamma^\alpha}{\gamma}\left(\frac{|\omega|}{\omega_p}\right)^{-5}\exp\left(-\frac{5}{4}\left(\frac{\omega}{\omega_p}\right)^{-4}\right) \tag{6–25}$$

where:

$$
\left.
\begin{aligned}
\alpha &= \exp\left(-\frac{1}{2}\left(\frac{|\omega|-\omega_p}{\sigma_f(\omega)\,\omega_p}\right)^2\right) \\[4pt]
\sigma_f(\omega) &= \begin{cases} 0.1, & |\omega| < \omega_p \\ 0.5, & |\omega| \geq \omega_p \end{cases} \\[4pt]
\beta &= 1.4017 \\[4pt]
\gamma &= 3.3 \\[4pt]
T_p &= \frac{2\pi}{\omega_p} = \sqrt{\frac{180 H_s}{g}}
\end{aligned}
\right\} \tag{6–26}
$$

H_s is the *significant wave height*, which is usually defined as the average of the highest third of the wave heights, T_p is the *peak period* and g is the *acceleration of gravity*. The values of β and γ in Eq.(6-26) ensure that $H_s = 4.0\sigma_\eta$, corresponding to Rayleigh-distributed wave heights. The significant wave height and the peak period are positively correlated, but in reality do not exactly follow the functional relation

---

[1]K. Hasselmann *et al.* Measurement of wind wave growth and swell decay during the joint North Sea project (JONSWAP). *Erganzungsheft zur Deutschen Hydrograph. Z.*, Reihe A, No. 12, Hamburg, 1973.

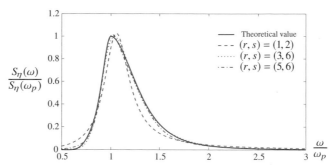

Fig. 6–5 Rational approximations to the auto-spectral density function (plotted as one-sided function) of the sea-surface elevation process.

in Eq. (6-26). However, the indicated relation has been applied in the ISO/NP 19902 code of practice for fixed offshore structures.

The one-sided auto-spectral density function corresponding to Eq. (6-25) has been plotted as the solid curve in Fig. 6-5. The abscissa has been normalized with respect to the angular peak frequency, and the ordinate has been normalized with respect to the peak value $S_\eta(\omega_p) = \beta \frac{5}{16} \exp\left(-\frac{5}{4}\right) \frac{H_s^2}{\omega_p}$ of the auto-spectral density.

A rational auto-spectral density function as given by Eqs. (3-58) and (3-54) is to be determined as an approximation to the double-sided auto-spectral density function $S_{\eta\eta}(\omega)$. Without loss of generality, the white noise input process is assumed to be a unit white noise, i.e., $S_0 = \frac{1}{2\pi}$. Then, from Eqs. (3-54), (3-55) and (3-58), $S_{\eta\eta}(\omega)$ can be approximated as:

$$S_{\eta\eta}(\omega) \simeq \frac{1}{2\pi} \left| \frac{p_0 z^r + p_1 z^{r-1} + \cdots + p_{r-1} z + p_r}{z^s + q_1 z^{s-1} + \cdots + q_{s-1} z + q_s} \right|^2, \quad z = i\omega \qquad (6\text{–}27)$$

The coefficients of the rational filter should be selected in such a way that the right-hand side of the equation gives the best fit to the left-hand side over a given range of angular frequencies. Further, the stability condition for the poles of the denominator polynomial should be fulfilled as shown in Eq. (3-57). This leads to the following constrained optimization problem:

$$\left.\begin{array}{l} \displaystyle \min_{\substack{p_0,\,\ldots,\,p_r \\ q_1,\,\ldots,\,q_r}} \int_{-\infty}^{\infty} \left(S_{\eta\eta}(\omega) - \frac{1}{2\pi} \frac{P(i\omega)}{Q(i\omega)} \frac{P(-i\omega)}{Q(-i\omega)} \right)^2 d\omega \\[4mm] \text{s.t} \\[2mm] \mathrm{Re}(z_j) < 0, \quad j = 1, \ldots, s \\[4mm] \displaystyle \frac{1}{2\pi} \frac{P(i\omega_p)}{Q(i\omega_p)} \frac{P(-i\omega_p)}{Q(-i\omega_p)} = S_{\eta\eta}(\omega_p) \end{array}\right\} \qquad (6\text{–}28)$$

The last constraint in Eq. (6-28) insures that the values of the rational auto-spectral density function and the target auto-spectral density function are identical at the peak angular frequency ω_p.

Table 6-1 Non-dimensional rational filter parameters for JONSWAP auto-spectral density function, $H_s = 7$ m.

r	s	$\bar{p}_0$	$\bar{p}_1$	$\bar{p}_2$	$\bar{p}_3$	$\bar{p}_4$	$\bar{p}_5$	$\bar{q}_1$	$\bar{q}_2$	$\bar{q}_3$	$\bar{q}_4$	$\bar{q}_5$	$\bar{q}_6$
1	2	0.7826	0.0000					0.3081	1.119				
2	4	0.7292	0.5605	0.0000				1.029	3.158	1.300	1.912		
3	4	0.2345	0.8134	0.1785	0.0000			1.017	2.964	1.216	1.707		
3	6	3.4667	0.5382	1.4353	0.0000			3.283	7.064	8.479	8.347	4.379	2.032
5	6	0.3905	0.0000	0.0000	1.2581	1.9101	0.0000	3.086	9.264	11.12	16.61	8.141	7.308

The solutions to the optimization problem are given in the form:

$$
\left.
\begin{aligned}
p_j &= \sqrt{S_{\eta\eta}(\omega_p)}\, \omega_p^{s-r+j}\, \bar{p}_j, &\quad j &= 0, \dots, r \\
q_j &= \omega_p^j\, \bar{q}_j &\quad , \quad j &= 1, \dots, s
\end{aligned}
\right\}
\tag{6–29}
$$

The optimal non-dimensional rational filter parameters $\bar{p}_j$ and $\bar{q}_j$ have been indicated in Table 6-1 for some low order rational filters. The reason for setting $\bar{p}_r = p_r = 0$ is to eliminate the apparent drift or slow convergence of the mean value μ_η as noticed in Fig. 6-3. Hence, p_r has not been included in the optimization problem in Eq. (6-28) for obtaining the results in Table 6-1.

The obtained rational auto-spectral density functions (one-sided) have been depicted in Fig. 6-5 for the cases $(r, s) = (1, 2), (r, s) = (3, 6)$ and $(r, s) = (5, 6)$. $(r, s) = (5, 6)$ gives a better approximation than $(r, s) = (3, 6)$, because more free parameters are available in the optimization problem. However, the difference is hardly visible in Fig. 6-5. On the other hand, the rational auto-spectral density function corresponding to $(r, s) = (1, 2)$ differs significantly from the target spectrum.

One further observation is that the theoretical auto-spectral density function Eq. (6-25) varies asymptotically as $\left(\frac{\omega}{\omega_p}\right)^{-5} \exp\left(-\frac{5}{4}\left(\frac{\omega}{\omega_p}\right)^{-4}\right)$ as $\omega \to \infty$, whereas the rational approximation varies as $\left(\frac{\omega}{\omega_p}\right)^{2r-2s}$, which represents a much weaker decay rate. If the tails of the spectrum are of importance to the dynamic response, the fit of the target and rational auto-spectral density functions at the crucial angular frequencies should be taken into consideration as a constraint in the optimization.

The filter differential equation in Eq. (3-157) driven by a unit white noise process $\{W(t), t \in R\}$ is written as an equivalent system of s first order differential equations as given by Eq. (3-168). Based on the solution stochastic vector process $\{\mathbf{Y}(t), t \in [t_0, \infty[\}$, the surface elevation is obtained from, cf. Eq. (3-171):

$$
\eta(t) = \mathbf{p}^T \mathbf{Y}(t)
\tag{6–30}
$$

The cut-off angular frequency ω_m in the broken line process is taken as $4.0\,\omega_p$. Then, the condition for the time interval (for linear interpolation) in Eq. (6-12) becomes:

$$
\Delta t < \frac{0.2455}{4.0\frac{2\pi}{T_P}} = \frac{1}{102.4}T_P
\tag{6–31}
$$

In the following $\Delta t = \frac{1}{105}T_P$ is applied. σ_W can thus be obtained from Eq. (6-13):

$$
\sigma_W = \sqrt{\frac{2\pi}{\frac{T_P}{105}}\frac{1}{2\pi}} = \frac{10.25}{\sqrt{T_P}}
\tag{6–32}
$$

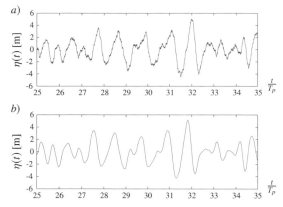

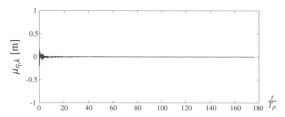

Fig. 6–6 Realizations of the sea-surface elevation process by solving linear stochastic differential equations. Rational approximation to JONSWAP wave spectrum, $H_s = 7$ m. a) $(r, s) = (5, 6)$. b) $(r, s) = (3, 6)$.

Fig. 6–7 Convergence study for $\mu_{\eta,k}$. Rational approximation to JONSWAP wave spectrum, $(r, s) = (3, 6)$, $H_s = 7$ m.

Fig. 6-6a and Fig. 6-6b show the realizations obtained for the same realization $\tilde{w}(t)$ of the broken line equivalent white noise process for rational filters of the order $(r, s) = (5, 6)$ and $(r, s) = (3, 6)$, respectively. The time-series have been obtained by numerical integration of the filter differential equation using the 4th order Runge-Kutta scheme with the time step $\Delta t_0 = \Delta t = \frac{T_p}{105}$. Obviously, the realizations are quite similar. However, small irregular fluctuations are present at the top of the realization in Fig. 6-6a, reflecting the fact that the response from the indicated filter to a white noise process is not differentiable. This is so, because the numerator polynomial is only one order lower than the denominator polynomial. The following analysis is carried out with the rational spectrum of the order $(r, s) = (3, 6)$.

$\eta_j = \eta(t_j)$, $t_j = j\Delta t_0$, $j = 0, 1, 2, \cdots$ is determined at the end of each time step Δt_0. Based on these sample values, estimations of the mean value and the standard deviation at the end of the kth time step are obtained by Eq. (6-24). Fig. 6-7 shows the convergence of $\mu_{\eta,k}$, and Fig. 6-8 shows the convergence of $\sigma_{\eta,k}$ normalized with respect to the exact value $\sigma_\eta = H_s/4.0$. It is seen that convergence has been achieved after approximately 160 peak wave periods.

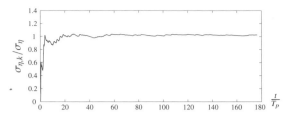

Fig. 6–8 Convergence study for $\sigma_{\eta,k}/\sigma_\eta$. Rational approximation to JONSWAP wave spectrum, $(r, s) = (3, 6)$, $H_s = 7$ m.

6.2   Simulation methods based on stochastic finite difference equations

Monte Carlo simulation based on numerical time integration may be time-consuming, especially if the time step needs to be limited due to stability or accuracy considerations. The idea in the present section is to reformulate the differential equation into an equivalent matrix finite difference equation. In principle, the devised difference equation will give the correct answer for arbitrary time steps.

The linear stochastic state vector differential equation is considered, cf. Eq. (3-209):

$$\left.\begin{aligned} \frac{d}{dt}\mathbf{Y}(t) &= \mathbf{A}\,\mathbf{Y}(t) + \mathbf{B}\,\mathbf{W}(t), \quad t > t_0 \\ \mathbf{Y}(t_0) &= \mathbf{Y}_0 \end{aligned}\right\} \tag{6–33}$$

where $\mathbf{A}$ and $\mathbf{B}$ are time-invariant real matrices of dimension $n \times n$ and $n \times m$, respectively. $\{\mathbf{Y}(t),\ t \in [t_0, \infty[\}$ is an n-dimensional stochastic state vector process with the initial value vector $\mathbf{Y}_0$, and $\{\mathbf{W}(t),\ t \in R\}$ is an m-dimensional stochastic load vector process of which the component processes are mutually independent unit white noise processes. Based on Eq. (3-210), the following property of $\{\mathbf{W}(t),\ t \in R\}$ holds:

$$E\left[\mathbf{W}(t)\mathbf{W}^T(t + \tau)\right] = \mathbf{I}\,\delta(\tau) \tag{6–34}$$

where $\mathbf{I}$ is an $m \times m$-dimensional identity matrix. If random, the initial value vector $\mathbf{Y}_0$ is assumed to be stochastically independent of the load process.

The solution to Eq. (6-33) can be written as:

$$\mathbf{Y}(t) = e^{\mathbf{A}(t-t_0)}\,\mathbf{Y}_0 + \int_{t_0}^{t} e^{\mathbf{A}(t-\tau)}\,\mathbf{B}\,\mathbf{W}(\tau)\,d\tau \tag{6–35}$$

where $e^{\mathbf{A}t}$ is the *exponential matrix function*, which is the fundamental solution matrix of the differential system in Eq. (6-33).[1] $e^{\mathbf{A}t}$ can be obtained as the solution to the initial value

---

[1] D.G. Zill and M.R. Cullen. *Differential Equations with Boundary-Value Problems, 7th Edition*. Brooks/Cole Publishing Company, Belmont, California, 2009.

problem:

$$\left.\begin{array}{l} \dfrac{d}{dt}\,e^{\mathbf{A}t} = \mathbf{A}\,e^{\mathbf{A}t}, \quad t > 0 \\[2mm] e^{\mathbf{A}0} = \mathbf{I} \end{array}\right\} \tag{6–36}$$

Example 6-4: Properties of the exponential matrix function

The exponential matrix function may be represented by the following matrix series expansion, as can be proved by insertion into Eq. (6-36):

$$e^{\mathbf{A}t} = \mathbf{I} + \mathbf{A}\,t + \frac{1}{2!}\,\mathbf{A}\mathbf{A}\,t^2 + \frac{1}{3!}\,\mathbf{A}\mathbf{A}\mathbf{A}\,t^3 + \cdots \tag{6–37}$$

The symbolic representation $e^{\mathbf{A}t}$ of the solution to Eq. (6-36) is based on the similarity of the series Eq. (6-37) to the *MacLaurin series* of the exponential function e^{at}. The convergence of the series Eq. (6-37) is guaranteed for arbitrary time intervals t. However, the series is only accurate for calculation of $e^{\mathbf{A}t}$ with small time intervals.

Consider the eigenvalue problem related to the system matrix $\mathbf{A}$:

$$\mathbf{A}\,\mathbf{\Psi}_j = \lambda_j\,\mathbf{\Psi}_j, \quad j = 1, \ldots, n \tag{6–38}$$

In general, the eigenvalues λ_j and the related eigenvectors $\mathbf{\Psi}_j$ are complex, and appear in complex conjugated pairs. Although not necessary, we shall assume for ease that the eigenvalues λ_j are simple, i.e., $\lambda_j \neq \lambda_k$ for $j \neq k$. Then, the n eigenvalue problems in Eq. (6-38) may be assembled in the matrix equation:

$$\mathbf{A}\,\mathbf{\Psi} = \mathbf{\Psi}\,\mathbf{\Lambda} \tag{6–39}$$

where:

$$\mathbf{\Lambda} = \begin{bmatrix} \lambda_1 & 0 & \cdots & 0 \\ 0 & \lambda_2 & \cdots & 0 \\ \vdots & \vdots & \ddots & \vdots \\ 0 & 0 & \cdots & \lambda_n \end{bmatrix} \tag{6–40}$$

$$\mathbf{\Psi} = \begin{bmatrix} \mathbf{\Psi}_1 \mathbf{\Psi}_2 \cdots \mathbf{\Psi}_n \end{bmatrix} \tag{6–41}$$

Multiplying Eq. (6-39) by $\mathbf{\Psi}^{-1}$, $\mathbf{A}$ can be obtained as:

$$\mathbf{A} = \mathbf{\Psi}\,\mathbf{\Lambda}\,\mathbf{\Psi}^{-1} \tag{6–42}$$

Obviously, Eq. (6-42) leads to the following representation:

$$\mathbf{A}\mathbf{A} = \mathbf{\Psi}\,\mathbf{\Lambda}\,\mathbf{\Psi}^{-1}\,\mathbf{\Psi}\,\mathbf{\Lambda}\,\mathbf{\Psi}^{-1} = \mathbf{\Psi}\,\mathbf{\Lambda}\,\mathbf{\Lambda}\,\mathbf{\Psi}^{-1} \tag{6–43}$$

Generalizing the result in Eq. (6-43), we have:

$$\mathbf{A}^j = \mathbf{A} \cdots \mathbf{A} = \mathbf{\Psi}\,\mathbf{\Lambda} \cdots \mathbf{\Lambda}\,\mathbf{\Psi}^{-1} = \mathbf{\Psi}\,\mathbf{\Lambda}^j\,\mathbf{\Psi}^{-1}, \quad j = 0, 1, 2, \ldots \tag{6–44}$$

where:

$$
\mathbf{\Lambda}^j = \mathbf{\Lambda} \cdots \mathbf{\Lambda} =
\begin{bmatrix}
\lambda_1^j & 0 & \cdots & 0 \\
0 & \lambda_2^j & \cdots & 0 \\
\vdots & \vdots & \ddots & \vdots \\
0 & 0 & \cdots & \lambda_n^j
\end{bmatrix}
\tag{6-45}
$$

Eq. (6-45) holds, because $\mathbf{\Lambda}$ is a diagonal matrix.

Insertion of Eq. (6-44) into Eq. (6-37) provides:

$$
e^{\mathbf{A}t} = \mathbf{\Psi}\, e^{\mathbf{\Lambda}t}\, \mathbf{\Psi}^{-1}
\tag{6-46}
$$

$$
e^{\mathbf{\Lambda}t} =
\begin{bmatrix}
e^{\lambda_1 t} & 0 & \cdots & 0 \\
0 & e^{\lambda_2 t} & \cdots & 0 \\
\vdots & \vdots & \ddots & \vdots \\
0 & 0 & \cdots & e^{\lambda_n t}
\end{bmatrix}
\tag{6-47}
$$

Eq. (6-47) is obtained by applying the MacLaurin expansion of the diagonal $e^{\lambda_j t}$ in $e^{\mathbf{\Lambda}t}$. Unlike Eq. (6-37), Eqs. (6-46) and (6-47) provide a solution of $e^{\mathbf{A}t}$ for arbitrary large values of t.

Inserting $t = t_1 + t_2$ into Eq. (6-46) provides:

$$
e^{\mathbf{A}(t_1+t_2)} = \mathbf{\Phi}\, e^{\mathbf{\Lambda}(t_1+t_2)}\, \mathbf{\Phi}^{-1} = \mathbf{\Phi}\, e^{\mathbf{\Lambda}t_1}\, \mathbf{\Phi}^{-1}\, \mathbf{\Phi}\, e^{\mathbf{\Lambda}t_2}\, \mathbf{\Phi}^{-1} = e^{\mathbf{A}t_1}\, e^{\mathbf{A}t_2}
\tag{6-48}
$$

where it has been used that $e^{\mathbf{\Lambda}(t_1+t_2)} = e^{\mathbf{\Lambda}t_1} e^{\mathbf{\Lambda}t_2}$, since the involved matrices are diagonal.

For $t_2 = -t_1 = -t$, Eq. (6-48) leads to the following property of $e^{\mathbf{A}}$:

$$
e^{\mathbf{A}0} = \mathbf{I} = e^{\mathbf{A}t}\, e^{-\mathbf{A}t} \qquad \Rightarrow
$$

$$
\left(e^{\mathbf{A}t} \right)^{-1} = e^{-\mathbf{A}t}
\tag{6-49}
$$

Finally, it follows from Eq. (6-37) that:

$$
\left(e^{\mathbf{A}t} \right)^T = e^{\mathbf{A}^T t}
\tag{6-50}
$$

Let $\mathbf{Y}_j = \mathbf{Y}(t_j)$ signify the solution given by Eq. (6-35) at the time $t_j = t_0 + j\Delta t$, $j = 0, 1, \ldots$, where Δt indicates the transition time step in the finite difference algorithm. Then, the solution at the next instant of time t_{j+1} can be written as:

$$
\mathbf{Y}_{j+1} = e^{\mathbf{A}\Delta t}\, \mathbf{Y}_j + \mathbf{R}_j, \quad j = 0, 1, \ldots
\tag{6-51}
$$

$$
\mathbf{R}_j = \int_{t_j}^{t_{j+1}} e^{\mathbf{A}(t_{j+1}-\tau)}\, \mathbf{B}\, \mathbf{W}(\tau)\, d\tau = e^{\mathbf{A}\Delta t} \int_0^{\Delta t} e^{-\mathbf{A}u}\, \mathbf{B}\, \mathbf{W}(u + t_j)\, du
$$

$$
= e^{\mathbf{A}\Delta t} \int_0^{\Delta t} e^{-\mathbf{A}u}\, \mathbf{B}\, \mathbf{W}(u)\, du = \mathbf{R}
\tag{6-52}
$$

The second last statement of Eq. (6-52) follows, since $\{\mathbf{W}(u),\ u \in R\}$ is stationary and hence invariant to a shift t_j in the index set. Eq. (6-52) shows that the stochastic vectors $\mathbf{R}_j$ are all identically distributed as $\mathbf{R}$. Moreover, $\mathbf{R}_j$ are mutually independent. These observations form the basis of the following simulation method, which has been proposed by Franklin.[1]

At the end of the jth time step, i.e., $t = t_j$, the realization $\mathbf{y}_j$ is assumed to be known. Next, an independent sample $\mathbf{r}_j$ of the stochastic vector $\mathbf{R}$ is generated. According to Eq. (6-51), the realization of the state vector at the time t_{j+1} is obtained as:

$$\mathbf{y}_{j+1} = e^{\mathbf{A}\Delta t}\,\mathbf{y}_j + \mathbf{r}_j \tag{6-53}$$

where the matrix $e^{\mathbf{A}\Delta t}$ is determined from Eq. (6-37) or Eq. (6-46). This method is accurate regardless of the length of the time step Δt, and only requires that samples of $\mathbf{R}$ can be generated. Since $\mathbf{R}$ is obtained as a stochastic integral of the Gaussian white noise stochastic process $\{\mathbf{W}(u),\ u \in [0, \Delta t]\}$, it follows that $\mathbf{R}$ becomes a normal stochastic vector, $\mathbf{R} \sim N(\boldsymbol{\mu}_{\mathbf{R}}, \mathbf{C}_{\mathbf{RR}})$. The mean value vector and the covariance matrix can be expressed as:

$$\boldsymbol{\mu}_{\mathbf{R}} = E[\mathbf{R}] = e^{\mathbf{A}\Delta t} \int_0^{\Delta t} e^{-\mathbf{A}u}\, \mathbf{B}\, E\big[\mathbf{W}(u)\big]\, du = \mathbf{0} \tag{6-54}$$

$$\mathbf{C}_{\mathbf{RR}} = e^{\mathbf{A}\Delta t} \int_0^{\Delta t}\int_0^{\Delta t} \left(e^{-\mathbf{A}u_1}\,\mathbf{B}\right) E\big[\mathbf{W}(u_1)\mathbf{W}^T(u_2)\big] \left(e^{-\mathbf{A}u_2}\,\mathbf{B}\right)^T du_1 du_2 \left(e^{\mathbf{A}\Delta t}\right)^T$$

$$= \int_0^{\Delta t} \left(e^{\mathbf{A}(\Delta t - u_2)}\,\mathbf{B}\right)\left(e^{\mathbf{A}(\Delta t - u_2)}\,\mathbf{B}\right)^T du_2 = \int_0^{\Delta t} \left(e^{\mathbf{A}u}\,\mathbf{B}\right)\left(e^{\mathbf{A}u}\,\mathbf{B}\right)^T du \tag{6-55}$$

Obviously, $\mathbf{C}_{\mathbf{RR}}$ can be obtained by numerical integration of Eq. (6-55). However, a more efficient alternative is usually used for obtaining $\mathbf{C}_{\mathbf{RR}}$. In Eq. (6-51), by setting $j = 0$ and assuming the initial value vector to be deterministic as $\mathbf{Y}_0 = \mathbf{0}$, we have $\mathbf{Y}_1 = \mathbf{Y}(\Delta t) = \mathbf{R}$. Therefore, $\mathbf{C}_{\mathbf{RR}} = \mathbf{C}_{\mathbf{Y}(\Delta t)\mathbf{Y}(\Delta t)}$ can be obtained as the solution at the time Δt of the covariance differential equation, cf. Eq. (3-211), with given initial value vector $\mathbf{C}_{\mathbf{RR}}(0)$:

$$\left.\begin{aligned} \frac{d}{dt}\mathbf{C}_{\mathbf{RR}}(t) &= \mathbf{A}\,\mathbf{C}_{\mathbf{RR}}(t) + \mathbf{C}_{\mathbf{RR}}(t)\,\mathbf{A}^T + \mathbf{B}\mathbf{B}^T, \quad t \in\,]0, \Delta t] \\ \mathbf{C}_{\mathbf{RR}}(0) &= \mathbf{0} \end{aligned}\right\} \tag{6-56}$$

From the obtained covariance matrix $\mathbf{C}_{\mathbf{RR}}$ of $\mathbf{R}$, a matrix $\mathbf{C}_{\mathbf{RR}}^{\frac{1}{2}}$ is determined with the property:

$$\mathbf{C}_{\mathbf{RR}} = \mathbf{C}_{\mathbf{RR}}^{\frac{1}{2}} \left(\mathbf{C}_{\mathbf{RR}}^{\frac{1}{2}}\right)^T \tag{6-57}$$

There are multiple solutions to Eq. (6-57). As an example, $\mathbf{C}_{\mathbf{RR}}^{\frac{1}{2}}$ can be selected as a lower triangular matrix by *Choleski decomposition* of the symmetric positive definite matrix $\mathbf{C}_{\mathbf{RR}}$.

---

[1]J. N. Franklin. Numerical simulation of stationary and non-stationary Gaussian random processes. *SIAM Review* 1965; **7** (1): 68-81.

Next, the following n-dimensional stochastic vector is introduced:

$$\mathbf{W} = \begin{bmatrix} W_1, \ldots, W_n \end{bmatrix}^T \tag{6–58}$$

W_j signifies mutually independent normal variables, all identically distributed as $W \sim N(0,1)$, so $E[\mathbf{W}] = \mathbf{0}$ and $E[\mathbf{W}\mathbf{W}^T] = \mathbf{I}$. Then, samples of $\mathbf{R} \sim N(\mathbf{0}, \mathbf{C}_{\mathbf{RR}})$ can be generated from:

$$\mathbf{R} = \mathbf{C}_{\mathbf{RR}}^{\frac{1}{2}} \mathbf{W} \tag{6–59}$$

This result follows, since the covariance matrix of the right-hand side of Eq. (6-59) becomes:

$$E\left[\mathbf{C}_{\mathbf{RR}}^{\frac{1}{2}} \mathbf{W}\mathbf{W}^T \left(\mathbf{C}_{\mathbf{RR}}^{\frac{1}{2}} \right)^T \right] = \mathbf{C}_{\mathbf{RR}}^{\frac{1}{2}} E\left[\mathbf{W}\mathbf{W}^T \right] \left(\mathbf{C}_{\mathbf{RR}}^{\frac{1}{2}} \right)^T = \mathbf{C}_{\mathbf{RR}} \tag{6–60}$$

The indicated method provides the exact solution for arbitrary time step Δt. Further, when $e^{\mathbf{A}\Delta t}$ and $\mathbf{C}_{\mathbf{RR}}^{1/2}$ are available after an initial calculation, the method is very fast. For this reason, it is considered the most effective method for Monte Carlo simulation of stochastic responses of a linear structural dynamic system exposed to Gaussian white noise processes.

Example 6-5: Simulation of realizations of the sea-surface elevation process based on stochastic finite difference equations

The auto-spectral density function $S_{\eta\eta}(\omega)$ of the sea-surface elevation process $\{\eta(t), t \in R\}$ is assumed to be described by the JONSWAP wave spectrum given by Eqs. (6-25) and (6-26).

A rational approximation of the order $(r,s) = (3,6)$ with filter parameters indicated in Table 6-1 is established for the target auto-spectral density function, as described in Example 6-3. Then, the filter differential equation in Eq. (3-168) is obtained with input process $\{X(t), t \in R\}$ as a unit white noise process $\{W(t), t \in R\}$. The system matrices $\mathbf{A}$ and $\mathbf{b}$ are given by Eqs. (3-169) and (3-170).

Realizations $\mathbf{Y}_j$ of the state vector process $\{\mathbf{Y}(t), t \in [0, \infty[\}$ can be generated by Eq. (6-51) at the times $t_j = j\,\Delta t$, $j = 1, 2 \ldots$, with the deterministic initial value $\mathbf{Y}_0 = \mathbf{0}$. The time step is taken as $\Delta t = \frac{T_p}{40}$. Finally, realizations η_j of the surface elevation process at the times t_j are obtained from, cf. Eq. (3-171):

$$\eta_j = \mathbf{p}^T \mathbf{Y}_j, \quad j = 0, 1, 2, \ldots \tag{6–61}$$

where $\mathbf{p}$ is given by Eq. (3-172).

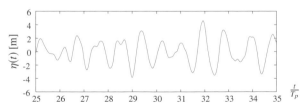

Fig. 6–9 Realization of the sea-surface elevation process based on stochastic finite difference equations. Rational approximation to JONSWAP wave spectrum, $(r,s) = (3,6)$, $H_s = 7$ m.

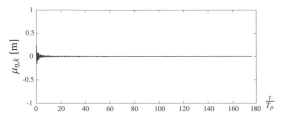

Fig. 6–10 Convergence study for $\mu_{\eta,k}$. Rational approximation to JONSWAP wave spectrum, $(r,s) = (3,6)$, $H_s = 7$ m.

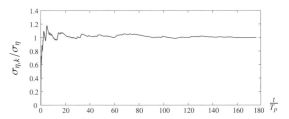

Fig. 6–11 Convergence study for $\sigma_{\eta,k}/\sigma_\eta$. Rational approximation to JONSWAP wave spectrum, $(r,s) = (3,6)$, $H_s = 7$ m.

Fig. 6-9 shows a realization of the surface elevation process for $H_s = 7$ m, using a rational auto-spectral density function of the order $(r,s) = (3,6)$.

Fig. 6-10 shows the convergence of μ_η, and Fig. 6-11 shows the convergence of $\sigma_{\eta,k}$ normalized with respect to the exact value $\sigma_\eta = H_s/4.0$. Again, convergence has been achieved after approximately 160 peak wave periods, same as in Figs. 6-7 and 6-8. However, the computational time for achieving the indicated accuracy is shorter than that in Example 6-3.

6.3 Almost periodic processes

Made up of a sum of harmonic processes, a process $\{\tilde{X}(t),\, t \in R\}$ is defined as:

$$\tilde{X}(t) = \sum_{j=1}^{J} R_j \cos\left(\omega_j t + \Phi_j\right) \tag{6–62}$$

where $\Phi_j \sim U(0, 2\pi)$, $R_j \sim R(\sigma_j^2)$, and ω_j are real constants. Further, the random variables $\Phi_j, R_j, j = 1, \ldots, J$, are all mutually independent. The mean value $\mu_{\tilde{X}}$, the auto-covariance function $\kappa_{\tilde{X}\tilde{X}}(\tau)$ and the auto-spectral density function $S_{\tilde{X}\tilde{X}}(\omega)$ are given by Eqs. (3-35), (3-36) and (3-37), respectively. The calibration of the process to a given stationary zero-mean Gaussian process $\{X(t),\, t \in R\}$ with the auto-spectral density function $S_{XX}(\omega)$ has been described in Fig. 3-5.

Example 6-6: Fourier-Stieltje transform of a stationary zero-mean Gaussian process

In Eq. (6-62), the frequency points are selected with equidistant spacing $\Delta\omega$, and ω_j can be written as:

$$\omega_j = j\,\Delta\omega, \quad j = 0, \ldots, J \tag{6–63}$$

Introducing the following definitions:

$$\left.\begin{array}{l} \omega_{-j} = -\omega_j \\ R_{-j} = R_j \\ \Phi_{-j} = -\Phi_j \end{array}\right\}, \quad j = 0, \ldots, J \tag{6–64}$$

Eq. (6-62) can thus be reformulated as:

$$\tilde{X}(t) = \sum_{j=-J}^{J} e^{i\omega_j t} \, \Delta F_j \tag{6–65}$$

where:

$$\Delta F_j = \begin{cases} 0 & , \quad j = 0 \\ \dfrac{1}{2} R_j \, e^{i\Phi_j}, & j = \pm 1, \pm 2, \ldots, \pm J \end{cases} \tag{6–66}$$

The symmetry property of ΔF_j follows from Eqs. (6-64) and (6-66):

$$\Delta F_j = \Delta F_{-j}^* \tag{6–67}$$

where ΔF_j^* indicates the complex conjugate of ΔF_j.

Recalling from Example 2-7 that $R_j \cos \Phi_j \sim N(0, \sigma_j^2)$, $R_j \sin \Phi_j \sim N(0, \sigma_j^2)$, and $R_j \cos \Phi_j$ and $R_j \sin \Phi_j$ are stochastically independent, we have:

$$E\big[\Delta F_j\big] = 0 \tag{6–68}$$

$$E\big[\Delta F_j \Delta F_k^*\big] = \begin{cases} 0 & , \quad j \neq k \\ E\Big[\big(\mathrm{Re}(\Delta F_j)\big)^2 + \big(\mathrm{Im}(\Delta F_j)\big)^2\Big] = \dfrac{1}{2}\sigma_j^2 = S_{XX}(\omega_j)\,\Delta\omega, & j = k \end{cases} \tag{6–69}$$

where Eq. (3-38) has been used.

In the limit as $J \to \infty$ and $\Delta\omega \to 0$, it follows from Eqs. (6-65), (6-68) and (6-69) that the target process $\{X(t),\, t \in R\}$ may be represented by the stochastic integral:

$$X(t) = \int_{-\infty}^{\infty} e^{i\omega t} \, dF(\omega) \tag{6–70}$$

where:

$$E\big[dF(\omega)\big] = 0 \tag{6–71}$$

$$E\big[dF(\omega_1)dF^*(\omega_2)\big] = \begin{cases} 0 & , \quad \omega_1 \neq \omega_2 \\ S_{XX}(\omega_1)\,d\omega_1, & \omega_1 = \omega_2 \end{cases} \tag{6–72}$$

Eqs. (6-70), (6-71) and (6-72) are known as the *Fourier-Stieltje spectral decomposition* of the zero-mean stationary Gaussian process $\{X(t), t \in R\}$. $\{F(\omega), \omega \in R\}$ is designated the spectral process. As a consequence of Eq. (6-72), the real and imaginary parts of this process have independent increments. Since the realizations are continuous, it follows from Example 3-8 that $\{\mathrm{Re}\big(F(\omega)\big), \omega \in R\}$

and $\{\mathrm{Im}(F(\omega)), \; \omega \in R\}$ are both Wiener processes (since the Wiener process is the only process with independent increments, which has continuous realizations).

If almost all realizations $x(t)$ of $\{X(t), \; t \in R\}$ have a finite number of discontinuities, have finite variation and are absolute integrable in the interval $] - \infty, \infty[$, the following *stochastic Fourier transform pair* becomes possible:[1]

$$
\left.
\begin{aligned}
X(t) &= \int_{-\infty}^{\infty} e^{i\omega t} X(\omega)\, d\omega \\
X(\omega) &= \frac{1}{2\pi} \int_{-\infty}^{\infty} e^{-i\omega t} X(t)\, dt
\end{aligned}
\right\}
\tag{6-73}
$$

where $\{X(\omega), \; \omega \in R\}$ is the stochastic Fourier transform of $\{X(t), \; t \in R\}$.

The indicated necessary conditions for Eq. (6-73) to be possible can be fulfilled by some non-stationary stochastic processes such as the stochastic earthquake process. Eq. (6-70) also apply to these cases, so the processes $\{F(\omega), \; \omega \in R\}$ and $\{X(\omega), \; \omega \in R\}$ are related as:

$$
X(\omega) = \frac{d}{d\omega} F(\omega)
\tag{6-74}
$$

However, for stationary processes, almost all realizations of $\{X(t), \; t \in R\}$ are not absolute integrable over $] - \infty, \infty[$. Eq. (6-73) is thus no longer valid. According to Eq. (6-74), $\{X(\omega), \; \omega \in R\}$ may formally be interpreted as a white noise process in this case.

Made up of a sum of *random phase processes*, a stochastic process $\{\tilde{X}(t), \; t \in R\}$ is defined as:

$$
\tilde{X}(t) = \sum_{j=1}^{J} a_j \cos\left(\omega_j t + \Phi_j\right)
\tag{6-75}
$$

Different from the Rayleigh-distributed random variables R_j in Eq. (6-62), the amplitudes $a_j, \; j = 1, \ldots, J$, are deterministic positive real constants. ω_j and Φ_j have the same meaning as in Eq. (6-62). The stochastic process $\{\tilde{X}_j(t), \; t \in R\}$, $\tilde{X}_j(t) = a_j \cos(\omega_j t + \Phi_j)$ is designated a random phase process, and Eq. (6-75) is made up of a finite sum of such processes. The following properties of the random phase process hold:

$$
E\left[\cos(\omega t + \Phi_j) \right] = \int_0^{2\pi} \cos(\omega t + \Phi_j) \frac{1}{2\pi}\, dx = 0
\tag{6-76}
$$

$$
E\left[\cos(\omega t_1 + \Phi_j) \cos(\omega t_2 + \Phi_k) \right] =
\begin{cases}
0 & , \quad j \neq k \\
\dfrac{1}{2} \cos\left(\omega(t_2 - t_1) \right), & j = k
\end{cases}
\tag{6-77}
$$

From Eqs. (6-76) and (6-77), the mean value function, the auto-covariance function and the auto-spectral density function of the process $\tilde{X}(t)$ defined in Eq. (6-75) can be obtained:

$$
\mu_{\tilde{X}} = 0
\tag{6-78}
$$

---

[1] E. Kreyszig. *Advanced Engineering Mathematics, 10th Edition.* John Wiley & Sons, Inc., Hoboken, New Jersey, 2011.

$$\kappa_{\tilde{X}\tilde{X}}(\tau) = \sum_{j=1}^{J} \frac{a_j^2}{2} \cos(\omega_j \tau) \tag{6–79}$$

$$S_{\tilde{X}\tilde{X}}(\omega) = \sum_{j=1}^{J} \frac{a_j^2}{4} \left(\delta(\omega - \omega_j) + \delta(\omega + \omega_j) \right) \tag{6–80}$$

The above results are identical to those of a sum of harmonic processes, as given by Eqs. (3-35), (3-36) and (3-37), as long as σ_j^2 is replaced with $\frac{a_j^2}{2}$.

The calibration of the parameters a_j follows the procedure for the sum of harmonic processes. With reference to Fig. 3-5, we have, cf. Eq. (3-38):

$$\frac{1}{2} \frac{a_j^2}{2} = \int_{\omega'_{j-1}}^{\omega'_j} S_{XX}(\omega) \, d\omega \quad \Rightarrow$$

$$a_j^2 = 4 \int_{\omega'_{j-1}}^{\omega'_j} S_{XX}(\omega) \, d\omega, \quad j = 1, 2, \ldots, J \tag{6–81}$$

The probability density function of $\tilde{X}_j(t)$ is given by Eq. (3-239) with $r = a_j$, which is far from a normal probability density function (as already indicated by Fig. 3-24b). Nevertheless, as a consequence of the central limit theorem, $\tilde{X}(t)$ will approach a normal distribution with the correct variance in the limit as $J \rightarrow \infty \wedge \max(\omega'_{j+1} - \omega'_j) \rightarrow 0$, as long as the amplitudes a_j are of comparable magnitudes. Further, the multi-dimensional generalization of this theorem insures that $\{\tilde{X}(t), t \in R\}$ approaches asymptotically to the target zero-mean stationary Gaussian process $\{X(t), t \in R\}$ under the same conditions.

Example 6-7: Simulation of realizations of the sea-surface elevation process using a sum of random phase processes

Same as in Examples 6-3 and 6-5, the auto-spectral density function $S_{\eta\eta}(\omega)$ of the sea-surface elevation process $\{\eta(t), t \in R\}$ is assumed to be described by the JONSWAP wave spectrum given by Eqs. (6-25) and (6-26).

Monte Carlo simulations are performed based on the process made up of a sum of random phase processes, cf. (6-75):

$$\eta(t) = \sum_{j=1}^{J} a_j \cos \left(\omega_j t + \Phi_j \right) \tag{6–82}$$

where a_j are determined from Eq. (6-81):

$$a_j = 2\sqrt{S_{\eta\eta}(\omega_j) \Delta\omega_j} \tag{6–83}$$

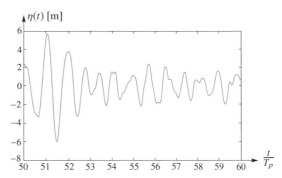

Fig. 6–12 Realization of the sea-surface elevation process using a sum of random phase processes. $H_s = 7$ m, $J = 400$, $\Delta t = \frac{T_p}{40}$.

Half of the J angular frequencies are distributed equidistantly in the interval $[0.5\,\omega_p, 1.5\,\omega_p]$. The other half is distributed equidistantly in the interval $]1.5\,\omega_p, 6.5\,\omega_p]$. The variance contributions outside the interval $[0.5\,\omega_p, 6.5\,\omega_p]$ are ignored. Correspondingly, we have:

$$
\Delta\omega_j = \begin{cases} \dfrac{\omega_p}{J/2}, & \omega_j \in [0.5\,\omega_p, 1.5\,\omega_p] \\[2ex] 5\dfrac{\omega_p}{J/2}, & \omega_j \in]1.5\,\omega_p, 6.5\,\omega_p] \end{cases} \tag{6–84}
$$

The following combinations of the number of frequencies J and time step Δt are considered: $(J, \Delta t) = \left(400, \frac{T_p}{40}\right)$, $(J, \Delta t) = \left(100, \frac{T_p}{40}\right)$ and $(J, \Delta t) = \left(400, \frac{T_p}{20}\right)$. In each of these three cases, the significant wave height $H_s = 7$ m has been used.

The standard deviation σ_η, the average wave period $\bar{T}$, the average of the highest third of the wave heights $\bar{H}_{1/3}$ and the probability density function of the 1st order $f_\eta(\eta)$ are to be determined by ergodic sampling.

Fig. 6-12 shows a segment of a typical realization of $\{\eta(t), t \in R\}$. At each time step $t_k = k\Delta t$, the sea-surface elevation $\eta_k = \eta(k\Delta t)$, $k = 0, 1, 2, \ldots$, is simulated using Eq. (6-82) with $(J, \Delta t) = \left(400, \frac{T_p}{40}\right)$.

The standard deviation $\sigma_{\eta,k}$ after k sampled values is estimated by Eq. (6-24). The convergence of $\sigma_{\eta,i}/\sigma_\eta$ as a function of the sampling length is shown in Fig. 6-13, where $\sigma_\eta = \frac{H_s}{4.0} = 1.75$ m is the exact standard deviation. It appears that reliable estimates of σ_η (with a relative error $\leq 2\%$) demand a sampling length of at least $300T_p$. Comparing with Examples 6-3 and 6-5, the computational time of the sample values η_j in this method is longer. Actually, the computational time for achieving the same accuracy turns out to be shortest for the Monte-Carlo simulation method based on stochastic finite difference equations described in Section 6.2. However, the advantage of the random phase method is that rational approximation of the target auto-spectral density function is not needed and simulations can be carried out directly using the JONSWAP spectrum in this example.

As shown in Fig. 6-14, the wave heights H_j are defined as the difference between a wave crest and the immediately preceding wave trough. After the kth wave height has been sampled, these sample values are renumbered and arranged as an order statistics, so that $H_1 > H_2 > \cdots > H_k$. Then, after k sampled

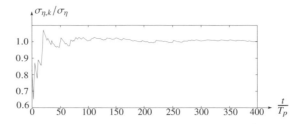

Fig. 6–13 Convergence study for $\sigma_{\eta,k}$. $H_s = 7$ m, $\Delta t = \frac{1}{40}$, $J = 400$.

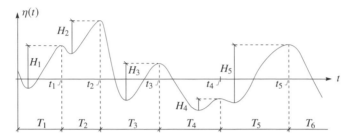

Fig. 6–14 Definition of the wave heights H_j and the wave periods T_j.

wave heights, the average of highest third of the wave heights $\bar{H}_{1/3,k}$ is estimated as:

$$\bar{H}_{1/3,k} = \frac{1}{\left[\frac{k}{3}\right] + 1} \sum_{j=1}^{\left[\frac{k}{3}\right]+1} H_j \qquad (6\text{–}85)$$

where $[x]$ indicates the integer part of x (e.g., $[2.6]=2$).

In Fig. 6-15, the convergence study for $\bar{H}_{1/3,k}/H_s$ is shown as a function of the number of sampled wave heights, k. It appears that at least 700 samples of the wave height are needed before a sufficient convergence is obtained. From the plot, it is seen that:

$$\lim_{k \to \infty} \frac{\bar{H}_{1/3,k}}{H_s} = 0.882 \qquad (6\text{–}86)$$

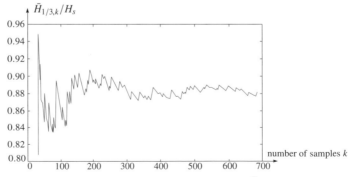

Fig. 6–15 Convergence study for $\bar{H}_{1/3,k}$. $H_s = 7$ m, $\Delta t = \frac{T_p}{40}$, $J = 400$.

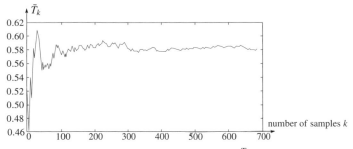

Fig. 6–16 Convergence study for $\bar{T}_k$. $H_s = 7$ m, $\Delta t = \frac{T_p}{40}$, $J = 400$.

Hence, the limit value is not 1.0. This means that the sampled wave heights H_j are not Rayleigh distributed.

Further, the wave periods T_j are identified as the interval between two successive wave crests, as shown in Fig. 6-14. After k sampled wave periods, the average wave period $\bar{T}_k$ is estimated as:

$$\bar{T}_k = \frac{1}{k} \sum_{j=1}^{k} T_j \tag{6–87}$$

The convergence of $\bar{T}_k$ as a function of the number of sampled wave periods k has been shown in Fig. 6-16.

Table 6-2 summarizes the simulation results obtained for various values of J and Δt. When Δt is increased, many small waves will not be observed. As indicated by the table, a doubling of Δt from $T_p/40$ to $T_p/20$ causes a drop in the number of observed waves from 689 to 666 for a same sampling length of $400T_p$. As a consequence, the final estimate $\bar{T}_\infty$ of $\bar{T}_j$ is increased correspondingly. In contrast, $\bar{H}_{1/3,k}$ and $\sigma_{\eta,k}$ are rather insensitive to the time step Δt. Reduction of the number J of frequencies has a similar effect as increasing Δt, as observed from the 2nd column of the table. The conclusion is that relatively many frequencies have to be used to ensure a stable estimate of $\bar{T}_\infty$. Moreover, it should also be noted that $\bar{T}_\infty$ implies the average interval between maxima, whereas T_p signifies the average interval between zero up-crossings. The deviation between these two quantities is a consequence of the broad-bandedness of the sea-surface elevation process.

Finally, the probability density function of the 1st order, $f_\eta(x)$, is determined by ergodic sampling, cf. Eq. (3-229). The class division of the sampling interval $]-\infty, \infty[$ is chosen as follows. The sub-intervals $[-3\sigma_\eta, -2\sigma_\eta[$ and $]2\sigma_\eta, 3\sigma_\eta]$ are divided into 4 classes of the width $\Delta x = \frac{\sigma_\eta}{4}$, the sub-intervals $[-2\sigma_\eta, -\sigma_\eta[$ and $]\sigma_\eta, 2\sigma_\eta]$ are divided into 10 classes of the width $\Delta x = \frac{\sigma_\eta}{10}$, and the sub-interval

Table 6-2 The effect of the number of frequencies J and the time step Δt on the simulation results. The sampling length is $400T_p$ in all cases.

	$(J, \Delta t) = \left(400, \frac{T_p}{40}\right)$	$(J, \Delta t) = \left(100, \frac{T_p}{40}\right)$	$(J, \Delta t) = \left(400, \frac{T_p}{20}\right)$
$H_{1/3,\infty}/H_s$	0.882	0.895	0.890
$\sigma_{\eta,\infty}/\sigma_\eta$	1.000	1.000	1.000
$\bar{T}_\infty/T_p$	0.581	0.662	0.602
Number of waves	689	640	666

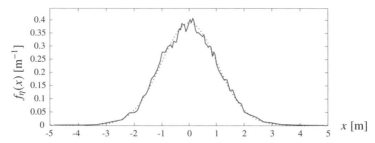

Fig. 6–17 Simulated and theoretical first order probability density function for the sea-surface elevation process $\{\eta(t),\ t \in R\}$. Solid line: simulated result. Dashed line: theoretical result.

$[-\sigma_\eta, \sigma_\eta]$ is divided into 40 classes of the width $\Delta x = \frac{\sigma_\eta}{20}$.

In the present case, totally n samples of the sea-surface elevation η are generated, depending on the sampling length $400\,T_p$ and the time step Δt:

$$n = \frac{400\,T_p}{\Delta t} = \frac{400\,T_p}{\frac{T_p}{40}} = 16000 \tag{6–88}$$

Assuming that Δn_j samples have been registered in the jth class interval, the probability density function $f_\eta(x)$ in the indicated interval can thus be estimated from, cf. Eq. (3-229):

$$f_\eta(x_j) = \frac{1}{\Delta x_j} \frac{\Delta n_j}{n} \tag{6–89}$$

where x_j and Δx_j are the center point and the width of the indicated jth class interval, respectively. The estimated probability density function has been shown in Fig. 6-17 together with its theoretical counterpart (normal distribution):

$$f_\eta(x) = \frac{1}{\sqrt{2\pi}\sigma_\eta} \exp\left(-\frac{1}{2}\frac{x^2}{\sigma_\eta^2}\right) \tag{6–90}$$

Next, let's consider a stationary zero-mean Gaussian N-dimensional stochastic vector process $\{\mathbf{X}(t),\ t \in R\}$, $\mathbf{X}(t) = \left[X_1(t),\ldots,X_N(t)\right]^T$. The correlation structure of the vector process is described by the cross-spectral density matrix function $\mathbf{S}_{\mathbf{XX}}(\omega)$, where the component in the jth row and the kth column indicates the cross-spectral density function $S_{X_j X_k}(\omega)$ of the component processes $\{X_j(t),\ t \in R\}$ and $\{X_k(t),\ t \in R\}$.

From Eqs. (3-24) and (3-25), the following properties of $\mathbf{S}_{\mathbf{XX}}(\omega)$ hold:

$$\mathbf{S}_{\mathbf{XX}}(\omega) = \mathbf{S}_{\mathbf{XX}}^{*T}(\omega) \tag{6–91}$$

$$\mathbf{S}_{\mathbf{XX}}(\omega) = \mathbf{S}_{\mathbf{XX}}^{*}(-\omega) \tag{6–92}$$

Eq. (6-91) qualifies $\mathbf{S}_{\mathbf{XX}}(\omega)$ as a *Hermitian matrix*. Hence, all eigenvalues of $\mathbf{S}_{\mathbf{XX}}(\omega)$ are real.

Further, it can be shown that $\mathbf{S_{XX}}(\omega)$ is positive definite (Bochner's theorem)[1], with the following property:

$$\forall \mathbf{a} \in C^N, \; |\mathbf{a}| > 0 \; : \; \mathbf{a}^* \, \mathbf{S_{XX}}(\omega) \, \mathbf{a}^T > 0 \qquad (6\text{--}93)$$

Eq. (6-93) implies that all eigenvalues of $\mathbf{S_{XX}}(\omega)$ are positive and real.

As a result, $\mathbf{S_{XX}}(\omega)$ can be factorized as:

$$\mathbf{S_{XX}}(\omega) = \mathbf{H}^*(\omega) \, \mathbf{H}^T(\omega) \qquad (6\text{--}94)$$

The factorization in Eq. (6-94) automatically fulfills Eq. (6-91). The symmetry condition in Eq. (6-92) implies that the following similar condition also needs to be fulfilled by $\mathbf{H}(\omega)$:

$$\mathbf{H}(\omega) = \mathbf{H}^*(-\omega) \qquad (6\text{--}95)$$

The concept of a *unitary matrix* is a complex generalization of orthonormal real matrices. A unitary matrix $\mathbf{A}$ is defined by the following property:

$$\mathbf{A}^{-1} = \mathbf{A}^{*T} \qquad (6\text{--}96)$$

For any unitary matrix $\mathbf{A}$, we have:

$$\mathbf{H}^*(\omega) \, \mathbf{H}^T(\omega) = \mathbf{H}^*(\omega) \, \mathbf{A}^{-1} \mathbf{A} \, \mathbf{H}^T(\omega) = \left(\mathbf{H}(\omega) \mathbf{A}^T \right)^* \left(\mathbf{H}(\omega) \mathbf{A}^T \right)^T \qquad (6\text{--}97)$$

Hence, the factorization in Eq. (6-94) is not unique. If $\mathbf{H}(\omega)$ is a solution, $\mathbf{H}(\omega)\mathbf{A}^T$ will be a solution as well. Usually, $\mathbf{H}(\omega)$ is determined as a lower triangular matrix by Cholesky decomposition.

The matrix $\mathbf{H}(\omega)$ forms the basis for simulating realizations of zero-mean Gaussian N-dimensional stochastic vector process. The formulations below are in essence due to Deodatis.[2]

The following sum of complex random phase processes is considered for generating realizations of the component process $\{X_j(t), \; t \in R\}$:

$$\tilde{X}_j(t) = \sum_{l=-J}^{J} \sum_{r=1}^{N} H_{jr}(\omega_l) \, \exp\left(i \, (\omega_l t + \Phi_{lr}) \right) \sqrt{\Delta\omega}, \quad j = 1, 2, \ldots, N \qquad (6\text{--}98)$$

where Φ_{lr} are mutually independent, uniformly distributed (from 0 to 2π) stochastic variables for arbitrary index values l and r, i.e., $\Phi_{lr} \sim U(0, 2\pi)$, fulfilling the symmetry property:

$$\Phi_{lr} = -\Phi_{-lr}, \quad l = 1, 2, \ldots, J \qquad (6\text{--}99)$$

Hence, merely J numbers of stochastically independent variables are entering the model for a given index value r.

---

[1] Y.K. Lin. *Probabilistic Theory of Structural Dynamics*. McGraw-Hill, New York, 1967.

[2] G. Deodatis. Simulation of ergodic multivariate stochastic processes. *Journal of Engineering Mechanics, ASCE* 1996; **122** (8): 778-787.

The discrete angular frequencies ω_l fulfill a similar symmetry property:

$$\omega_l = -\omega_{-l} = l\,\Delta\omega, \quad l = 1, 2, \ldots, J \tag{6–100}$$

Eqs. (6-99) and (6-100) in combination with the symmetry property Eq. (6-95) insure that $\tilde{X}_j(t)$ becomes a real stochastic variable, so Eq. (6-98) can be written as:

$$\tilde{X}_j(t) = 2\,\mathrm{Re}\left(\sum_{l=1}^{J}\sum_{r=1}^{N} H_{jr}(\omega_l)\,\exp\left(i\,(\omega_l t + \Phi_{lr})\right)\sqrt{\Delta\omega}\right) \tag{6–101}$$

From Eq. (6-76) and (6-77), the following expectations are obtained:

$$E\left[\exp\left(i\,(\omega_l t + \Phi_{lr})\right)\right] = E\left[\cos(\omega_l t + \Phi_{lr}) + i\,\sin(\omega_l t + \Phi_{lr})\right] = 0 \tag{6–102}$$

$$E\left[\exp\left(-i(\omega_l t_1 + \Phi_{lr})\right)\,\exp\left(i(\omega_m t_2 + \Phi_{ms})\right)\right] = \begin{cases} 0 & , \quad l \neq m \vee r \neq s \\ \mathrm{e}^{i\,\omega_l\,(t_2-t_1)}, & \quad l = m \wedge r = s \end{cases} \tag{6–103}$$

Then, the mean value function and the cross-covariance function become:

$$\mu_{\tilde{X}_j} = \sum_{l=-J}^{J}\sum_{r=1}^{N} H_{jr}(\omega_l)\,E\left[\exp\left(i\,(\omega_l t + \Phi_{lr})\right)\right]\sqrt{\Delta\omega} = 0 \tag{6–104}$$

$$\kappa_{\tilde{X}_j\tilde{X}_k}(t_1, t_2) = E\left[\tilde{X}_j(t_1)\tilde{X}_k(t_2)\right] = E\left[\tilde{X}_j^*(t_1)\tilde{X}_k(t_2)\right]$$

$$= \sum_{l=-J}^{J}\sum_{m=-J}^{J}\sum_{r=1}^{N}\sum_{s=1}^{N} H_{jr}^*(\omega_l)\,H_{ks}(\omega_m)\,E\left[\exp\left(-i\,(\omega_l t_1 + \Phi_{lr})\right)\,\exp\left(i\,(\omega_m t_2 + \Phi_{ms})\right)\right]\Delta\omega$$

$$= \sum_{l=-J}^{J}\sum_{r=1}^{N} \mathrm{e}^{i\,\omega_l\,(t_2-t_1)}\,H_{jr}^*(\omega_l)\,H_{kr}(\omega_l)\,\Delta\omega$$

$$= \sum_{l=-J}^{J} \mathrm{e}^{i\,\omega_l\,\tau}\,S_{X_jX_k}(\omega_l)\,\Delta\omega, \quad \tau = t_2 - t_1 \tag{6–105}$$

where the last statement of Eq. (6-105) follows from Eq. (6-94).

The right-hand side of Eq. (6-105) converges to the target cross-covariance function $\kappa_{X_jX_k}(\tau)$ as $J \to \infty \wedge \Delta\omega \to 0$, cf. Eq. (3-23).

From the multi-dimensional generalization of the central limit theorem and from Eq. (6-102), it follows that Eq. (6-98) approaches a zero-mean Gaussian vector process as $J \to \infty$. This demonstrates the applicability of Eq. (6-101) for Monte Carlo simulation of realizations of stationary zero-mean Gaussian vector process.

Finally, it is noticed that the summations in Eq. (6-101) may be calculated by FFT (Fast Fourrier Transform), which will reduce the calculation time significantly.

Example 6-8: Stochastic modelling of turbulent wind field on a chimney by means of a sum of random phase processes

In this example, the turbulent wind field on a chimney is simulated by means of a sum of random phase processes. The height of the chimney is h and the diameter of its cross-section is D, as shown in Fig. 6-18.

The *instantaneous wind velocity* $V(z, t)$ at the height z above ground level in the direction of the mean wind may be written as:

$$V(z, t) = \bar{v}(z) + v(z, t) \tag{6–106}$$

where $\bar{v}(z)$ denotes the *mean wind velocity* given by:

$$\bar{v}(z) = E\big[V(z, t)\big] \tag{6–107}$$

The mean wind velocity is assumed to follow a logarithmic profile:

$$\bar{v}(z) = \bar{v}_{10} \frac{\ln \frac{z}{z_0}}{\ln \frac{z_{10}}{z_0}} \tag{6–108}$$

where $\bar{v}_{10}$ indicates the mean wind velocity at the height $z_{10} = 10$ m. z_0 is a measure of the surface irregularity at the position of the structure.

$v(z, t)$ in Eq. (6-106) indicates the *turbulence* at the height z at the time t. Due to Eq. (6-107), the mean value of $v(z, t)$ is zero per definition. We shall assume that the turbulence is fully correlated over the diameter D of the chimney. Then, the turbulence may be modelled by a zero-mean Gaussian stochastic process $\{v(z, t), (z, t) \in [z_0, h] \times R\}$, which is assumed to be stationary in the time index parameter t, and non-homogeneous in the spatial index parameter z.

The auto-spectral density function of the turbulence is known to follow an $\omega^{-5/3}$ relationship in the so-called inertial sub-range of the universal equilibrium range of the turbulence spectrum. Many civil engineering structures have their lowest angular frequency placed in this range, for which reason the

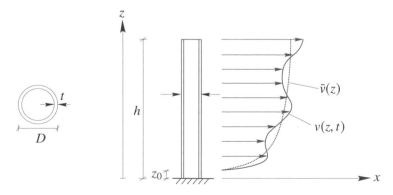

Fig. 6–18 A steel chimney with constant cross-section in turbulent wind field.

analytical expression for the auto-spectral density function should display the indicated angular frequency relationship. This is indeed the case for the *Simiu spectrum*, which can be written in the form:[1]

$$S_{vv}(\omega, z) = \frac{1}{3\,\omega_0(z)} \frac{\sigma_v^2(z)}{\left(1 + \frac{|\omega|}{\omega_0(z)}\right)^{5/3}} \tag{6–109}$$

where:

$$\omega_0(z) = \alpha\,\frac{\bar{v}(z)}{z}, \quad \alpha = \frac{2\pi}{50} \tag{6–110}$$

$$\sigma_v(z) = I\,\bar{v}(z) \tag{6–111}$$

I signifies the *turbulence intensity*, which is assumed to be independent of z in the present case. As shown in Eq. (6-111), I is equivalent to the variational coefficient of $V(z,t)$, cf. Eq. (2-52).

The factor $\frac{1}{3\,\omega_0(z)}$ is merely a normalization term. For arbitrary $\omega_0(z)$, using substitution $x = \frac{|\omega|}{\omega_0(z)}$, we have:

$$\int_{-\infty}^{\infty} S_{vv}(\omega, z)\, d\omega = \sigma_v^2(z)\,\frac{2}{3}\int_{0}^{\infty} \frac{1}{(1+x)^{5/3}}\, dx = \sigma_v^2(z) \tag{6–112}$$

The turbulence processes along the chimney are simulated at N discrete numbers of heights $z_j = z_0 + j\Delta z$, $j = 1, 2, \ldots, N$, with the interval:

$$\Delta z = \frac{h - z_0}{N} \tag{6–113}$$

Therefore, the stochastic turbulence field is reduced to a stationary zero-mean Gaussian N-dimensional stochastic vector process $\{\mathbf{v}(t),\ t \in R\}$, $\mathbf{v}(t) = \begin{bmatrix}v_1(t), \ldots, v_N(t)\end{bmatrix}^T$. The correlation structure of this vector process is described by the cross-spectral density matrix function $\mathbf{S_{vv}}(\omega)$, where the component in the jth row and the kth column indicates the cross-spectral density function $S_{v_j v_k}(\omega)$ of the component processes $\{v_j(t),\ t \in R\}$ and $\{v_k(t),\ t \in R\}$. $\{v_j(t),\ t \in R\}$ and $\{v_k(t),\ t \in R\}$ denote the stationary zero-mean Gaussian turbulence processes at the heights z_j and z_k, respectively. They can be described by the corresponding auto-spectral density functions $S_{v_j v_j}(\omega)$ and $S_{v_k v_k}(\omega)$, which are obtained from Eq. (6-109) for $z = z_j$ and z_k, respectively. The cross-spectral density function $S_{v_j v_k}(\omega)$ may be expressed as:

$$S_{v_j v_k}(\omega) = \sqrt{S_{v_j v_j}(\omega)}\,\sqrt{S_{v_k v_k}(\omega)}\,\mathrm{Coh}_{v_j v_k}(\omega)\,e^{i\Phi_{v_j v_k}(\omega)} \tag{6–114}$$

where $\mathrm{Coh}_{v_j v_k}(\omega)$ and $\Phi_{v_j v_k}(\omega)$ signify the so-called *coherence spectrum* and the *phase spectrum*, respectively. The following properties need to be fulfilled for $\mathrm{Coh}_{v_j v_k}(\omega)$ and $\Phi_{v_j v_k}(\omega)$:

$$\left.\begin{aligned}
\mathrm{Coh}_{v_j v_j}(\omega) &= \mathrm{Coh}_{v_k v_k}(\omega) = 1 \\
\mathrm{Coh}_{v_j v_k}(\omega) &= \mathrm{Coh}_{v_j v_k}^*(\omega) = \mathrm{Coh}_{v_k v_j}(\omega) \\
\mathrm{Coh}_{v_j v_k}(\omega) &= \mathrm{Coh}_{v_j v_k}(-\omega)
\end{aligned}\right\} \tag{6–115}$$

[1] E. Simiu. Wind spectra and dynamic alongwind response. *Journal of the Structural Division*, ASCE 1974; **100** (9): 1897-1910.

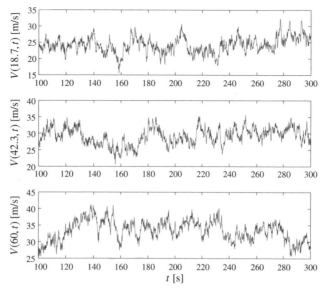

Fig. 6–19 Simulated realizations of the instantaneous wind velocity at the heights $z = 18.7$ m, $z = 42.3$ m, $z = 60$ m. $z_0 = 1$ m, $I = 0.1$, $h = 60$ m, $\bar{v}_{10} = 20$ m/s, $N = 20$.

$$
\left.
\begin{aligned}
\Phi_{v_j v_j}(\omega) &= \Phi_{v_k v_k}(\omega) &= 0 \\
\Phi_{v_j v_k}(\omega) &= -\Phi_{v_k v_j}(\omega) \\
\Phi_{v_j v_k}(\omega) &= -\Phi_{v_j v_k}(-\omega)
\end{aligned}
\right\} \tag{6–116}
$$

As a result, $S_{v_j v_k}(\omega)$ fulfills the Hermitian properties, cf. Eqs. (3-24) and (3-25):

$$
\left.
\begin{aligned}
S_{v_j v_k}(\omega) &= S^*_{v_k v_j}(\omega) \\
S_{v_j v_k}(\omega) &= S^*_{v_j v_k}(-\omega)
\end{aligned}
\right\} \tag{6–117}
$$

Traditionally, the coherence spectrum and the phase spectrum are given by the following expressions [1,2], compatible to the requirements in Eqs. (6-115) and (6-116):

$$
\mathrm{Coh}_{v_j v_k}(\omega) = \exp\left(-\beta \frac{|z_j - z_k|}{\bar{v}_{jk}} |\omega|\right), \quad \beta = \frac{10}{2\pi} \tag{6–118}
$$

$$
\Phi_{v_j v_k}(\omega) = \gamma \frac{z_j - z_k}{\bar{v}_{jk}} \omega, \quad \gamma = 1.2 \tag{6–119}
$$

$$
\bar{v}_{jk} = \frac{1}{2}\left(\bar{v}(z_j) + \bar{v}(z_k)\right) \tag{6–120}
$$

With the given cross-spectral density matrix function $\mathbf{S}_{vv}(\omega)$, Monte Carlo simulations of the N-dimensional turbulence are obtained by Eq. (6-101). Fig. 6-19 shows the obtained realizations of the instantaneous wind velocity at the heights $z = 18.7$ m, $z = 42.3$ m and $z = 60$ m.

[1]A.G. Davenport. Gust loading factors. *Journal of the Structural Division, ASCE* 1967; **93** (3): 11-34.

[2]M. Shiotani and Y. Iwatani. Correlation of the wind velocities in relation to the gust loadings. *Proceedings of the 3rd International Conference on Wind Effects on Buildings and Structures*, Tokyo, 1971.

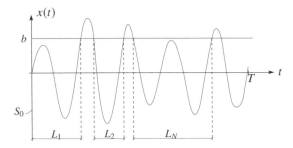

Fig. 6–20 Ergodic sampling of interval lengths spent in the safe domain before an out-crossing.

6.4 Simulation of the first-passage time probability density function

In this section, it is demonstrated how the probability density function of the first-passage time can be determined by Monte Carlo simulation using Eq. (5-17).

Let the safe domain be time-invariant and $\{X(t), t \in R\}$ be a stationary stochastic process possessing sufficient ergodicity properties, so the probability distribution function $F_L(l)$ and the expected value $E[L]$ of the time interval L spent in the safe domain, as well as the probability $P(X(0) \in S_0)$, can all be determined by ergodic sampling from a single realization $x(t)$ of the length T (sufficiently long). In this realization, totally N out-crossings from the safe domain are observed, and the corresponding interval lengths spent in the safe domain before the out-crossings are measured as $L_1, \ldots, L_N$, as shown in Fig. 6-20. Let $N_{\leq l}$ denote the number of intervals in the sample, for which $L_j \leq l$. Then, $F_L(l)$, $E[L]$ and $P(X(0) \in S)$ are estimated from the following statistics:

$$F_L(l) \simeq \frac{N_{\leq l}}{N} \tag{6–121}$$

$$E[L] \simeq \frac{1}{N} \sum_{j=1}^{N} L_j \tag{6–122}$$

$$P(X(0) \in S_0) \simeq \frac{1}{T} \sum_{j=1}^{N} L_j \tag{6–123}$$

Insertion of Eqs. (6-121) and (6-122) into Eq. (5-17) provides robust estimation of the first-passage probability density function, because the sampling has been related to a probability distribution function $F_L(l)$, and not directly to the probability density function $f_T(t)$ itself. Curve c in Fig. 5-12 has been determined in this way.

The indicated method for the determination of $f_T(t)$ can immediately be generalized to an ergodic stationary vector process $\{\mathbf{X}(t), t \in R\}$ with time-invariant safe domains. For both the scalar and the vector cases, the sample size should encompass at least $N=1000$ out-crossings. Hence, the length of the realization of the considered ergodic process should at least be $T = 1000\frac{1}{f_1}$.

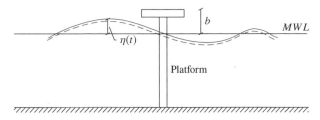

Fig. 6–21 A monopile platform in irregular sea-state.

Example 6-9: Exceeding probability of critical level for the sea-surface elevation process

Fig. 6-21 shows a monopile platform in irregular sea-state. The distance from the *mean water level* (MWL) to the deck of the platform is b. The elevation of the surface is modelled as a zero-mean stationary Gaussian process $\{\eta(t),\ t \in R\}$ with the auto-spectral density function $S_{\eta\eta}(\omega)$ given by the JONSWAP wave spectrum in Eqs. (6-25) and (6-26).

The significant wave height is chosen to be $H_s = 7$ m, from which the standard deviation of the sea-surface elevation process becomes $\sigma_\eta = \frac{H_S}{4.0} = 1.75$ m. The problem is to determine the probability that the platform will be flooded for a time duration of $15\,T_p$, when $b = 2.0\sigma_\eta = 3.5$ m. T_p indicates the peak period of the JONSWAP spectrum, and is related to H_s as indicated in Eq. (6-26). The initial value $\eta(0)$ of the sea-state at the start of the considered interval $[0, 15T_p]$ is assumed to be unknown. Hence, the reliability problem has random initial start, cf. Eq. (5-7).

Based on a realization of the sea-surface elevation with $N = 985$ out-crossings, the first-passage probability density function is obtained and illustrated in Fig. 6-22. Next, the following quantities are computed:

$$P\big(\eta(0) \in S_0\big) = P\big(\eta(0) < b\big) = \Phi\Big(\frac{b}{\sigma_\eta}\Big) = \Phi(2.0) = 0.9773 \tag{6–124}$$

$$F_T(15T_p) = \int_0^{15T_p} f_T(\tau)\,d\tau = 0.9148 \tag{6–125}$$

Finally, the requested failure probability (exceeding probability) follows from Eq. (5-7):

$$P_f\big([0, 15T_p]\big) = 1 - 0.9773\,(1 - 0.9148) = 0.9167 \tag{6–126}$$

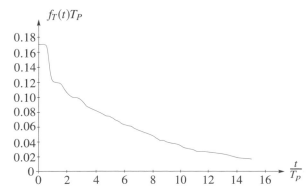

Fig. 6–22 First-passage probability density function of the sea-surface elevation process. $H_s = 7$ m, $b = 3.5$ m.

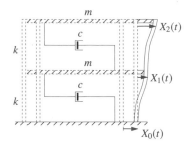

Fig. 6–23 Two-storey planar shear frame exposed to a horizontal ground surface motion.

Example 6-10: Failure probability of a two-storey planar shear frame subjected to a horizontal ground surface motion modelled as a Gaussian white noise

Fig. 6-23 shows a linear elastic two-storey shear frame, which is exposed to a horizontal ground surface motion $\ddot{X}_0(t)$. The displacements of the storey masses relative to the ground surface are denoted $X_1(t)$ and $X_2(t)$, respectively. Both storeys have the same mass m, and both inter-storeys have the same linear shear stiffness k. The dissipation in the shear columns during relative motion of the storeys is modelled by a linear viscous damping element with the damping constant c.

The equations of motion are given by Eq. (4-149):

$$\ddot{\mathbf{X}}(t) + 2\zeta_0\omega_0\,\mathbf{k}\,\dot{\mathbf{X}}(t) + \omega_0^2\,\mathbf{k}\,\mathbf{X}(t) = \mathbf{b}\,\ddot{X}_0(t) \tag{6–127}$$

where:

$$\mathbf{X}(t) = \begin{bmatrix} X_1(t) \\ X_2(t) \end{bmatrix}, \quad \mathbf{b} = -\begin{bmatrix} 1 \\ 1 \end{bmatrix}, \quad \mathbf{k} = \begin{bmatrix} 2 & -1 \\ -1 & 1 \end{bmatrix} \tag{6–128}$$

ω_0 and ζ_0 are given by Eq. (4-151).

The angular eigenfrequencies and the modal damping ratios become, cf. Eqs. (4-152) and (4-155):

$$\left.\begin{array}{ll} \omega_1 = \omega_0\sqrt{2 - 2\cos\left(\dfrac{\pi}{5}\right)} = 0.6180\,\omega_0, & \omega_2 = \omega_0\sqrt{2 - 2\cos\left(\dfrac{3\pi}{5}\right)} = 1.6180\,\omega_0 \\[3mm] \zeta_1 = \zeta_0\sqrt{2 - 2\cos\left(\dfrac{\pi}{5}\right)} = 0.6180\,\zeta_0, & \zeta_2 = \zeta_0\sqrt{2 - 2\cos\left(\dfrac{3\pi}{5}\right)} = 1.6180\,\zeta_0 \end{array}\right\} \tag{6–129}$$

The undamped eigenmodes become, cf. Eq. (4-153):

$$\mathbf{\Phi}_1 = \begin{bmatrix} \sin(\frac{\pi}{5}) \\ \sin(\frac{3\pi}{5}) \end{bmatrix} = \begin{bmatrix} 0.5878 \\ 0.9511 \end{bmatrix}, \quad \mathbf{\Phi}_2 = \begin{bmatrix} \sin(\frac{2\pi}{5}) \\ \sin(\frac{6\pi}{5}) \end{bmatrix} = \begin{bmatrix} 0.9511 \\ -0.5878 \end{bmatrix} \tag{6–130}$$

The modal masses become, cf. Eq. (4-154):

$$m_1 = m_2 = \frac{5}{4}\,m \tag{6–131}$$

The frequency response functions $H_1(z)$ and $H_2(z)$ of the inter-storey displacements $Y_1(t) = X_1(t)$ and $Y_2(t) = X_2(t) - X_1(t)$ due to a harmonically varying ground surface acceleration $\ddot{X}_0(t)$ can be written as, cf. Eq. (4-157):

$$H_k(z) = \frac{P_k(z)}{Q(z)} = \sum_{j=1}^{2} \frac{c_{jk}}{z^2 + 2\zeta_j\omega_j z + \omega_j^2}, \quad z = i\omega \tag{6–132}$$

where from Eqs. (4-156) and (4-158), we have:

$$c_{jk} = \frac{1}{m_j} \mathbf{a}_k^T \, \mathbf{\Phi}_j \mathbf{\Phi}_j^T \, \mathbf{b}, \quad \mathbf{a}_1 = \begin{bmatrix} 1 \\ 0 \end{bmatrix}, \quad \mathbf{a}_2 = \begin{bmatrix} -1 \\ 1 \end{bmatrix} \quad \Rightarrow$$

$$c_{11} = -0.7236 \frac{1}{m}, \quad c_{21} = -0.2764 \frac{1}{m}, \quad c_{12} = -0.4472 \frac{1}{m}, \quad c_{22} = 0.4472 \frac{1}{m} \tag{6–133}$$

On the right hand side of Eq. (6-132), upon bringing the terms within the summation on a common fraction line, the parameters of the polynomial $P_k(z)$ and $Q(z)$ may be identified. The poles of $Q(z)$ are:

$$\left.\begin{matrix} z_1 \\ z_2 \end{matrix}\right\} = \omega_1 \left(-\zeta_1 \pm i\sqrt{1 - \zeta_1^2} \right) \tag{6–134}$$

$$\left.\begin{matrix} z_3 \\ z_4 \end{matrix}\right\} = \omega_2 \left(-\zeta_2 \pm i\sqrt{1 - \zeta_2^2} \right) \tag{6–135}$$

The ground surface acceleration process $\{\ddot{X}_0(t), \, t \in R\}$ is modelled as a stationary Gaussian white noise process with the auto-spectral density S_0. The stationary standard deviations σ_{Y_1} and σ_{Y_2} of the inter-storey displacement stochastic processes $\{Y_1(t), \, t \in R\}$ and $\{Y_2(t), \, t \in R\}$, which are the physical quantities causing stresses in the structure, are obtained from Eq. (3-59) for $\tau = 0$, using the rational frequency response functions defined by Eq. (6-132).

The modal damping ratios are chosen as $\zeta_1 = 0.01 \Rightarrow \zeta_2 = \frac{1.6180}{0.6180} 0.01 = 0.02468$.

The structure is assumed to be safe if the inter-storey displacements $Y_1(t)$ and $Y_2(t)$ are simultaneously in the safe domain as shown in Fig. 6-24, i.e.:

$$S = \{(y_1, y_2) | -b_1 < y_1 < b_1 \land -b_2 < y_2 < b_2\} \tag{6–136}$$

where the barrier levels b_1 and b_2 are chosen as $b_1 = 2.0\sigma_{Y_1}$ and $b_2 = 2.0\sigma_{Y_2}$.

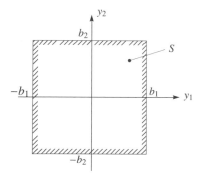

Fig. 6–24 Safe domain for the two-storey planar shear frame.

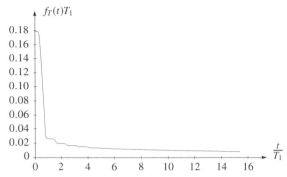

Fig. 6–25 First-passage probability density function for a two-storey shear frame. $\zeta_1 = 0.01$, $b_1 = 2.0\sigma_{Y_1}$, $b_2 = 2.0\sigma_{Y_2}$.

The probability of failure during an interval of the length $15\,T_1$ is requested, where $T_1 = \frac{2\pi}{\omega_1}$ is the lowest undamped eigenperiod as determined from Eq. (6-129). The initial value $\mathbf{Y}(0)$ at the start of the considered interval $[0, 15T_1]$ is assumed to be unknown. Hence, the reliability problem has random initial start, cf. Eq. (5-7).

A realization $\mathbf{y}(t)$ of the relative displacement vector process $\{\mathbf{Y}(t), t \in R\}$ is obtained through numerical integration of Eq. (6-127). This necessitates that $\{\ddot{X}_0(t), t \in R\}$ is replaced by an equivalent white noise process, where the broken line process described in Section 6.1 has been selected.

The first-passage time probability density function, based on a realization $\mathbf{y}(t)$ with $N = 15804$ out-crossings from S, is shown in Fig. 6-25. The following results are also computed:

$$P\big(\mathbf{Y}(0) \in S\big) = 0.9471 \tag{6–137}$$

$$E[L] = 18.69\,T_1 \tag{6–138}$$

$$F_T(15\,T_1) = \int_0^{15T_1} f_T(\tau)\,d\tau = 0.5243 \tag{6–139}$$

Finally, the requested failure probability follows from Eq. (5-7):

$$P_f\big([0, 15T_1]\big) = 1 - 0.9471\,(1 - 0.5243) = 0.5494 \tag{6–140}$$

APPENDICES

Appendix A: Calculation of the auto-covariance function for a stochastic process with rational auto-spectral density function

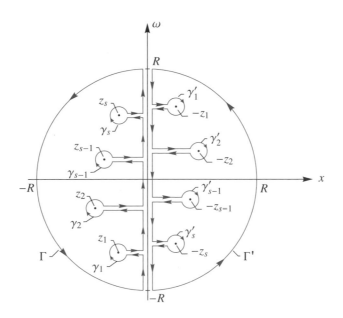

Fig. A-1. Circulation paths.

The auto-covariance function of a stochastic process with rational auto-spectral density function follows from Eqs. (3-23), (3-54) and (3-58):

$$\kappa_{XX}(\tau) = \int_{-\infty}^{\infty} e^{i\omega t} \frac{P(i\omega)P(-i\omega)}{Q(i\omega)Q(-i\omega)} S_0 \, d\omega \tag{A-1}$$

where $P(i\omega)$ and $Q(i\omega)$ are given by Eqs. (3-55) and (3-56), respectively.

The following complex function is considered:

$$f(z) = e^{z\tau} \frac{P(z)P(-z)}{Q(z)Q(-z)} S_0, \quad z = x + i\omega \tag{A-2}$$

All roots $z_1, \ldots, z_s$ of $Q(z) = 0$ are assumed to have negative real parts. Consequently, all roots $-z_1, \ldots, -z_s$ of $Q(-z) = 0$ have positive real parts. From this it is seen that $f(z)$ has no poles within the contour Γ (in Fig. A-1) which encircles all simple poles $z_1, \ldots, , z_s$ in the half plane $\text{Re}(z) < 0$, or within the contour Γ' which encircles all simple poles $-z_1, \ldots, -z_2$ in the half plane $\text{Re}(z) > 0$. Hence, $f(z)$ is analytical in the interior of Γ and Γ'. Then, the following expressions follow from *Cauchy's integration theorem*:[2]

$$\oint_{\Gamma} f(z)\, dz = 0 \tag{A-3}$$

$$\oint_{\Gamma'} f(z)\, dz = 0 \tag{A-4}$$

In Eqs. (A-3), (A-4) and below, all circulation integrals are assumed to be performed in the positive, counterclockwise direction.

Initially, it is assumed that $\tau > 0$, and the circulation integral along Γ in Eq. (A-3) is considered. The absolute value of $e^{z\tau}$ along this semicircle thus becomes:

$$\left| e^{z\tau} \right| = \left| e^{\tau |z| e^{i \arg(z)}} \right| = \left| e^{\tau R \left(\cos(\arg(z)) + i \sin(\arg(z)) \right)} \right| = e^{\tau R \cos(\arg(z))} \tag{A-5}$$

$\cos\left(\arg(z)\right) < 0$ along the semicircle. Since $\tau > 0$, it follows that $|e^{z\tau}| \to 0$ as $R \to \infty$, and the integral in Eq. (A-3) vanishes due to the exponentially decay in Eq. (A-5).

Along the ω-axis, we have $z = i\omega$. Further, the line integrals forth and back to the circulations γ_j cancel pairwise. Hence, Eq. (A-3) reduces to:

$$\int_{-\infty}^{\infty} f(i\omega)\, d(i\omega) - \sum_{j=1}^{s} \oint_{\gamma_j} f(z)\, dz = 0 \tag{A-6}$$

where γ_j signifies a circulation path along a circle with its center at z_j and with infinite small radius r_j. The negative sign in front of the sum in Eq. (A-6) is because the circulation along γ_j in Fig. A-1 should be performed clockwise, while the sign in Eq. (A-6) assumes a counterclockwise contour integration.

[2]E. Kreyszig. *Advanced Engineering Mathematics, 10th Edition*. John Wiley & Sons, Inc., Hoboken, New Jersey, 2011.

Along γ_j, we have $z - z_j = r_j e^{i\theta}$, $dz = i r_j e^{i\theta} d\theta$. Further, $P(z) \simeq P(z_j)$, $e^{z\tau} \simeq e^{z_j \tau}$, $z - z_k \simeq z_j - z_k$, $k \neq j$. Therefore, the following expression follows from Eqs. (A-2) and (3-56):

$$\oint_{\gamma_j} f(z)\, dz = \lim_{r_j \to 0} \int_0^{2\pi} e^{z_j \tau} \frac{P(z_j)P(-z_j)}{\displaystyle\prod_{\substack{k=1 \\ k \neq j}}^{s}(z_k - z_j) \prod_{k=1}^{s}(-z_k - z_j)} \frac{i r_j\, e^{i\theta}}{r_j\, e^{i\theta}} S_0\, d\theta$$

$$= i\, 2\pi\, S_0 \frac{e^{z_j \tau}}{z_j} \frac{P(z_j)P(-z_j)}{\displaystyle\prod_{\substack{k=1 \\ k \neq j}}^{s}(z_k^2 - z_j^2)} \tag{A-7}$$

Finally, from Eqs. (A-1), (A-2), (A-6) and (A-7), the expression for the auto-covariance function becomes:

$$\kappa_{XX}(\tau) = -\pi S_0 \sum_{j=1}^{s} \frac{e^{z_j \tau}}{z_j} \frac{P(z_j)P(-z_j)}{\displaystyle\prod_{\substack{k=1 \\ k \neq j}}^{s}(z_k^2 - z_j^2)} \tag{A-8}$$

For $\tau = 0$, $e^{z\tau} = 1$. In this case, the integral along the semicircle only vanishes if $s > r$. On the other hand, if this is the case, Eq. (A-8) is also valid for $\tau = 0$.

For $\tau < 0$, the integration is performed along the circulation Γ' given by Eq. (A-4). However, the result can immediatly be obtained from the symmetry condition in Eq. (3-22). Consequently, the auto-covariance function must have the form:

$$\kappa_{XX}(\tau) = -\pi S_0 \sum_{j=1}^{s} \frac{e^{z_j |\tau|}}{z_j} \frac{P(z_j)P(-z_j)}{\displaystyle\prod_{\substack{k=1 \\ k \neq j}}^{s}(z_k^2 - z_j^2)} \tag{A-9}$$

Eq. (3-59) is thus proved.

Appendix B: Impulse response function for a system with rational frequency response function

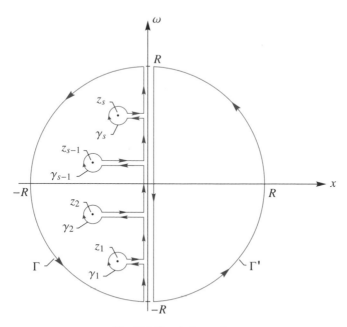

Fig. B-1 Circulation paths.

The impulse response function follows from Eqs. (3-54) and (3-99):

$$h(t) = \frac{1}{2\pi} \int_{-\infty}^{\infty} e^{i\omega t} \frac{P(i\omega)}{Q(i\omega)} \, d\omega \tag{B-1}$$

where $P(i\omega)$ and $Q(i\omega)$ are given by Eqs. (3-55) and (3-56), respectively.

The following complex function is considered:

$$f(z) = \frac{1}{2\pi} e^{zt} \frac{P(z)}{Q(z)}, \quad z = x + i\omega \tag{B-2}$$

All roots, $z_1, \ldots, z_s$, of $Q(z) = 0$, are assumed to have negative real parts. The roots have been indicated in Fig. B-1.

Initially, it is assumed that $t > 0$. With arguments identical to those in Appendix A, it follows that the contour integral along the semicircle vanishes. Further, the integrals forth and back from the circles with circumference γ_j vanish mutually. $f(z)$ is analytical in the interior of the contour Γ. Then, Cauchy's integration theorem leads to the following equation, cf. Eq. (A-6):

$$\int_{-\infty}^{\infty} f(i\omega) \, d(i\omega) - \sum_{j=1}^{s} \oint_{\gamma_j} f(z) \, dz = 0 \tag{B-3}$$

Along the circumference γ_j with radius $r_j \to 0$, we have:

$$\oint_{\gamma_j} f(z)\,dz = \frac{1}{2\pi} \lim_{r_j \to 0} \int_0^{2\pi} e^{z_j t}\, \frac{P(z_j)}{\prod\limits_{\substack{k=1 \\ k \neq j}}^{s}(z_k - z_j)}\, \frac{i r_j\, e^{i\theta}}{r_j\, e^{i\theta}}\, d\theta = i\, e^{z_j t}\, \frac{P(z_j)}{\prod\limits_{\substack{k=1 \\ k \neq j}}^{s}(z_k - z_j)} \tag{B-4}$$

It follows from Eqs. (B-1), (B-2), (B-3) and (B-4) that:

$$h(t) = \sum_{j=1}^{s} e^{z_j t}\, \frac{P(z_j)}{\prod\limits_{\substack{k=1 \\ k \neq j}}^{s}(z_k - z_j)}, \quad t > 0 \tag{B-5}$$

Next, it is assumed that $t < 0$. The assumption of all poles having a negative real part implies that $f(z)$ is analytical in the interior of the contour Γ' shown in Fig. B-1. From Cauchy's integration theorem, we have:

$$\oint_{\Gamma'} f(z)\,dz = 0 \tag{B-6}$$

$\cos\left(\arg(z)\right) > 0$ along this semicircle. Since $t < 0$, $\lim\limits_{R \to \infty} |e^{zt}| = 0$, cf. Eq. (A-5). Consequently, the part of the contour integral in Eq. (B-6) along the semicircle vanishes, and Eq. (B-6) reduces to:

$$\int_{-\infty}^{\infty} f(i\omega)\,d(i\omega) = 0, \quad t < 0 \tag{B-7}$$

Then, it follows from Eqs. (B-1) and (B-2) that:

$$h(t) = 0, \quad t < 0 \tag{B-8}$$

The result in Eq. (B-8) is obtained because no poles are present within Γ'. Hence, the assumption that all roots of $Q(z) = 0$ have negative real parts is tantamount to the causality condition in Eq. (3-91). As seen from Eq. (B-5), the obtained impulse response function will also be absolutely integrable over any finite or infinite time interval, implying the system is stable, cf. Eq. (3-92). Therefore, the assumption that all roots of $Q(z) = 0$ have negative real parts is the necessary and sufficient condition for a system with rational frequency response function to be causal and stable.

Appendix C: Expected value of a combined stochastic variable generated by a normally distributed stochastic vector

It is assumed that $\mathbf{X} \sim N(\boldsymbol{\mu}_{\mathbf{X}}, \mathbf{C_{XX}})$, $\mathbf{X} = [X_1, \ldots, X_n]^T$. $\boldsymbol{\mu}_{\mathbf{X}} = [\mu_{X_1}, \ldots, \mu_{X_n}]$ is the mean value vector and $\mathbf{C_{XX}}$ is the covariance matrix with elements $C_{X_j X_k}$.

Considering an arbitrary, combined stochastic variable $g(\mathbf{X})$, the following equation can be obtained:

$$E\left[(X_j - \mu_{X_j}) g(\mathbf{X})\right] = \sum_{k=1}^{n} C_{X_j X_k} E\left[\frac{\partial g(\mathbf{X})}{\partial x_k}\right] \tag{C-1}$$

Eq. (C-1) is proved below in a few steps ending with Eq. (C-5).

The probability density function of $\mathbf{X}$ is given by Eq. (2-15). For an arbitrarily state variable x_j, we have:

$$\int_{-\infty}^{\infty} \frac{\partial}{\partial x_j} \left(g(\mathbf{x}) f_{\mathbf{X}}(\mathbf{x})\right) dx_j = \left[g(\mathbf{x}) f_{\mathbf{X}}(\mathbf{x})\right]_{x_j=-\infty}^{x_j=\infty} = 0 \quad \Rightarrow$$

$$\int_{-\infty}^{\infty} \frac{\partial}{\partial x_j} \left(g(\mathbf{x})\right) f_{\mathbf{X}}(\mathbf{x}) dx_j = -\int_{-\infty}^{\infty} g(\mathbf{x}) \frac{\partial}{\partial x_j} \left(f_{\mathbf{X}}(\mathbf{x})\right) dx_j \tag{C-2}$$

Next, Eq. (C-2) is integrated over the remaining $(n-1)$ state variables $x_1, \ldots, x_{j-1}, x_{j+1}, \ldots, x_n$:

$$\int_{-\infty}^{\infty} \cdots \int_{-\infty}^{\infty} \frac{\partial}{\partial x_j} \left(g(\mathbf{x})\right) f_{\mathbf{X}}(\mathbf{x}) dx_1 \cdots dx_n = -\int_{-\infty}^{\infty} \cdots \int_{-\infty}^{\infty} g(\mathbf{x}) \frac{\partial}{\partial x_j} \left(f_{\mathbf{X}}(\mathbf{x})\right) dx_1 \cdots dx_n$$

$$\tag{C-3}$$

Using the symmetry of the inverse covariance matrix, $C_{X_j X_k}^{-1} = C_{X_k X_j}^{-1}$, it follows from Eq. (2-15) that:

$$\frac{\partial}{\partial x_j} \left(f_{\mathbf{X}}(\mathbf{x})\right) = -\sum_{k=1}^{n} C_{X_j X_k}^{-1} (x_k - \mu_{X_k}) f_{\mathbf{X}}(\mathbf{x}) \tag{C-4}$$

Eq. (C-4) is then inserted into Eq. (C-3). After multiplication with $\mathbf{C_{XX}}$ on both sides, the following result is obtained:

$$\int_{-\infty}^{\infty} \cdots \int_{-\infty}^{\infty} (x_j - \mu_{X_i}) g(\mathbf{x}) f_{\mathbf{X}}(\mathbf{x}) dx_1 \cdots dx_n = \sum_{k=1}^{n} C_{X_j X_k} \int_{-\infty}^{\infty} \cdots \int_{-\infty}^{\infty} \frac{\partial g(\mathbf{x})}{\partial x_k} f_{\mathbf{X}}(\mathbf{x}) dx_1 \cdots dx_n$$

$$\tag{C-5}$$

which is identical to Eq. (C-1).

A special case of $g(\mathbf{X})$ is considered:

$$g(\mathbf{X}) = \left(X_{j_2} - \mu_{X_{j_2}}\right)\left(X_{j_3} - \mu_{X_{j_3}}\right), \quad j_2, j_3 = 1, \ldots, n \tag{C-6}$$

Insertion of Eq. (C-6) into Eq. (C-1), the 3rd order joint central moments of a normally distributed n-dimensional stochastic vector can be obtained:

$$E\left[\left(X_{j_1} - \mu_{X_{j_1}}\right)\left(X_{j_2} - \mu_{X_{j_2}}\right)\left(X_{j_3} - \mu_{X_{j_3}}\right)\right]$$

$$= \sum_{j=1}^{n} C_{X_{j_1}X_j}\left(E\left[\delta_{j_2 j}\left(X_{j_3} - \mu_{X_{j_3}}\right)\right] + E\left[\delta_{j_3 j}\left(X_{j_2} - \mu_{X_{j_2}}\right)\right]\right) = 0 \tag{C-7}$$

where δ_{ij} signifies Kronecker's delta.

Next, consider another special case of $g(\mathbf{X})$:

$$g(\mathbf{X}) = \left(X_{j_2} - \mu_{X_{j_2}}\right)\left(X_{j_3} - \mu_{X_{j_3}}\right)\left(X_{j_4} - \mu_{X_{j_4}}\right), \quad j_2, j_3, j_4 = 1, \ldots, n \tag{C-8}$$

Insertion of Eq. (C-8) into Eq. (C-1) provides the following result for the 4th order joint central moment:

$$E\left[\left(X_{j_1} - \mu_{X_{j_1}}\right)\left(X_{j_2} - \mu_{X_{j_2}}\right)\left(X_{j_3} - \mu_{X_{j_3}}\right))\left(X_{j_4} - \mu_{X_{j_4}}\right)\right]$$

$$= \sum_{j=1}^{n} C_{X_{j_1}X_j}\left(E\left[\delta_{j_2 j}\left(X_{j_3} - \mu_{X_{j_3}}\right)\left(X_{j_4} - \mu_{X_{j_4}}\right)\right] + E\left[\delta_{j_3 j}\left(X_{j_2} - \mu_{X_{j_2}}\right)\left(X_{j_4} - \mu_{X_{j_4}}\right)\right]\right.$$

$$\left. + E\left[\delta_{j_4 j}\left(X_{j_2} - \mu_{X_{j_2}}\right)\left(X_{j_3} - \mu_{X_{j_3}}\right)\right]\right)$$

$$= C_{X_{j_1}X_{j_2}}C_{X_{j_3}X_{j_4}} + C_{X_{j_1}X_{j_3}}C_{X_{j_2}X_{j_4}} + C_{X_{j_1}X_{j_4}}C_{X_{j_2}X_{j_3}} \tag{C-9}$$

Joint central moments of arbitrary order can be evaluated in the same manner. Take the one-dimensional case for example, $X \sim N(\mu_X, \sigma_X^2)$, with $g(X) = (X - \mu)^{m-1}$. Then, the use of Eq. (C-1) provides:

$$E\left[(X - \mu_X)^m\right] = \sigma_X^2 (m - 1) E\left[(X - \mu)^{m-2}\right], \quad m = 3, 4, \ldots \tag{C-10}$$

From the recursive formula in Eq. (C-10), the following result can be obtained:

$$E\left[(X - \mu_X)^m\right] = \begin{cases} 0 & , \quad m \text{ odd} \\ \sigma_X^2 (m - 1)(m - 3) \cdots 3 \cdot 1, & m \text{ even} \end{cases} \tag{C-11}$$

INDEX

A

acceleration of gravity, 178
acceleration process, 58, 61, 86
acceleration vector process, 98
almost periodic processes, 187
auto-correlation coefficient function, 29, 32
auto-covariance function, 28, 32
auto-spectral density function, 33, 34, 37
auxiliary process, 63, 64

B

band-limited white noise, 42, 166
bandwidth parameter, 41, 42, 143
Belyaev's formula, 142
Bernoulli-Euler beam theory, 10, 89, 123
Box-Müller transformation, 21
broad-banded process, 39
broken line process, 173

C

Cauchy's integration theorem, 206, 208
Caughey damping model, 101
causality, 9, 49
central limit theorem, 36, 190
combined stochastic variable, 16, 24
complex conjugate, 34
conditional mean value, 24, 26
conditional probability density function, 23
conditional variance, 24
continuous parameter stochastic process, 27
continuous systems, 85, 126
correlation coefficient, 23
correlation length, 32, 35, 37
coupled first order differential equations, 65

covariance, 22
covariance matrix, 16, 22
Cramér and Leadbetter envelope process, 155
cross-covariance function, 31, 33
cross-spectral density function, 33

D

d'Alembert's principle, 89
damped angular eigenfrequency, 86, 103
damping coefficient, 66, 85, 126
damping matrix, 90, 99
damping ratio, 41, 67, 86
decoupling condition, 127
derivative process, 56, 175
derivative process of the nth order, 57
derivative vector process of the nth order, 57
determinism, 9
dimensionality paradox in reliability theory, 152
discrete parameter stochastic process, 27
displacement process, 58, 61, 85
displacement vector process, 98, 104
double-sided auto-spectral density function, 40
Duhamel's integral, 11, 86
dynamic variance, 97

E

eigenvalue problem, 19, 127, 183
envelope process, 153
equivalent white noise approximation, 95, 108
equivalent white noise process, 172
ergodic in the auto-covariance function, 76, 79
ergodic in the mean value, 76, 79
ergodic process, 76